Latin America

Cultures in Conflict

Robert C. Williamson

LATIN AMERICA: CULTURES IN CONFLICT

First published in 2006 by
PALGRAVE MACMILLAN™
175 Fifth Avenue, New York, N.Y. 10010 and
Houndmills, Basingstoke, Hampshire, England RG21 6XS
Companies and representatives throughout the world.

PALGRAVE MACMILLAN is the global academic imprint of the Palgrave Macmillan division of St. Martin's Press, LLC and of Palgrave Macmillan Ltd. Macmillan® is a registered trademark in the United States, United Kingdom and other countries. Palgrave is a registered trademark in the European Union and other countries.

ISBN-13: 978-1-4039-6885-2 (hardback: alk. paper)
ISBN-13: 978-1-4039-6886-9 (pbk.: alk. paper)
ISBN-10: 1-4039-6885-3 (hardback: alk. paper)
ISBN-10: 1-4039-6886-1 (pbk.: alk. paper)

Library of Congress Cataloging-in-Publication Data

Williamson, Robert Clifford
    Latin America : cultures in conflict / Robert C. Williamson.
        p. cm.
    Includes bibliographical references and index.
    ISBN 1-4039-6885-3 (hardback : alk. paper)
    ISBN 1-4039-6886-1 (pbk. : alk. paper)
        1. Social change—Latin America. 2. Social conflict—Latin America.
    3. Social institutions—Latin America. 4. Latin America—Social
    conditions—1982–. I. Title.

HN110.5.A8W485 2006
306.098—dc22   2005049543

A catalogue record for this book is available from the British Library.

Design by Newgen Imaging Systems (P) Ltd., Chennai, India.

First edition: May 2006

10 9 8 7 6 5 4 3 2 1

Printed in the United States of America.

Transferred to digital printing in 2007.

Contents

Tables, Maps, and Figures

Tables

Maps

Figures

Preface

This work represents a continuation of my previous volume *Latin American Socities in Transition* (Praeger, 1997), which dealt with sociopolitical problems—demography, rural society, ethnicity, stratification, and urbanization. The first half of the present book analyzes the geographical, anthropological, historical, and especially the sociological bases of today's cultures and focuses on several conflict issues—political, economic, ethnic, class, and gender, among others. More than my previous treatment it offers a theoretical approach to the questions of social structure, values, identity, modernity, the arts, and societal change, as viewed by a range of observers, including writers and political figures as well as social scientists.

The second half of the book examines in detail the socializing institutions—family, education, religion, communication including mass media, science, and technology. The final chapter looks at the future of the society as well as placing Latin American society, especially its prospects for change, in a conceptual setting. Among the theories examined are conflict (including neo-Marxist), structural functionalism, and world systems, with reference to Weber, Collins, Giddens, Foucault, Habermas, and Inglehart, to mention several social scientists. Among the sources are the research studies of sociologists and social anthropologists as well as economists, political scientists, historians, and representatives of literature and the arts. I also draw on my own research in Central America, Chile, and Colombia supported by the Fulbright and other agencies. In order to avoid the unconscious bias of any foreign observer I include Latin American researchers along with North Americans and Europeans.

Of course, any analysis of Latin America must consider the complexity of this area in all its cultural and regional variations. Inevitably this text emphasizes certain larger nations like Brazil, Mexico, and Argentina more than smaller ones like Ecuador or Haiti, but none of the twenty republics will be ignored.

Like most social scientists I am reluctant to make predictions, yet my background convinces me that changes in the twentieth century, slow and incomplete though they were, did effect considerable societal transformation. I would surmise that the twenty-first century will involve an accelerated rate of innovations, with both positive and negative effects. At least, this book brings together a number of generalizations that may help the reader to understand the present dynamics and the potential for further development.

Robert C. Williamson

Acknowledgments

In writing this book I am grateful for the contribution of a number of agencies and individuals. The Fulbright Commission, Social Science Research Council, and Lehigh University Office of Research enabled me to experience and investigate Latin America in considerable depth. I am also indebted to both the reference and technical staff of the Lehigh University libraries. I am also most appreciative to a number of individuals for reading portions of the manuscript, including Anna Adams and Hannah Stewart-Gambino. My greatest gratitude is to my wife Virginia for her patience and help and to the staff of Palgrave Macmillan.

Chapter One

Latin America: Change and Diversity

Even to those who have never been to its shores, the mention of Latin America conjures up a series of images and easy generalizations. Many foreigners, notably North Americans, see backwardness, dishonesty, disorder, even violence. The reality is far more complex. Indeed, Latin America is undergoing a rapid transformation. For instance, at the time of my first visit to Bogotá, Colombia, some forty years ago the city of two million had a number of inviting restaurants, an occasional concert, and no end of American movies. In 1993 when I returned to teach at the University of the Andes, skyscrapers towered over this city of six million, with bookstores, florist shops, a symphony orchestra, restaurants with a variety of cuisines, especially in the fashionable northern reaches of the city. As a film buff I continue to revel in the cinema clubs' offerings from nearly every continent. Yet, there is a continuity from my visits decades earlier. The *chozas* (huts) of corrugated tin and wood still occupy interstitial areas, but are less evident than they were a generation earlier. Even more than earlier years, many areas of the city are at the visitor's risk, as are some interstitial areas of North American cities. Yet, an upwardly mobile middle class is ever more visible.

Contrasts and complexities still characterize the twenty republics that we usually identify as Latin America, both within and between nations, as is true for most areas of the world. Visitors to the United States can hardly escape the differences between Vermont and Texas, Manhattan and Des Moines, not to mention disparities in social class, belief systems, lifestyles, including a number of discontinuities comparable to the negative images they see south of the Rio Grande, for instance, violence and corruption—commodities hardly absent in the United States. Probably Latin America offers even more variations than does the United States. Social stratification is hardly less acute in the United States where the upper 1 percent of the population owns over 40 percent of the country's wealth. However, poverty is less extensive and severe there than in Latin America.

Most striking about Latin America is its sheer size and growth. This area is almost three times the size of the United States. It is after Africa the most rapidly growing area of the world. With a half billion people and a gradual but uneven rise in the standard of living it offers a widening market for the twenty-first century. Also, Latin America is salient in the markets of Europe,

Asia, and especially North America. Already, roughly twenty-five million Latin Americans reside in the United States, which is now the fifth largest Spanish-speaking nation in the world (after Mexico, Spain, Colombia, and Argentina).

Further, in this note of superlatives one cannot forget that Brazil is exceeded in size only by Russia, China, Canada, and the United States (if Alaska is included). With approximately one hundred and ninety million inhabitants it is the fifth largest nation after China, India, United States, and Indonesia along with being the world's ninth (between Italy and Mexico) industrial power as measured by GNP (gross national product). More people speak Portuguese than French, German, or Italian as a first language. In view of Brazil's ever-growing role on the world stage why are not more colleges offering its language? One more example of cultural lag.

Again, the excitement of Latin America is in its diversity. First of all, its landscapes display beguiling vistas from the cordillera stretching from Mexico to Patagonia, a coastline varying from Brazil's palm-studded beaches to Chile's fjords, in addition to Iguassú, the world's largest waterfall, and Venezuela's Angel Falls, the tallest in the Americas. Beyond these scenic wonders are the people living highly diverse lives—peasants laboring on Andean mountainsides, workers in coal, tin, and copper mines, fishermen on Lake Titicaca and in the Caribbean, servants, clerks, and stockbrokers in Latin America's some thirty cities of over a million population.

The Human Variable

Most evident in Latin America is the diverse origins of its people—indigenous, European, African, and mestizo—with variations in each of those segments, both physical and cultural, all influenced by region, education, and social class. Latin America's phenomenal growth is shown in figure 1.1.

First was the indigenous population. It is difficult to secure an accurate estimate of the original indigenous population; the figures range between 80 and 150 million, but barely a third of the pre-Columbian population survived until the seventeenth century as a result of European diseases for which the natives had no immunity—measles, influenza, and small pox. Obliterated by disease and strenuous work, and relegated to marginal lands spurned by their European conquerors, the indigenous were reduced to the lowest rung of the caste structure in the colonial period. All this becomes more remarkable for Mesoamerica and the Andes, where the Spanish encountered—and destroyed—civilizations comparable in several respects to the cultural level of the Iberian Peninsula in the sixteenth century. Fragments of these civilizations still remain in diet, housing, language, music, ceremonies, and other folkways. In four countries, Bolivia, Ecuador, Guatemala, and Peru, a third to more than half of the population are Indians and adhere in varying degrees to their ancient culture in work,

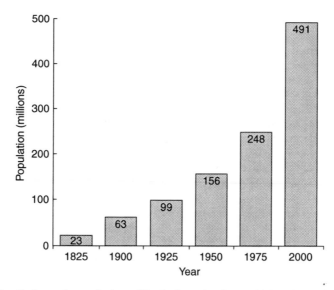

Figure 1.1 Estimated population of Latin America from 1825 to 2000 (in millions).

Source: *América Latina y Desarrollo Social* (Santiago: Centro para el Desarrollo Económico y Social, 1965) and World Fact Sheet, United Nations, 2003.

dress, language, and belief systems. In still other countries indigenous groupings persist to this day, though threatened by incursions from the national culture.

As the indigenous population along with Africans was historically considered nonpersons they were spared the inquisition. This was of small consolation as they remained in this submerged position until the twentieth century when revolutionary movements, at least in Mexico and Bolivia, gave them a degree of status and glorification, more ceremonial than real. Of course, several countries, possibly Argentina more than others, ignore their indigenous and African past. Miscegenation over the last few centuries conveniently permits this deception. In the last third of the nineteenth century, the Indian population was reduced tenfold in the "cone" countries (Argentina, Chile, and Uruguay), largely by intermixture, outmigration, privation, and extermination. Afro-Argentines were a fourth of the Buenos Aires population in the eighteenth century, yet by 1887 only 2 percent survived.[1] European immigrants also "whitened" the country.

No less than North America, Latin America is peopled and shaped by immigrants, not only from Spain and Portugal but also from Britain, France, Germany, Holland, Italy, and Poland, among other nations. Brazil probably represents the most vibrant mixture. Yet, to Mesoamerica and to all of South America came a stream of Europeans and Asians. Most obscene in this saga of immigrants was the forced importation of Africans in the early sixteenth century as slaves, first in the sugar plantations of Brazil and the Caribbean but, soon after, sought for labor in every Latin American republic—and in the U.S. South (note inset 1a).

INSET 1A   SLAVERY: THE CURSE OF THE AMERICAS

The history of slavery remains, along with the holocaust of Indians, the other tragedy of the colonial and early national periods. The slave trade brought nearly ten million Africans to the Americas in addition to those sent to Europe. Because of the horrible conditions on the ships, a fifth of the slaves-to-be died in transit. About four million landed in Brazil, with another five million in the remainder of the hemisphere. The high ratio of males hardly favored natural increase; consequently the slave trade continued over three centuries. Despite privation and hardship, Africans in Brazil enjoyed more rights such as marriage and manumission than did the Anglo-Protestant South of the United States. Besides being able to retain remnants of their African past, they intermittently organized rebellions, the most dramatic being the revolt in Haiti resulting in an independent republic in 1804.

After Britain outlawed the slave trade in 1804, several regions, notably those finding slavery less crucial such as Mexico and the cone countries (Argentina, Chile, and Uruguay) "ended" slavery. Cuba and Brazil were the last, 1886 and 1888, respectively.[2] This emancipation hastened miscegenation and the blending of African Americans into the population in most of Latin America.

Mestizos (the mixture of European and Indian) account for over 60 percent of the population in several countries from Mexico through Central America to Colombia, Venezuela, and Chile. A rich vocabulary designates the categories of mixtures, whether the combination is white with African or Indian, African with indigenous, or other variations. As we shall see in chapter 4 the status hierarchy became more diffuse as the colonial tradition gave way to broader groupings in this century. For instance, mestizos moved upward as a result of the Mexican Revolution (1910–20), and the indigenous obtained at least a sentimental and ceremonial niche after the revolution, Similarly, both mestizo and indigenous reached a marginal degree of bargaining power in the Bolivian Revolution of 1952. Even so, throughout Latin America, as the truism goes, the whiter the skin the higher the socioeconomic status. The varying dimensions of demography and economic levels are shown in table 1.1.

Not only do Latin American nations differ in ethnic and socioeconomic composition, they also differ in population density, El Salvador being the most crowded, and Bolivia having the sparsest population. These differences derive from the physical environment—Bolivia's inhospitable physical environment limits growth, whereas El Salvador's is favored with a fertile soil. Also, its small size (only slightly larger than New Jersey) results in a heavy outmigration, leading to considerable friction with its neighbor Honduras. Even more significant is the differing rates of growth between the twenty republics. Most stable populations are the more industrialized and European nations—Argentina, Chile, and Uruguay—with low rates of growth. Intermediate are countries undergoing economic expansion—Mexico, Costa Rica, Colombia, Peru, and Brazil, among others. The third group includes countries such as Guatemala, Honduras, Nicaragua, and Ecuador, which remain largely rural and agricultural. A rapid growth rate

Table 1.1   Population and per capita income for Latin America in 2000

Country	Population (in thousands)	GNP per capita ($U.S.)*
Argentina	37,215	10,880
Bolivia	8,139	2,460
Brazil	1,731,791	7,710
Chile	14,667	9,820
Colombia	40,037	6,370
Costa Rica	3,259	8,480
Cuba	11,131	3,174
Dominican Rep.	8,262	6,640
Ecuador	12,782	3,582
El Salvador	5,925	4,262
Guatemala	12,670	4,080
Haiti	6,922	1,160
Honduras	5,862	4,130
Mexico	102,027	8,970
Nicaragua	4,851	2,170
Panama	2,281	7,781
Paraguay	5,580	4,610
Peru	27,166	5,010
Uruguay	3,337	7,830
Venezuela	23,590	5,380

* Income data are subject to currency exchange fluctuations, therefore not too meaningful in regard to purchasing power.

Source: "Cultural Liberty in Today's Diverse World," *Human Development Report*, United Nations Development Program, 2004.

characterized the 1950s to the 1970s, when better health standards permitted increased longevity, Beginning with the 1980s nearly all countries have lower fertility, if at different rates. The causes for this decline run the gamut of the spread of contraceptive methods, education, urbanization, and rising mobility aspirations. Latin America still has larger growth rate than most of the world, but lags Africa.

Diversity, whether in culture or in population, makes the concept of a *Latin* America all the more questionable. When the area was open to exploration and exploitation in the sixteenth century, the Vatican settled the dividing line between Spain and Portugal in the Treaty of Tordesillos, implying an Iberic America, but the label of "Latin" gradually took root as it covered both regions in addition to the French occupation of Haiti, conveniently ignoring native, African, Asian, and other ingredients. Moreover, broad categorizations such as Latin America, South America, and Third World are only abstractions. We are trained to think of continents or nations as rather sacred entities, but in reality these designations become almost mythical.[3] Residents of Buenos Aires feel closer to Paris or Madrid than to Asunción; Afro-Brazilians may identify more with Nigeria or New Orleans than with Uruguay or even São Paulo. Even the smaller nations such as El Salvador or Ecuador represent complex entities with cultures and subcultures according to ethnic, class, and regional variations. When we turn to Latin America's largest nation the differentiation becomes far greater. The European or white population is 54 percent, *paardo or* brown 38.5, African or black 6, and indigenous, Asian, or undeclared 1.5. Subtle distinctions exist, for instance, among the dozens of term used are *Branco fino* (fine white), to *Negro retinto* (very dark black). Still terms include *branco de Bahia, mulatto, mulatto claro, mulatto escuro, cabo verdde, cabra,* and *Preto* or *criollo,* more or less based on a white to black continuum. Terms also refer to the white Indian mixture *(mameluco)* and black-Indian *(cafuso).*[4] The ambivalence about race in many Latin American countries arises from the superiority the Europeans feel and the reality that the national culture is enriched by the contribution of indigenous and African. Music derives from African and indigenous sources. Also, for many Brazilians, magical rites are a part of the national heritage. The tradition or stereotype of "black powers" of healing, sexual prowess, and musical ability are often found in the middle class—an index of the nuancing of racism in various Latin American regions.[5] As we shall see in chapters 4 and 5, these identities are interlaced with socioeconomic considerations and related symbols of status. To repeat, generalizations and stereotypes, such as the underdeveloped or developing world, are only marginally useful—one cannot place Colombia, India, and Turkey in the same package.[6]

Historic Roots

The indigenous history of this hemisphere reached a climax in the civilizations of Mexico and Peru. A number of Amerind groups still exert their influence in Latin America, from the Araucanians in Chile, Guaraní in Paraguay, Tupani in Brazil to the Motolones in Colombia, and the Seri in northern Mexico, to mention a few of the more conspicuous societies. In reality, 386 indigenous groups constitute nearly a million-and-half people—in Brazil alone are 116 entities with a total population of two hundred and twenty-five thousand. The major Indian societies of Mesoamerica and the northern Andes made the greatest impact. In particular, Indians in Guatemala, Ecuador, Peru, and Bolivia reflect the great empires of centuries ago.

In attempting to look at Latin America's history, as with any part of the world, one has to consider the usual caveats. As with social science in general, the observer attempts to be value-free but she or he can never completely overcome her or his biases. It is difficult to view a contemporary culture objectively, even more so for one in the past. History itself is "an ideological construct," constantly reanalyzed by individuals who seldom escape their own value sets.[7] Marxist or anti-Marxist views, empathy or detachment, macro or micro approaches, all can color the analysis. Even in a multivolume work the author or editor is selective; hence in this section only a rough sketch is possible. Despite these pitfalls a historical introduction is called for.

The Peopling of the Americas

The first human beings entered the Americas across the Bering Strait over twenty thousand years ago. Definitive evidence was first provided in 1927 when a fossilized human being was discovered at Folsom, Nevada, in deposits of the upper Pleistocene age. Investigations using radioactive carbon techniques similarly supported this date. The last Ice Age reportedly ended approximately ten thousand to twenty thousand years ago, providing further confirmation. More recent finds place the date beyond twenty-four thousand years, especially as the level of cultural development and variety of indigenous societies suggest a relatively early date. Whether or not later migrations of peoples from the Eurasian continent to the New World occurred in prehistoric times is a moot point. Similarities in domesticated plants as well as other parallelisms point to transoceanic influences. However, the possibility of extensive overseas travel is dubious. Also, cultural dissimilarity surpasses similarity. No record of these migrations exists in the extensive mythology of Oceanic and Asian peoples. Yet, the culture of Easter Island, fifteen hundred miles from the coast of Chile, has unmistakable Polynesian relationships.

Cultural Variations in the New World

At the time of Columbus's arrival in 1492, in what was later to be named Latin America, scores of distinct cultures existed. Many of these continue, if fragmented, into the present. For one, in the Caribbean area was the Arawak culture, organized in village-like chiefdoms, of which only remnants remain today. Other groupings include the Amazonian tropical forest peoples who exist in relative isolation, even though Europeans and Africans in this area adopted a number of items of the indigenous material culture. Still different are the nomadic tribes of Patagonia, sedentary farmers of central Chile, and warrior Araucanians of southern Chile, who were not subdued until the 1880s and live today in a reservation-like atmosphere with marginal integration into Chilean society, as determined by the political climate—the Pinochet dictatorship (1973–89) reversing previous reforms. In the Ecuadorian Amazon basin is another contrast, the Shuar (previously labeled as Jívaro or "wild"), who in the twentieth century gave up tribal warfare and headhunting for a productive economy with stress on education. Consequently, any attempt to classify the indigenous is fraught with oversimplification culturally—and linguistically—as over a hundred language groupings have been identified.[8] A number of indigenous peoples, especially in Amazonia, adopt Western culture as well as retaining their own arts and crafts in a cultural mélange that includes computers and airplanes.

Traditionally, interest focuses on the civilizations of the highlands and lowlands of Mesoamerica and the central Andes. Despite physical destruction by the conquistadores, their cultural features left a permanent heritage.

Table 1.2   Average population growth rates
in percentages for selected countries, 1961–99

Country	1961–70	1991–99
Argentina	1.5	1.2
Bolivia	2.4	2.1
Brazil	2.8	1.2
Chile	2.2	1.3
Colombia	3.0	1.7
Costa Rica	3.4	2.5
Ecuador	3.2	1.9
Haiti	2.1	1.5
Honduras	3.1	2.5
Mexico	3.3	1.5
Peru	2.9	1.6
Uruguay	1.0	0.6
Venezuela	3.5	1.9
Latin America	*2.8*	*1.5*

Source: Latin American Center, *Statistical Abstract of Latin
America* (Los Angeles: University of California, 2002), p. 525.

Descendents of these peoples still terrace the mountainsides; their foodstuffs,
textiles, and building styles left a perceptible impact on Western society.
Unfortunately, however, the intensified modernization in the late twentieth
century disturbed the way of life for many indigenous, especially the
younger generation, as in Mayan areas of Mesoamerica. Forced settlements,
mobility, including contacts with other communities, and missionary influences led to a different cultural orientation and perhaps most serious, less
sensitivity to the forest ecology of plant–animal interrelationships, As Latin
America continues its population growth (table 1.2) ecological problems
will become even more severe.

On the whole, Latin America's diversity seems more salient than its
similarities, as it is shaped by nationality, region, and a variety of
"subcultures"—ethnicity, class, gender, lifestyle, and the arts. We examine
the institutional structure, notably the economy and the state. Other social
institutions will be the focus of later chapters.

The Conceptual Framework

In an analysis of any complex topic some kind of theoretical mooring is
essential. For nearly a century behavioral scientists have been devising theoretical approaches in order to "explain" or "understand" social phenomena,
including society itself. At the risk of gross oversimplification, one might
divide these theoretical approaches into two camps: one is functionalism and
related approaches while the other is a number of models labeled as variants

of conflict—Marxism, neo-Marxism, critical theory, poststructuralism. Other variants include systems theory, ethnomethodology, feminism, and so on. Occasionally theories overlap, even between the two camps, as they deal with micro (short-term or immediate) and macro (total or long-range) problems on stability and change. Latin American social scientists gravitate toward both rubrics, but increasingly toward the conflict school—criticism and skepticism, in which the bases of social knowledge are questioned. Intellectuals tend to be critical of the status quo. In this context neo-Marxism has intermittent support.

Functionalism (sometimes referred to as structural functionalism) derives from several sociologists, particularly Talcot Partsons, revolving around a number of concepts—roles, values, personalities, and action systems—offering an explanation of the social system. The theory provides a plausible, if intricate, explanation of society. Other approaches, such as exchange theory and symbolic interactionism add to our understanding of how a society and its institutions function. The difficulty is that many societies are not as neat and orderly as the Parsonian model oriented to the mid-twentieth century suggests.

Latin America, among other regions of the world, is in a state of flux with instability and conflict. Violence is endemic in several nations—Colombia, Guatemala, Peru, and Venezuela, among others. As future chapters reveal, rivalries continue between classes, ethnic groups, regions, belief systems—the urban and the rural. Gender relations, economic stresses, political clashes, labor relations represent schism, or at least disarray. It is not to say that conflict resolution and cohesiveness are absent, but tensions lie underneath. All countries compete for scarce resources; differential goals divide individuals and groups. Certain areas of Latin America display a particularly disturbing profile of conflict. During the 1970s and 1980s well over half of the nations were caught in authoritarian governments, most notoriously Argentina and Chile, with reigns of terror. Even Colombia, a nominally democratic regime, continues to be torn apart by a violent struggle between government troops, a counter insurgency, occasionally aided by drug warlords.

Central America displays a tragic cycle of civil war—reactionary governments in concert with the economic oligarchy against peasants and workers. In her book *Paradise in Ashes*, Beatriz Manz describes the life, death, and rebirth of a Guatemalan enclave, one of many communities destroyed in the late 1970s and early 1980s by government troops in a war against Mayan peasants.[9] Similarly, El Salvador and Nicaragua were caught in a bloody struggle between peasants and workers for survival. But conflict and violence in Latin America go far beyond political and economic struggles. For instance, another study of Guatemalan society points to the pervasiveness of conflict and violence in family relations and other interpersonal encounters among the urban poor, including the role of alcohol and drugs in personal breakdown and their effects on social groupings such as crime and delinquency.[10] In other words, conflict cuts across various dimensions of social life. These problems are especially acute in many areas of Latin America, notably the lower classes, which is over two-thirds of the population in most countries. Consequently, patterns of conflict will be a recurrent theme in this text. Finally, a continuing challenge is how to resolve or reduce these disequilibria.

Chapter Two

Geography: Barriers and Challenges

The geography of Latin America is no less complex than its peoples. This chapter examines the physical structure, including the topography of plains, valleys, mountains, and river systems, focusing on the implications of these features on human habitation. In this context, geographic variables determine cultural and economic regionalism. Further, the area's resources, both natural and human, are analyzed for their contribution to the economy in the world setting of the twenty-first century. In other words, geographic factors are always changing, even more so in an era of globalization.

The Physical Environment

The geography of Latin America might be approached in a series of superlatives, yet many of these imply paradoxes or challenges. For instance, the Andes represent the longest mountain range on this planet, even though large areas forbid cultivation, complicating the welfare, if not the survival, of its inhabitants, as in the Bolivian plateau. The Amazon is the largest river system in the world. The area is also the largest tropical forest, but is now suffering from soil destruction and human exploitation, seriously reducing its capacity to produce ozone. Moreover, the continent has immense resources; yet their remoteness along with awesomely high mountain passes, poor transport facilities, including the lack of natural harbors, remain a formidable obstacle. One more contradiction is the juxtaposition of extensive iron reserves; yet forming less than 2 percent of the world's coal deposits.[1] These various impediments could be solved with technological breakthroughs, but financial barriers are evident.

The land is used differently by its occupants—Amerinds, Europeans, Africans, and assorted mixtures. This diversity of images prompts one geographer to think of the physical environment as

1. *divine*—animistic view of the mountains, rivers, and the earth itself;
2. *prey*—plundering and exploitation of its resources, mineral and human;
3. *illusion*—visitors from Columbus to the expeditions of scientist Alexander von Humboldt at the end of the eighteenth century saw the area as an exotic Garden of Eden; or
4. *obstacle*—the impediments of jungles, mountain ranges, and volcanoes.[2]

These physical features give rise to natural areas. However, sociocultural change surmounts physical barriers, at least within limits.

Geographically, it is desirable to look at Latin America in the context of its development from a colonial past shaped by mercantilist Spain to a quasi-colonial status in the nineteenth and early twentieth centuries to the global economy of today. Consequently, a characterization of Latin American geography might be in three divisions:

1. A core of urban industrial areas with communication networks, an agriculture serving urban markets, and above-average education and income levels as in southeast Brazil, the Puebla-Mexico City-Guadalajara axis, similar core areas in Argentina, Chile, and other relatively developed nations.
2. Stagnant regions usually marked by primary economic activities (agriculture and mining), which may have considerable population, but have been bypassed by more dynamic centers. Southern Mexico, the south central sierra of Peru, and northeast Brazil belong to this rubric.
3. Sparsely populated regions where potential resources have not yet been exploited—for better or for worse—as in the Brazilian west.[3]

Surface Features, Tectonics, and Natural Disasters

In Middle America the tectonic or geologic systems extend from the United States to central Mexico, where they give way to volcanoes, especially south and east of Mexico City. Southern Mexico and Central America also offer a complex system of volcanic structures relating to the islands of the Caribbean. Map 2.1 of Central and South America give an idea of the intricate physical structure.

The major feature of South America is the cordillera or Andes, which stretch for over 4,000 miles from Colombia to Patagonia. This folded and volcanic chain begins in Colombia as a three-pronged chain (the Western, Central, and Eastern Cordillera), then becomes two parallel ranges down the western edge of the continent to Chile, where they form a single chain. In the northern portion the range is 100–200 miles in width but increases to 400 miles in southern Peru and Bolivia. Over a dozen peaks are above 20,000 feet, Mt. Aconcagua being the highest peak in the hemisphere at 22,834 feet. More than half of Mexicans, Guatemalans, Colombians, Ecuadorians, and Bolivians live in the intermountain valleys and plateaus. Altitude accounts for a wide distribution of temperature and precipitation, providing a diversity of crops and assuring the population of moderate temperatures in contrast to the heat and humidity of the coast and lowlands.

These same tectonic forces—the entire intercontinental system—are responsible for volcanic and seismic activity, both ancient and recent. A cross-section of the Andean region is seen in figure 2.1. For instance,

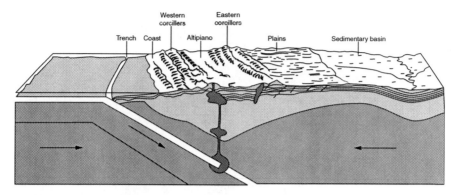

Figure 2.1   Cross-section through South America from the Pacific to the Amazon basin at the latitude of southern Peru.

Source: Simon Collier, Thomas E. Skidmore, and Harold Blakeman (eds), *The Cambridge Encyclopedia of Latin America and the Caribbean* (New York: Cambridge University Press, 1992) p. 15.

Paricutín arose out of a cornfield in central Mexico in 1943, El Chichón exploded in 1981, with the killing of approximately two thousand people. Most savage of all was the 1985 eruption of Nevado del Ruíz in western Colombia, which led to the loss of twenty-five thousand lives. Earthquakes pan the entire Pacific Rim. The 1939 Chile earthquake, Chillán, cost nearly twenty thousand lives; the 1960 Chile quake offshore was less destructive, but was followed by a *tsunami* (tidal wave) over much of the Pacific area. Mesoamerica is also highly unstable; the most costly seismic events were the 1972 Managua and 1985 Mexico City quakes—both killed an estimated ten thousand people.

Eastern South America is marked by extensive highlands in Venezuela and Brazil. More than half of Brazil is a highland or plateau with an escarpment of almost ten thousand feet just west of Rio, providing the majestic background to beaches such as Cococabana and Victoria. Over half of Brazil is a highland, generally at an elevation of two thousand feet or higher—the setting of Brasilia and São Paulo (the second largest Latin American city after Mexico City).

Plateaus, Valleys, and Plains

The Peruvian–Bolivian *altiplano* is, outside Tibet, the largest high plateau area of the world. At over 12,000 feet, La Paz is the world's highest capital; nor are Quito at 9,300 and Bogotá at 8,720 far behind. In Bolivia and south central Peru half the population live at 10,000 feet or above. Most newcomers to the *altiplano* experience *siroche* (dizziness and nausea) for a day or so after their arrival. The early Spanish settlers complained of sterility, and even today this towering plateau is associated with a high mortality rate because of low food supply, pulmonary disorders, and dependence on coca leaves in order to still the hunger.

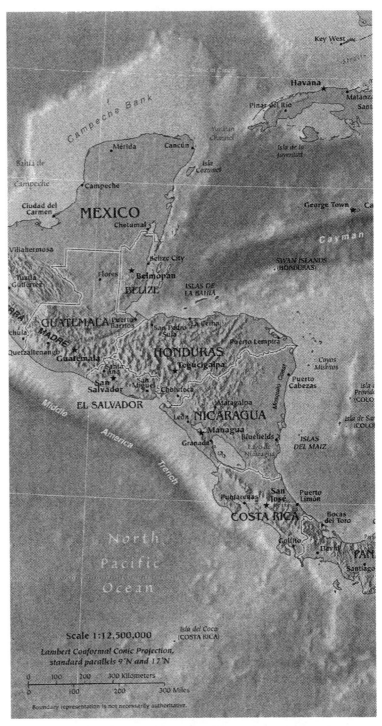

Map 2.1 Natural regions of Central America and the Caribbean.

On the other hand, the intermountain valleys are livable. The *conquistadores* sought these areas not only to avoid the heat and disease-carrying insects of the lowlands, but because the highlands were centers of the Aztecs in Mexico, Chibchas in Colombia, and Incas in Peru. Several major cities are located in these mountain valleys. Mexico City's estimated nineteen million habitants live in the Valle de México, roughly twenty by fifty miles at nearly seventy-three hundred feet altitude. Other cities such as Guadalajara, Puebla, and Oaxaca occupy a similar setting. Most major cities of Colombia are located well above sea level, with an even more rugged topography than found in Mexico. Today—as in the era of the Aztec, Chibcha, and Inca empires—peasants work the terraces and river valleys, where soil and rainfall permit a rich agriculture.

The advent of disease control and air conditioning eventually made the sea coast and lowlands more appealing. In reality, the coast of Brazil continually drew most settlers to that country until well into the nineteenth century. Also, the forts on the Spanish Main such as Cartagena, Colombia, protected the interior from foreign invasion.

Beyond the mountains are the plains. In time the humid pampa of Argentina yielded far greater riches than the mines of Mexico, Peru, and Bolivia, but as a much later development. Further south, the pampa, nearly as large as the American Midwest, is endowed with the rich chernozem or prairie black soil. Serious development of the area began with immigration in the latter part of the nineteenth century and the building of railroads. By the early 1900s refrigerated ships permitted Argentina to transport a third of the total meat export of the world. Immigration principally from Italy and Spain continued into the twentieth century. Englishmen, Frenchmen, Germans, Swiss, Poles, Yugoslavs, Russians, and people of other nationalities came in large numbers. Jewish refugees from Russia and Poland established eleven colonies, the first in the 1890s—a movement later augmented by German refugees during the Third Reich (1933–45). This colonization is, aside from Israel, a rare case of Jewish agricultural societies. Immigration from Europe declined during World War I but rose markedly in the early 1920s when the United States restricted the entry of migrants from southern and eastern Europe.

The pampa extends into both Uruguay and the state of Rio Grande do Sul in southern Brazil. Here as in Argentina the *bola*-swinging gaucho is dressed in wide pantaloons, a brightly colored poncho, wide hat, and drinks *maté yerba*. As a legend of the pampa, he has virtually disappeared, outnumbered by immigrants from a score of nations and closed in upon by the civilization the gaucho sought to evade.

Except for the pampa, cultivated plains are less evident in Latin America than in Europe or North America. However, the *llanos* (plains) of eastern Colombia and western Venezuela are vast cattle lands; intensive cultivation and industrialization would seem to be at least a generation away. For reasons of soil, climate, and other ecological factors the almost interminable Amazon plain is only partially ready for development. Even so, agriculture flourishes in its flood plain, which accounts for 2 percent of the basin, an area nearly equal to a major western European nation. Moreover, several broad river valleys with large flat surfaces, such as the Cauca Valley in Colombia, are highly productive.

Adjacent to the Amazon selva is the *sertão*, the northeast of Brazil, which suffers from marginal soil, low rainfall, and maldistribution of land. Extreme drought from 1992 to 1994 pushed even more than the usual number of migrants to São Paulo and other areas of the southeast. A cultural geographer would look at Brazil and see the ravages of climate, land tenure, economic cycles, along with the inevitable migration. Migration from the *sertão* surpasses all other population movements in that country.[4] This severe dislocation demands economic reform and organizational skills from both federal and regional resources.[5] Geographic disparities plague many nations, but the northeast is a particular source of perpetual anguish for Brazilians.

River Systems and Water Resources

Although not as long as the Nile, the Amazon river system carries the greatest flow of water in the world. Its basin includes not only northern Brazil but portions of five countries (Venezuela, Colombia, Ecuador, Peru, and Paraguay) as well, a total watershed almost as large as the continental United States Ocean liners sail into Manuas, nearly a thousand miles from the mouth of the Amazon. Smaller ships proceed another thousand miles upstream to Iquitos, Peru. Tides swell the river five hundred miles from the Atlantic Ocean. Also, the thick growth of the selva together with the leeching of tropical reddish latosols hinders full economic utilization of the Amazon basin. The area awaits further work by agronomists and sanitary engineers. In the meantime, settlement continues as seen in inset 2a.

INSET 2A   SURVIVAL ON THE AMAZON FRONTIER

Among the settlements on the Amazon frontier is Santa Terezinha, a village of nineteen hundred inhabitants, forming a U-shaped configuration with a *praça* (square) on the Rua do Comercio and on adjoining *ruas* (streets) are the rice cleaning plant, old Catholic mission, new Catholic church, Pentecostal church, school, and health clinic. In addition are stores, bars, poolrooms, and boarding houses.

Beyond the town are a few vast *fazendas* (farm or ranch) and numerous small farms with an average size of 100 hectares or 250 acres—relatively large as compared to peasant holdings in areas of better soil or greater rainfall. Migrants procured these plots by squatters' rights (*posse*). The land is usually used for pasture, yet crops of beans and manioc are major projects. Most of the farmers depend on *roça* (a slash and burn process), typical of tropical areas by which old plantings give way to new growth. Often the clearings are performed by communal work parties (*multirão*), but that tradition is less popular than it was.

Although some homesteaders obtain title to their land, most do not. Consequently they are at the mercy of the large *fazandeiros* (landowners), who occasionally react violently. Indeed, in 1972 the *fazandeiros* tried to bulldoze the health clinic. Other problems for the homesteaders are marketing, lack of access to credit, agricultural technology, and inadequate infrastructure such as roads. Natural disasters are another source of anxiety, Community development by the government, state or federal, is a distant goal.

In reality, *fazendas* often constitute agroindustrial bureaucracies. Besides their regular staff they employ wage labor on contract or peons who work on a seasonal basis for minimal pay (minus any debts incurred for equipment or fertilizers, as peons may have their own plots). Pay depends on the contractor's resources and integrity. The worker has little legal recourse for nonpaid wages. Moreover, at least a tenth of the squatters had one or more evictions.

Men in landless families are especially prone to irregular wage labor, the women work as domestic help seamstresses, laundresses, cooks, and the like. Many peons are single and fund shelter in boarding houses or drift about the bordellos. The more settled families look down on the peons, who may move on to new settings, usually in the hope of finding a boomtown. The hierarchy of labor, from independent farmer to regularly employed, with the peon at the bottom, determines social relations, including marital candidacy. Landowners generally oppose marriage of their daughter to peons.

Coping strategies are hard pressed. Most men move from job to job, digging, construction, road building as well as agricultural tasks. Skilled work pays better, but opportunities are scarce. Economic pressures require two or more jobs. Even storekeepers may seek part-time work in agriculture; farmers may strive to work on a *fazenda*, Economic instability greatly increases the migration rate. A series of push and pull factors are in evidence: climatic conditions, kinship, educational possibilities, and of course the lure or rumor of work elsewhere. The gap always remains between hope and reality.[6]

The second largest river system of South America is the Plata–Paraná–Paraguay, which drains an area almost half the size of the United States. The river provides the major artery for products of a vast hinterland in Argentina, Brazil, Paraguay, and Uruguay. The flow of water has been accompanied by a flow of migrants into both settled areas and tropical frontiers. The peopling of this region was among the most variegated of the New World: Jesuit missionaries in the early colonial period, followed by a score of cultural groups, notably Germans, Molokoners, and other religious dissenters from Eastern Europe. Today the Japanese are the largest immigrant group in Paraguay and like the Germans before them continue to transform a frontier. Highways, particularly in Brazil, are gradually superceding the Paraná River as the major artery to the outside world.

The Magdalena–Cauca river system in Colombia, the Orinoco in Venezuela, and the São Francisco in Brazil are all partly navigable and are major stimuli to agricultural expansion in their respective areas. As railroads are poorly developed in Latin America, airlines, trucks, and buses complement the river traffic.

An irony of these great river systems is the seeming inability to divert the flow into human use by more widespread irrigation and hydroelectric development. Latin America suffers from excess water in one region and deficiency in another. For dry areas evaporation usually exceeds precipitation. On that point, the lower the annual rainfall the more irregular it is. Most disturbing is the high rate of pollution (as measured by several indices including fecal coliform counts) in rivers, as it is twice as high as in most areas of the world.[7] Most developing countries are far from reaching the WHO (World Health Organization) norm of forty liters of clean water per inhabitant per day. Only the cone countries and Costa Rica offer satisfactory levels of water purification.

Climate: Contrasts but Few Extremes

Perhaps as much or more than surface configuration, temperature and rainfall shaped the early course of empire and the potential of human habitation. Again, we are struck by enormous climatic variations in Latin America, especially as altitude causes marked contrasts in a relatively small area. However, in some respects, temperature fluctuations are less severe than in Anglo-America.

Latin America's tropical location sets the general pattern of its climate. As over two-thirds of the area lies between the Tropics of Cancer and Capricorn, it has high annual temperatures but not the extremes as compared to other continents. It is rather the monotonous lack of relief than extreme heat of the equatorial zone that is unattractive. Actually the hottest temperatures of the hemisphere occur in the middle latitudes of the northern hemisphere during the summer, that is, the North American Southwest. Although the average temperature is over 80 degrees for each month in the Amazon basin and a few coastal lowlands, only a limited area of Latin America has temperatures of 110 degrees, namely Paraguay, northern Argentina, and northwestern Mexico. In comparison, much of the outer Midwest, from Texas to South Dakota, is subject to these temperatures.

Despite the advantage of a long growing season, life in the tropics means a loss of efficiency and the dismal prospect of watching the sun set nearly the same time each day throughout the year. Still, mountains and high plateaus offer an escape from the heat by changing altitude. *Bogotanos* can descend eight thousand feet in a couple of hours to thaw out at resorts on the Magdalena River, and Mexicans flee the seventy-three hundred feet level of their capital for Cuernavaca, all to change air pressure as well as temperature. As another example, Quito at nine thousand feet and Quayaquil at sea level enjoy twelve hours of daylight every day of the year, but because of altitude a twenty-four degree difference in temperatures separates the two cities with little variation from January to July—only one degree in Quito! In tropical zones the labels "summer" and "winter," if used at all, refer respectively to the dry and the rainy seasons.

Latin America has a smaller land mass than does North America, at least at the higher latitudes. Consequently a continental winter, as compared to the United States or Canada, is unknown. Still, Buenos Aires experiences the same winter temperature and a somewhat lower summer temperature as compared to Charleston, South Carolina. On the whole, in southern Brazil, Uruguay, and the Argentine pampa prevails a climate similar to the southeast United States.

The gradation of climate on the West Coast of South America repeats with some variation the pattern of the West Coast of North America. The equatorial area with its selva in Colombia and Ecuador rapidly gives way to desert conditions in Peru and northern Chile. This fifteen hundred miles of desert stretches further in latitude than does the desert coastal area of northwest Mexico. Moreover, the extremely cold Humboldt Current brings fog but no rain to the coast. On the other hand, central Chile has a Mediterranean type of climate, topography, and vegetation seemingly identical to California. Further south the climate becomes more like the Pacific Northwest and British Columbia with its fjorded coast, Tierra del Fuego being at the same latitude as Juneau, Alaska. Patagonia, in the rain shadow of the Andes, is one of the few examples in the world of an upper-latitude desert and steppe adjacent to an ocean and located on the east side of a continent.

Again, altitude forms the climatic character for Middle America and the Andean region. For Ecuador and Peru it is common to speak of three regions as based on topography and climate: the *costa, sierra,* and *montaña,* as one moves from the coast to the Andes to the Amazon basin. Three zones are conventionally designated for the Latin American tropics:

1. *tierra caliente* (hot country), below 4,000 feet, has temperatures throughout the year above seventy-five degrees (twenty degree centigrade). Sugar, bananas, pineapples, and cotton are the usual crops.
2. *tierra templada* (temperate zone), 4,000–6,500 feet, has average annual temperatures below seventy degrees and is the zone of coffee, various fruits, and cotton.
3. *tierra fria* (cold country), 6,500–10,000 feet, with temperatures fifty-five to sixty-five degrees—wheat, potatoes, and beans, in other words, products comparable to northern climes.

Corn and certain other crops are adjustable to almost any elevation. On the other hand, the forbidding *páramo* exists above the tierra fria. In these sparsely vegetated meadows little cultivation is possible, and they serve primarily as pasture land for the alpaca, vicuña, and llama. On the whole, divergence characterizes this vertical landscape and the combination of high mountains and infertile jungles make for crises in food production. The necessity to import cereals in Bolivia, Ecuador, and Peru places a strain on the national budget.[8]

The dramatic effects of altitude are demonstrated in northern Colombia, where the Sierra Nevada has a permanent snow line at fourteen thousand feet in elevation, directly above the tropical rainforest, yet on the map the two zones are only several miles distant. In further contrasts a few hundred miles to the east lies the rainless desert of the Guajira Peninsula. Rainfall differs greatly with varying altitude, as the eastern side of the Peruvian Andes demonstrates. A heavy rainfall falls at a high elevation as compared to four thousand feet below producing a striking transition from tropical forest to cactus. From the human side, differences in altitude and precipitation are related not only to crop diversity, but to clothing, building styles, and general habitability.

Despite a generally milder climatic picture as compared to the northern hemisphere, Latin America can suffer from severe meteorological events. Hurricanes frequently ravage Mesoamerica and the Caribbean with fatal effects. Some ten thousand persons were killed in Honduras and Nicaragua in the 1998 hurricane bringing floods and mudslides to whole villages, followed by the 1999 floods in coastal Venezuela and the island of Hispanola (Dominican Republic and Haiti) in 2004. Similarly, great destruction occurs with torrential rains, as in the 1920s when whole archeological zones in coastal Peru were washed away by what was decades later identified as El Niño. Similarly, El Niño ravished the ordinarily parched coast of Peru in 1994 and 1998.

Regionalism and Resources

In Latin America, mountains, rain forests, and deserts pose serious obstacles to development. Economic investments and extensive technology are struggling in the area to support present and future populations. The problem will become a crisis with the predicted near doubling of population over the next few generations. Latin Americans, like other inhabitants of the planet, often have a parochial perspective. Depending on the time and place, this outlook assumes local, regional, or nationalist lines. The isolation of large areas led to this provincial attitude.

Territory and Settlement Patterns

Except for several regions Latin America is at present not crowded. A half-billion people occupy an area five times the size of Europe (excluding Russia), which has a population of over three hundred million. Inevitably, the population is highly scattered; Bolivia has an average density of one inhabitant per square kilometer, as compared to El Salvador with twenty. In most countries several clusters constitute the greater part of the population. Often these agglomerations arise from historic, economic, and geographic factors. Certain regions are marked by intense cultivation, as in the highlands of Guatemala, Antioquia in Colombia, central Chile, and sugar plantations along Brazil's northeast coast. Also, oases exist in comparatively arid regions, examples being Tucamán, Argentina or Saltillo, Mexico. In rare instances, an intensive agriculture permits dense populations on both sides of a national border, as between Guatemala and El Salvador and between Colombia and Venezuela. The more familiar pattern is one of a wide dispersion of the population toward the frontier of the country.

This demographic sparseness combined with the cluster type of settlement points to the question of territoriality. The "total national territory," or the actual boundaries of the nation, can be contrasted with "effective national territory," which alone has demographic or economic significance.[9] This disparity is a factor in the history of many territorial disputes in Latin America. Moreover, structuring of the effective area can be critical in the integrity of a country. For instance, the unwieldy length of Chile does not prove to be a serious barrier since 90 percent of its population lives within four hundred miles north and south of Santiago. National sovereignty is further vindicated by the towering Andes along its border.

Transport: Accelerated Change

Previous to the advent of railroads and steamboats, settlement of the vast territory was extremely slow. Even today, the transportation system, however improved with massive road construction and airline expansion, remains a problem. Its railroads are relatively underdeveloped, not surprising in view of sparse settlement and the enormous cost of railroad building. For instance, the northern Andes are pierced only by the Peruvian Central Railroad with its standard gauge between Lima and Cerro de Pasco, its summit being at nearly sixteen thousand feet above sea level. More than in other areas of the world, railway construction moved ahead despite enormous costs because of the anticipated gain of bringing mineral and other riches to ports bound for overseas—all to the advantage of Europe and North America.[10]

Brazil, Chile, and Mexico each have over seven thousand miles of railways, or over half the entire railroad mileage of Latin America. As elsewhere, Mexico turns mainly to motor traffic, but railroads still transport a large portion of the freight (map 2.2). With 31 percent of Latin Amercia's railroads, Argentina could be compared with the world's more advanced countries. However, as with the Americas in general, from the 1970s onward railroads gave way to other forms of transport. As in the United States, passenger travel is more restrictive than it was in the mid-1900s.

In contrast to rail is the advance in road travel. Road mileage has grown ninefold since 1960—twelve hundred thousand miles in Brazil and one hundred and seventy thousand in Mexico. Ecuador alone has over nineteen thousand miles of paved roads. The Pan American Highway, stretching from the Rio Grande to Panama, is paved except for a gap of some two hundred miles between Panama and Colombia. From there the highway is traversable to southern Chile, east to Buenos Aires, and finally to Rio de Janeiro and beyond. Amazonia and Mato Grosso, once almost impenetrable, are crossed by two major highways—not without cost to their ecology and indigenous population—a reminder of the destruction North Americans brought to their frontier. In order to facilitate trade and commerce, the cone countries are especially eager to integrate their road network, as they did a century ago with rail. Not only trucks but buses ply highways in all the countries. The 1990s brought on a vast improvement in the advent of luxury buses between cities and between countries.

Another revolution is the rise of automobile manufacturing and assembly plants, which first appeared in the 1960s. Formerly only the upper and upper–middle classes could afford the luxury of a car. By 1990 one Mexican in twelve, one Brazilian in eight, and one Argentine in seven owned a car. Airways, established in the 1920s, markedly expanded after World War II. In view of the shortage of railways, air travel plays an even more important role in uniting a nation than in North America or Europe. Air travel is critical in linking the capital with other cities and towns, usually in a radial fashion rather than by connecting outlying centers with each other.

Resources: Assets and Liabilities

At least a third of the Latin American economy is its agriculture. Approximately 30 percent of its workers still till the land as compared to 3 percent in the United States. Historically the Aztec, Mayan, and Inca civilizations were based on cultivation of the land, and this tradition remains. It is a surprise to many Europeans and North Americans to realize that over half the food on their tables is of New World origin—corn, potatoes, tomatoes, among other products, in addition to tobacco and coffee. Other exotic items include cacao in Ecuador, *maté* in Paraguay, in addition to other resources such as mahogany and various precious woods in Central America and the Amazon. Moreover, Europeans contributed their products—wheat, rice, cattle, sheep, and vineyards. Exportation of beef from the Plata and coffee from Brazil, Colombia, and Central America brought to those countries more wealth than did the gold and silver of colonial times. Yet, rich as the land may be, few Latin Americans gain equitably from its resources. As in colonial times, minerals and farm products are still traded for manufactured goods from abroad.

The best terrain is assigned to the cultivation of coffee, bananas, sugar, and other export crops. Except for the pampa, cereals and dairy production are poorly developed. Production is far less efficient in labor input than in Anglo-America, and less intensive than in Europe and Asia. Fertilizers, crop rotation, irrigation of the drier lands, and lack of mechanization are perennial obstacles. Millions of subsistence farmers struggle to survive against drought and floods.

Still more tragic is the peon's depressed social status, the extremely low economic return, the tenancy situation, and acute land parcelization. In most countries two desperate extremes of land ownership persist: latifundia and minifundia—vast estates and subsistence plots.

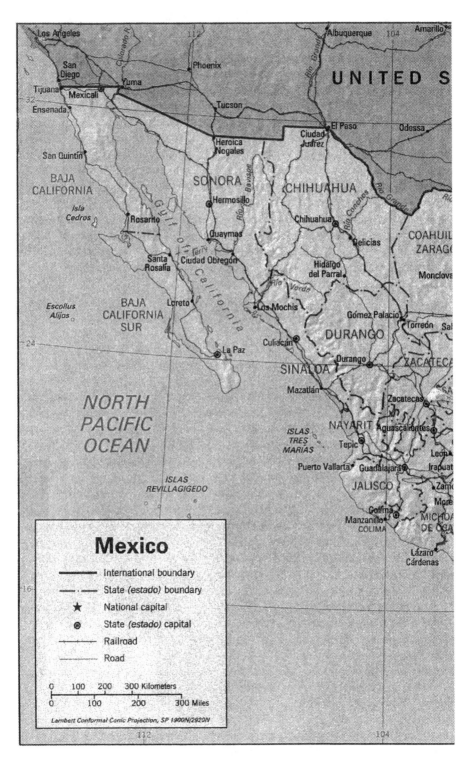

Map 2.2 Transport arteries of Mexico.

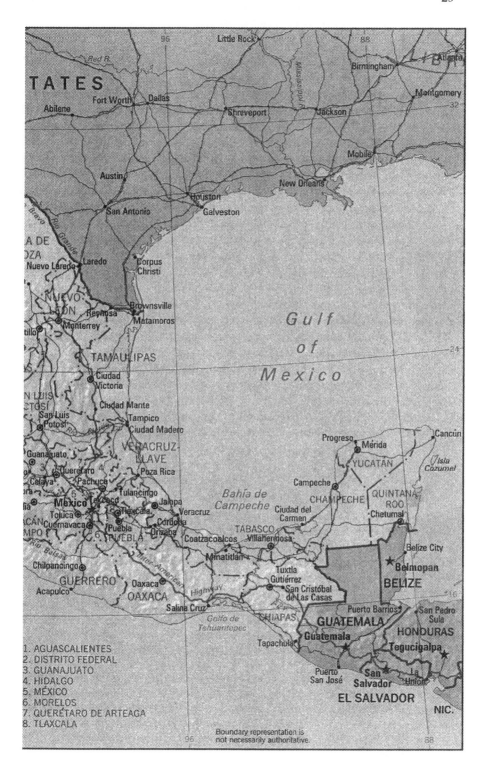

1. AGUASCALIENTES
2. DISTRITO FEDERAL
3. GUANAJUATO
4. HIDALGO
5. MÉXICO
6. MORELOS
7. QUERÉTARO DE ARTEAGA
8. TLAXCALA

Boundary representation is
not necessarily authoritative.

Forests, Ecosystems, and Global Change

Over the last century, population growth, increased agriculture, and industrialization have altered our planet's ecosystem. For example, the *conquistadores* brought sheep grazing to Mexico and the Andean countries resulting in environmental degradation—deterioration of vegetation, soil, and the water regime.[11] Ecological devastation appears to increase geometrically in our global corporate economy of the twenty-first century. Further, this impending tragedy appears to surface in the world's consciousness only in selected quarters. In reality, both preliterate and developing societies over several centuries destroyed forests in order to provide pasture and agriculture—not to mention housing and newsprint. Today the less developed and populated areas of the earth are called upon to restore an ecological balance. As this process, notably pollution by hydrocarbons and other chemicals, moves on a logarithm the ozone level is seriously affected. Ironically, the United States and other excessively polluting countries are lecturing Latin America on the need to preserve its forests. In comparison with areas such as the Amazon, Mexico has made reasonable progress in preserving its forests. The *ejido* system with its *campesino* (peasant) cooperatives reserves a sizeable portion of the forests from commercial lumbering. Although the balance sheet is not ideal, at least peasant organizations along with intellectuals and enlightened public officials realize the long-term consequences of forest neglect and profit making.[12]

Latin America contains immense rain forests, notably in Central America and Brazil. As indicated earlier, the Amazon rain forest, the world's largest, is rapidly being cut back for highways, settlements, questionable farming pursuits, and exploitation of its hardwoods. However, irregularities occur (inset 2b).

INSET 2B   INDIGENOUS RIGHTS VERSUS VESTED INTERESTS: MAPUCHES IN CHILE

The Mapuche Indians have occupied a large area of south central Chile, long before the arrival of the conquistadores. In 2004 a new crisis flared up in their struggle for autonomy when they defied the government over the claims to their forests. This Mapuche of over a million were subdued by a European population in the 1880s and assigned to reservations similar to the U.S. "solution."

Over the twentieth century they saw Chileans usurp their forests for both farmland (often going to immigrants from Germany, Italy, and Switzerland) and for commercial lumbering with exports of lumber to Europe and the United States. Beginning in the late 1990s the Mapuches have destroyed lumbering operations and burned several farmhouses.

As a result, officials resurrected a law from the Pinochet dictatorship (1973–89) in order to prosecute eighteen Mapuches, including several leaders. Human rights groups protested what they label as a violation of justice, Chilean president Ricardo Lagos himself raising questions about the treatment of the indigenous. The Mapuche struggle is one more episode in the worldwide campaign for rational ecology and the struggle of "have-nots" versus the "haves."[13]

Deforestation occurs at the annual rate of forty thousand square kilometers per year in South America, mostly in the Amazon—probably a loss of 18 percent in the last quarter of the twentieth century. Landowners simply ignored the law that a given portion of the property should remain as forest. The most serious ecological damage occurred during Brazil's military dictatorship (1964–85). Even after the restoration of democracy a system of tax credits permitted mining and agribusiness to expand. Much of ecological movement can be attributed to Carlos Menges, who agitated for ending the destruction by the lumbering and other industries with governmental connivance. His murder dramatized the need for serious action as represented by the 1994 international conference on conserving the Amazon rainforest. As a result, environmental protection networks and nongovernmental organizations (NGOs) are active in protecting the fragile ecology and the tragically endangered indigenous population.[14] In order to rectify the damage from lumbering, mining, and other commercial enterprises, ecologists urge various solutions, including intensive research such as the "Biological Dynamics of Forest Fragments" project.[15] In this context, because of the loss of flora and fauna, including many fish species, habitat conservation is essential. Reserve areas are being established but hardly at a rate to keep pace with destruction. In order to protect the soil, agrodiversity is also necessary. A mix of native and exotic plants is permissible to expand food production. However, importation of certain alien plants, particularly if it is in the floodplain and if chemicals are used, is risky. Foremost is preservation of the forest.[16] Otherwise the result is less oxygen for the urban populations and insufficient water vapor for the rainfall potential over the hemisphere.

Deforestation in other parts of Latin America is even worse—Haiti has only 2 percent of its original forests intact. Costa Rica lost 40 percent of its forests; in the 1970s the country had possibly the highest rate of deforestation in the world. However, by the 1980s a laissez-faire attitude gave way to interventionism, with finally a hybrid approach as necessitated by economic realities, including public debt.[17] Costa Rica is committed to effective planning with the development of national forests and parks. Tropical rain forests are incredibly endowed with fauna as well as flora—Costa Rica, the size of Ohio, has more species of birds than all of the United States Ecotourism in Costa Rica permits an economic return as well as preservation of its national treasures such as the quetzal bird.[18]

Southern Mexico offers another example of the dire effects of human deprivation of ecosystems as well as the intricacies of global change. As in the Amazon there is the ecological impact of oversettlement, ill-fated agricultural schemes, and the invasion of multinational companies. Sustainability of the ecosystem became ever more questionable with deforestation.[19] Public awareness of the problem arrives slowly. For instance, according to a survey of Chiapas, Mexico, inhabitants are marginally conscious of fewer wild animals, climatic changes, and more frequent floods, but less than 7 percent mention deforestation. Blame is placed principally on the government for its lack of foresight and its encouragement of developers. Few residents or governmental authorities see the long-range implications of deforestation and perceive the situation as local concerns rather than as the national or global future.

The case of Argentina, to cite another disaster area, shows a lengthy history of environmental degradation. From wealthy land barons to recent industrialists, abuse of the landscape continues. It may well be that the poor farmers have by lack of crop rotation, inability to buy soil nutrients, and faulty irrigation systems caused the greatest destruction. Guilt actually rests with a range of factors—*campesinos* desperate to survive, cattle barons, state-owned industries, transnational entrepreneurs. Interestingly, what attempts were made to reverse this trend met with resistance by the military dictatorship (1976–83), which persecuted peasant organizations along with other perceived threats to the status quo.[20]

The picture is not altogether hopeless. For instance, into the Ecuadorian Amazon a resettlement project moved refugees from the 1887 earthquake zone. Also, the government regulated land use in order to protect the forest and provide biodiversity. Although the end result is not yet certain, the report stresses the need for establishing protective areas.[21] Further, developmental strategies must involve the indigenous population in the protection of the natural environment.[22]

Reforestation is a major concern on all continents. Chile serves as an example. Since the 1930s the government has encouraged the marketing of wood products with an uneven policy of replacement. Despite the harsh, neoliberal economic policy of the Pinochet dictatorship (1973–89), the government adopted stringent policies against the multinational wood and paper industry.[23] Today, replacement is a key goal, with over a hundred thousand hectares planted every year. Chile's democratic regime is committed to ensure a healthy environment concerning pollution and protection of natural resources, particularly its forests.[24]

Subsoil Resources and Their Potential

Besides agriculture, mineral resources are the area's major means of contributing to the world market. The search for fortunes in precious metals brought Spaniards and Portuguese to the New World, and mining in various forms still constitutes a principal, if declining, medium of securing foreign capital. This dependence on one or two essential commodities is precarious for a given nation. A drop of a few cents per pound in the world copper price can cost Chile millions of dollars as can an equivalent fall in the coffee price for several countries. Also, synthetic production can be fatal, as demonstrated by the history of nitrates in the 1920s when Chile was no longer the almost exclusive supplier of that chemical. As another example, exploitation of rubber in the Amazon basin has a bleak history—competition from Malaya early in the twentieth century and introduction of synthetic rubber during World War II.

Most impressive is the wealth of resources lying under the soil of Latin America. Fortunately more than unfortunately, this abundance has only begun to be tapped. Brazil and Venezuela have two of the world's largest iron deposits, and huge reserves are believed to exist in all the Andean countries. Bolivia has an oil bed presumably surpassing that of Venezuela, development only recently underway, as with oil and gas reserves on the eastern side of the Andes in Peru. In Zipaquirá, only twenty miles from Bogotá, are salt deposits several hundred feet in thickness. In regard to power resources, Chile has only partially developed its hydroelectric potential. Brazil with its escarpments near its coast has the fourth largest electrical capacity of the world.

For the most part, these riches lie dormant because of seemingly insurmountable barriers—lack of capital, entrepreneurial skills, and technical personnel, poor endowment in coal, transport, among other variables. However, steel plants are in operation in several countries, Volta Redondo in Brazil being the largest. In addition to the problem of overland transport, the continent is deficient in natural harbors. Most West Coast ports are artificial. Even Valparaiso as a port depends on a breakwater for protection against storms. Other harbors are even more hazardous. The East Coast is slightly better supplied, yet the dazzling harbor of Rio de Janeiro is separated from the interior by an escarpment. The giant port of Buenos Aires must be intermittently dredged, as with other Argentine harbors. And as the third largest urban center of Latin Americas, Buenos Aires finds itself poorly supplied in fuel and power resources. Since the air age these problems are less serious; still, certain products can only be shipped by land and sea.

Globalization and Geopolitics

Any discussion of Latin American resources must be placed in the context of the economic and political setting. For one thing, we must remember that there are several Latin Americas. Besides the specific geographic and cultural contexts are marked economic differentials. This diversity is a product of sociopolitical developments, differential freedom from colonialism, and a complex infrastructure (especially educational opportunities as well as local resources). Economic philosophies are also relevant, as we shall see in more detail in chapter 4.

From the beginning of the twenty-first century a new cultural and socioeconomic framework is changing our geographic concepts. The meaning of time and space is redefined. Cyberspace, rapid communication, and transport, all make older notions of space and territoriality inadequate. We still consider the nation state as central but it is cut through with cross-national alliances. The individual's sense of self as bound to a given territory is now replaced by a host of identities, as will become clear in chapter 5. We no longer think of a two-dimensional grid, but of multiple networks—economic, political, professional, recreational, to mention a few.[25] In other words, the idea of the Global Village implies both micro and macro dimensions and assumes an uneven equilibrium between larger and smaller states.[26] For instance, the United States assumes a geopolitical approach as the world superpower, as in its embargo placed on Cuba with resulting food shortages for that island's people.[27] The U.S. geopolitical outlook with a cavalier attitude is only intermittently directed to Latin America, as will become evident in chapter 11.

Even before globalization our political–economic scheme could be approached as a *world system* in which three areas are identified: core, semi-periphery, and periphery.[28] Historically, Latin America remains between the periphery and semi-periphery, but even within Latin America certain core areas are salient, for instance, Mexico City, São Paulo, and Buenos Aires, with provincial cities as semi-periphery, and rural territory at the periphery, even though São Paulo is rapidly becoming a core area in the world system. The capitalist superpowers in Europe, North America, and more recently Asia are in a continuous rivalry, although a global economic network is changing the texture of this conflict. The formal imperialism of the colonial powers has given way to an informal imperialism, based on unequal exchange.[29] The concept of free trade, as in the North American Free Trade Association (NAFTA), eases the flow of goods and services, but the core areas, for instance the United States, import clothing made in sweatshops around the Third World, including Latin America, selling the items at twenty–thirty times what the worker receives. Consequently, opposition to globalization surfaces throughout the world as violent demonstrations beginning at a 1999 conference in Seattle, later spreading to a Genoa meeting and to other cities. In its desperate attempts to move beyond colonialism, Latin America first turned to ISI (import substitution by industrialization) as early as the 1920s, but could never overcome its disadvantage of supplying primary resources to the capitalist world. Through most of the twentieth century Latin America struggled against the yoke of economic colonialism, but was never able to overcome its disadvantaged position. With the fiscal collapse of the 1980s political and economic elites sought refuge in the global market, driven by neoliberalism. However, this globalization disrupts the local economy and represents colonialism under another guise.[30] Not least, global economy pushes nations into industrial expansion and exploitation of natural resources at a rate for which they are not prepared and with questionable ecological consequences.[31]

Especially difficult is the adjustment that the indigenous must make to global processes. Traditional cultural traits from land utilization to language and handcrafts have fallen to the needs of standardization and mass production. Protest movements appear in several countries, most dramatically in Colombia and Peru.[32] Beyond the economic effects of globalization are other institutional aspects, as suggested in inset 2c. At first glance, the world would seem to be more homogenized, except for the underclass who fall beyond or outside the global market and U.S. cultural imperialism.[33] Or one might rephrase the question: to what degree is society and culture becoming standardized, polarized, or hybridized? For instance, in the 1970s and 1980s, Latin America depended on the United States for 25 percent of its imports, whereas the United States in return imported less than 5 percent. By the 1990s Brazil and Mexico, for example, were approaching equity in their trade with the United States.

INSET 2c   TERRITORY AND THE NATION-STATE

What does globalization mean to the nation state? Nationalism has been a major cultural and political force from the late nineteenth century to the present. Also, as national identity is regarded as sacred, globalization makes for anxiety, if not panic, in many sectors.[34] The military, for one, is sensitive to national security, both internally and externally, and often intervenes—violating civil rights and international norms of conduct—on the pretext of "national security."[35] Anyone who crosses a national boundary in Latin America knows the sensitivity surrounding the concept of national territory. One variation is the U.S.–Mexican boundary with its fluidity of entry, at least in the southern direction. Over half the *maquiladora* plants (assembly-line operations employing cheap labor) were established in or near Mexican border cities, but more recently have spread to other parts of the country.

As another aspect of economic globalization, deterritorialization means spatial reorganization and for Latin America it represents the twilight of colonization. Space or territory translates into the commodity of power. This last century, as with centuries before, saw a race for conquest, military and nonmilitary, throughout the planet. As a result, a score of border disputes arose in Latin America—Argentina–Chile, Ecuador–Peru, El Salvador–Honduras, to mention a few. In this context we return to Amazonia for an instructive scenario of territorial development. In Brazil, because of the sugar trade, settlement was confined to the northeast coast and the immediate interior, but industrial and agricultural growth turned to the south. By the time of Getulio Vargas's Estado Novo (1937–45), more grandiose aims surfaced. In 1935 appeared GeraldoTravaassos's book *Projeção Continental do Brasil* with its call for expansion into the Amazon in order that the country should achieve new greatness. Development moved slowly until the military government (1964–85) expanded highway construction followed by economic exploitation.[36] Of the other nations coveting the Amazon, Ecuador and Peru were at loggerheads for decades in the attempt to control their portion of the basin. The boundary dispute was finally settled in 1999. The question of space and the nation state is now engulfed in an even larger domain. The global now supercedes traditional international norms.[37] Export-led development replaces economic nationalism. Regional and interregional pacts such as NAFTA are transforming the local and national economy throughout the region with ambiguous results, as will appear in later chapters. Industrialization and modernization remain in the vocabulary but they lead to multiple power relationships. All this brought on an acute expression of people power during the 1990s, notably in Chiapas, Mexico.

A Note of Caution

We must be cautious about applying any ready-made model of modernization to Latin America's social and economic development. Too often we think of these countries as feudalistic structures to be changed by the introduction of industrialism ushered in by economic reforms, copious amounts of foreign capital, and technical aid, One aspect of economic development is rampant urbanization—no less inevitable in the context of globalization and cyberspace than it was with earlier industrialization.

Latin American leaders still talk glibly of forming a technological elite who are to promote a "modern," global capitalistic state. This bias often bypasses other societal values. Social and personal distress, such as unemployment and labor turnover, follow globalization. More subtle effects also occur. For instance, psychologists report personality strain among students in U.S.-sponsored college programs when introduced to entrepreneurship, management techniques, and strict time limits as opposed to intellectual pursuits and other traditional values as an end in themselves. This problem is declining as modernization of the economy is almost universally accepted.

Most Latin American leaders still regard economic progress as their primary need. The determination to end colonialism and North American domination may eventually be realized by following a sequence of development not uniquely different from what occurred in Western Europe and the United States during the nineteenth and twentieth centuries. Industrial maturity, to whatever degree it is desirable, can be achieved in less time than what it took Europe and North America. However, the global environment will mean different options and challenges. Again, one must be wary of superimposing an extraneous model for development. Each nation, hopefully in concert with its neighbors, should find its own solution.

Chapter Three

An Unfolding Heritage

We have just seen the kaleidoscope that Latin America represents geographically. Its history and ethnic structure are no less diversified. The development of Ibero-America grew out of Indian cultures upon which Spanish and Portuguese feudalism was superimposed in the colonial period. Notwithstanding the importance of other influences, the mix of indigenous and Iberian institutions remains of fundamental significance into the present. The following pages reveal Latin American history beginning with the colonial and national periods; they are examined in order to understand the remarkable changes of the twentieth century.

The Conquest and Its Aftermath

First of all, one may ask how some six hundred men under Cortés could have conquered a society of several million Indians residing in central Mexico, or how less than two hundred men under Pizarro were able to subjugate the highly organized Inca social order. Likewise, the disorganization caused by imprisonment and murder of the Inca emperor proved disastrous to the survival of that empire. A fundamental problem was the over-centralization in New World empires. By the time the Incas reorganized to offer effective opposition, the Spanish were already in control of crucial areas of the empire. Somewhat the same tragedy had occurred in Mexico. Cortés' status was aided by the Toltec mythology that a supernatural agent of light complexion would appear some time.

The Spanish social structure, reflected in its colonial conquests and aspirations, had vague similarities to the cultures of Mexico and Peru. Among other features were a caste-like division of the society, an economy based on agriculture rather than on manufacturing, preoccupation with military aggrandizement, and unyielding religious zeal. Yet, these similarities should not conceal the immense differences. Spain was a coalition of feudal and privately owned estates on which a nobility and a system of royal patronage were superimposed on landless peasants, in contrast to the communal type of agriculture the conquistadores encountered. The new lands provided a repository of crops for both internal consumption and the Iberian market. However, these agricultural concerns were minor compared to the lust for gold and silver that prompted the conquest of the New World. The directions and routes of the Spanish and Portuguese are shown in map 3.1.

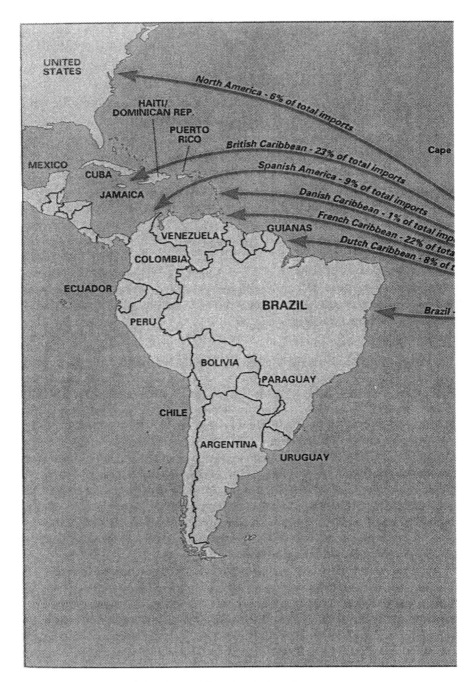

Map 3.1 Patterns of the slave trade to Latin America.

Source: Simon Collier, Thomas E. Skidmore, and Harold Blakemore, *The Cambridge Encyclopedia of Latin America and the Caribbean* (New York: Cambridge University Press, 1992), p. 139.

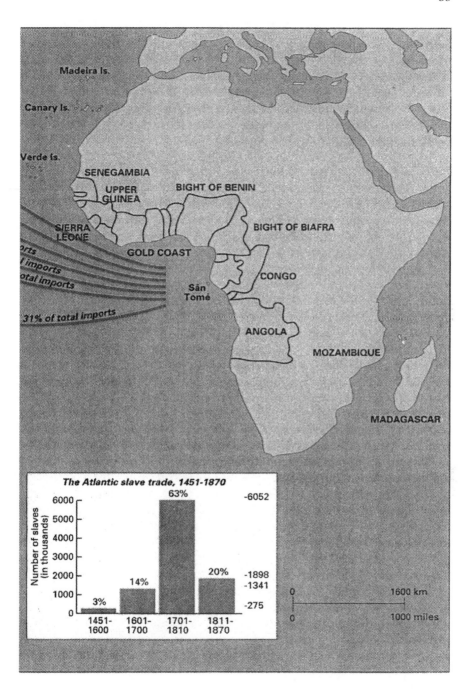

Madeira Is.

Canary Is.

Verde Is.

SENEGAMBIA

UPPER GUINEA

BIGHT OF BENIN

SIERRA LEONE

BIGHT OF BIAFRA

GOLD COAST

CONGO

Sân Tomé

ANGOLA

MOZAMBIQUE

MADAGASCAR

rts

l imports

otal imports

31% of total imports

The Atlantic slave trade, 1451-1870

Number of slaves (in thousands)

63% -6052

20% -1898
-1341

14%

3% -275

1451-1600   1601-1700   1701-1810   1811-1870

0   1600 km

0   1000 miles

Although both societies—indigenous and European—were stratified, the contours changed drastically after the Conquest. The rigidly organized Aztec socioeconomic system was only marginally articulated with the Spanish colonial world. A new caste structure based on ethnicity and social origin emerged. The various mixtures constituted the *castas* with an elaborate nomenclature to identify the specific ratio of ethnic origin, a structure that endured through the eighteenth century. This hierarchical system became ever more glaring with the slave trade placing Africans in nearly every region of the hemisphere with an especially heavy concentration in Brazil, the Caribbean and the southern United States.

In the indigenous empires religion provided the symbolic props for integrating the society. The Catholicism of the Spanish warriors, specifically the clergy and aristocracy who followed in their wake, permitted no compromise. The church rationalized a societal structure for the benefit of Christians. Still, concessions did occur. The Spanish outlawed worship of the sun god and chief deities of the Inca, but allowed them to continue their rituals in regard to lesser deities and sacred places. In fact, Roman Catholic rituals and beliefs could readily blend with the magical practices of indigenous peoples from southern Mexico to Bolivia, as visitors observe at Chichicastenango, Guatemala, or Cuzco, Peru, today.

The most tragic impact of the Spanish conquest was the decimation of the Indian population. It is estimated that of the eleven million Indians in Mexico in 1519, only two million remained in 1607. The population of the central Inca realm was reduced from approximately 3.5 million to 1.5 million within forty years after Pizarro's conquest. For all of Latin America estimates of population loss between 1492 and 1570 vary between 50 and 90 percent.[1] As noted in chapter 1, the principal cause of this population decline was the spread of European diseases, for which the natives had no immunity. Clusters of the indigenous today are shown in table 3.1.

With the introduction of Western agricultural features—cattle and the horse—lands previously cultivated were converted to pasturage. European crops were introduced, principally wheat, vegetables, and fruits. Moreover, Indians were forced into the least desirable lands, a pattern continuing to this day. Forced into hard labor, especially in the mines, under intolerable conditions, the natives barely survived. In certain areas the indigenous who resisted encroachment of Europeans were exterminated. However, as compared to Anglo-American colonies, the system integrated the indigenous, if indirectly and incompletely, into the total society, as illustrated by the miscegenation between Europeans and Indians. Further, cultural continuity was another factor, as a number of institutions testify to a similarity between indigenous and Hispanic. For example, the pueblos of the eighteenth-century Mexican highlands operated as highly stratified landed estates dominated by a few elite families often referred to as *caciques*—the basis of the colonial social structure.[2]

Spanish exploitation stood in contrast to English domination in North America, however unfavorable the effects of both on the native population.

Table 3.1 Indigenous population based on 1980 estimates

Country	Population	Percentage of total population
Argentina	360,000	1.1
Bolivia	4,150,000	56.8
Brazil	225,000	0.2
Chile	550,000	3.2
Colombia	300,000	0.9
Costa Rica	36,000	0.9
Ecuador	3,100,000	29.5
El Salvador	1,000	0.0
Guatemala	3,900,000	43.8
Honduras	110,000	2.1
Mexico	12,000,000	14.2
Nicaragua	49,000	1.2
Panama	99,000	4.1
Paraguay	80,000	1.9
Peru	9,100,000	40.8
Venezuela	10,000	0.8

Source: Adapted from *Statistical Abstract of Latin America* (Los Angeles: University of California, 2002), p. 171.

The Anglo-Saxons arrived in North America as intact families (at least committed to that goal with little miscegenation) dedicated to the ideal of individual religious freedom under a civil regime, if theocratic in character. This restless society, continually expanded by new recruits, had even more disastrous effects on the native population than did the conquistadores to the South.

The legacy of the Indians is incalculable. Beyond the New World food items are medicines and herbs. Not only the products survived, but the Indian village market of Mexico today is a direct carryover from Aztec and other cultures. To some extent, Hispanic American culture can be described as an integration of Old and New Worlds. Building styles are such an instance. Whereas rural housing remained predominantly indigenous, preference in the city was for the Spanish style of joining one house to the next and orientation around a central patio. The pre-Colombian pattern was almost invariably detached dwellings. Because of its dubious functionality and healthiness, many Latin Americans question the conglomerated urban style of the Spanish.

Occasionally a cultural synthesis was reached; yet in one sense the two worlds never really came together. The reservation system for the indigenous in Brazil, Chile, Argentina, and in the United States is a reminder of intransigence on both sides, but mainly vengeance of the conquerors. The *ladino* in Guatemala and the *cholo* in Peru—each region has its term for mestizo—live side-by-side but have little interaction with the indigenous. Upwardly mobile Indians make the transition to the Western lifestyle—wearing shoes and speaking Spanish—but discrimination remains.

The Roots of Ibero-American Culture

In order to understand the complexities of contemporary Latin American society the knowledge of its history since the conquest is indispensable. Although Indian—and to some extent African—cultures were of far-reaching influence, it was Spanish and Portuguese institutions that fashioned the structure of Hispanic American society over a period of three centuries of colonial rule. However, the roots go back to the history of Spain itself. Certain conditions made for a unique development, not least being Spain's peripheral position on the European continent. The Pyrenees acted as a barrier against European ideas, and six centuries of Moorish influence had its effect. Tight governmental and ecclesiastical controls were other inhibiting factors. Latin America was even more isolated than Spain from the stimulus of West European civilization. This remoteness continued through the nineteenth century. Inevitably, the region's history was to proceed along different lines from what developed in North America.

Spain at the Time of the Conquest

The Iberian Peninsula has a varied ethnic history. More than two thousand years ago, successive waves of Phoenicians, Greeks, and Carthaginians moved in from the Eastern Mediterranean, and Celts invaded from the north. The Romans, of course, left a definitive stamp on Spanish culture, followed by the Visigoths, a Germanic people recently converted to Christianity, who established a landed aristocracy. Another movement of the same period was the migration of Jews from the Middle East, settling in the few urban centers. Most critical in Spanish history was the Moorish invasion in the eighth century and their unchallenged occupation of almost the entire peninsula for three hundred years, climaxed by the high level of Arab culture in the tenth century. Their heritage is visible in architectural, linguistic, and institutional influences. Arabic overtones colored the Spanish value structure, from its mysticism to the inferior status of women.[3]

As the Spanish began the reconquest of their realm in the eleventh century, commerce increased, guilds expanded, towns and cities assumed new importance, especially during the Crusades. With the establishment of a *cortés* (parliament) reforms could be initiated, although power resided in the king and his advisors. The marriage of Isabella of Castille and Ferdinand of Aragon in 1469 unified Spain. Moreover, the last barrier to national unity was removed when Córdoba fell in 1492, the Caliph withdrawing to North Africa. After the defeat of the Moors, Ferdinand and Isabel consolidated their power even more fully, expelling the remaining Moors and Jews from the country except for those willing to be converted to Christianity. The Inquisition was inaugurated in order to purify the faith, protecting the social order from all heresy. Several decades later the Inquisition was installed in Lima in 1570, Mexico City in 1571, and Cartagena in 1620.

Charles I ruled as king in the sixteenth century and upon his accession to the Hapsburg dynasty in 1519 he became Charles V as ruler of what remained of the Holy Roman Empire—the same year as Cortés' invasion of Aztec Mexico. With these developments Spain's political power reigned supreme for more than a century. In this golden age of Spain, art and literature flourished, but by the end of the seventeenth century this glittering empire was on the wane. Portugal's career was almost equally impressive, notably in the discoveries of Henry the Navigator during the early fifteenth century, and of Vasco da Gama following Columbus's voyages. Although Portugal's decline set in earlier than Spain's, the achievements of these two powers gave to the New World a language and an institutional structure on which to build.

Colonial Socioeconomic Institutions

As Spanish power declined, the colonies were forced to furnish wealth for the crown and its military apparatus. Spain's control over its colonies was more unilateral than Britain's in its trans-Atlantic domain. The court in Madrid generally displayed a self-conscious power and the conquistadores differed greatly from the religious dissenters who went to the Atlantic seaboard a century later. Yet if we look at the seventeenth-century world, it is questionable how qualitatively different were the motives and methods of the Spanish from those of the English, Dutch, and French in their commercial and imperial enterprises.

The organization of colonial life in Spanish America was unmistakably set by the power structure. For example, Nueva España (Mexico and Central America) was to be a source of wealth and glorification of the mother country, the king representing *la suma del poder* (the essence of power). Even the conquistadores had to give a fifth of their booty to the monarch. The *audiencia*, a primary means of royal control, functioned as an administrative court. Later the viceroy superseded the power of the audiencia, but in either case the king was the final purveyor of power in the colonies. By the seventeenth century, as viceroys became more corrupt, power was allotted to smaller regions with the introduction of the *intendencia*. Specifically, as *municipalidades* directed civic life in Spanish America, government was largely relegated to the *cabildo* (city council) in contrast to the rural-oriented township concept of North America. Whereas in New England the city developed as a by-product of the rural hinterland, the municipality was the basic unit in Latin America. For example, Pizarro founded Lima as "The City of Kings" and Pedro de Valdivia established Santiago as a *municipalidad*, both in a comparative wilderness.

The *regidores* (councilors) of the *cabildo*, as with all positions of power in the society, were drawn from *peninsulares* (Spanish born), although occasionally by the late seventeenth century a Spaniard born in the New World was appointed. Some measure of democracy appeared in the *cabildo abierto* (open council or town meeting), where citizen members could voice their opinion on current issues. Still, veto power rested with the *audiencia* or viceroy and finally with the king. The king and his representatives almost invariably regarded the *criollo* (creole or American born) to be of inferior status. This discrimination further intensified after the Bourbons took over the Spanish court in 1763. Moreover, beyond the usual parameters of hierarchy, the Hispanic view of interpersonal deference affected architecture (inset 3a).

INSET 3A   THE COLONIAL MIND: SPACE AND SOCIAL DISTANCE

In any society the elites define their goals, including the social structure itself. For the Spanish colonists these decisions focused on class and ethnicity. This was not always simple as a large number of the Spanish males married mestizo or indigenous women.

As towns emerged, often built on remnants of previous cultures, whether Aztec, Inca, or other cultures, the ethnic element was seldom absent, especially as the indigenous population was still present. The ideal was the Spanish norm with a central plaza, and if possible, a grid street pattern, if not prevented by topography or contours of tan ancient site. The provincial city of Cuenca, Ecuador, illustrates a not atypical example of the evolvement of architecture in order to provide the proper dimensions of social relations. Blank walls on the street protect the family from public view; however, the balcony on the second floor permits the owner's surveillance of the outside world. The house was, and is, oriented around a central patio, or among the wealthier families two or three patios, the rear one being for the use of indigenous servants.

Arrangement of rooms defines ingress of servants and under what conditions and hours in order to enhance the privacy of family members. As the seventeenth century gave way to the eighteenth, social life, including the accoutrements of the elite, became more ritualized. The dining room is turned into the nexus of interpersonal relationships. It is also the major area of interaction between lord and servant. Both the dimensions and elaborateness of living quarters were conditioned by the gender and economic roles of the household head. Since business was mostly conducted in the home, separating domestic from commercial functions became arbitrary. Moreover, women (usually widows because of the practice of men choosing wives younger than themselves) had differing ideas on the use of space and its embellishment.[4]

The Economic Fabric

Since the conquest and colonization of the Americas remained primarily a means of economic aggrandizement, the societal framework is best understood in this perspective. The major areas of economic activity are themselves revealing. Mining was the chief endeavor for at least the first two centuries, production continuing at a high level throughout the colonial era. By 1800 mining in Latin America had reached six billion pesos (at least the equivalent of that many dollars today). The relative importance of various semi-precious stones, metals, and minerals varied. Gold first dominated the scene, particularly in Peru, followed by silver in Bolivia and Mexico, copper, zinc, and tin being a much later development. A number of mining towns flourished—Zacatecas, Guanajuato, and Taxco in Mexico were all dazzling centers. Most extravagant of all was Potosí in Bolivia, the "boom town" par excellence of the Americas. The fortunes from these mines yielded enormous wealth to Spain, but in the end led only to a ruinous course of inflation as involvement in European wars made for eventual economic decline.

In the long run, agriculture, with New and Old World crops in varying regional patterns, was a far more profitable venture. Most dedicated to agricultural development were the Jesuits, not only in their vast system of missions in the Plata area, but in Mexico, Peru, and other areas with livestock, grains, and vegetables.[5] From the conquest onward, land ownership was the privilege of the chosen few. Titles to land were prodigiously conferred to the *adelantes* (chieftains) such as Pedro Alvarado, who conquered Central America and Pedro de Valdivia who "settled" Chile. Ever more, immense holdings fell to the church. The indigenous, of course, were targeted for labor. The skills and resources of the Incas, for instance, merged with the cultural assets of the Spanish, notably domesticated animals and the wheel.

Among the major diffusions was the introduction of sugar to the Caribbean and Brazil, and a variety of cereals to the plateaus and highlands of nearly all Latin America. Nearly everywhere Indians retained their diet of maize, potatoes, manioc, tomatoes, and cocoa. New World stables still led the total output, making impressive inroads into the European market of the sixteenth and seventeenth centuries. After the overproduction of sugar in the 1670s, agriculture throughout Hispanic America took on a new momentum, especially as the temperate climate of the southern third of South America became an auspicious environment for traditional European agriculture. Cattle were introduced to the pampa before the end of the sixteenth century; by the end of the eighteenth century, export of preserved or salted meat was a reality.

Slight as it was, manufacturing, primarily for local consumption, gained in output throughout the colonial period. Crafts could compete in the European market if they conformed to the mercantilism of the Spanish court. Textile factories led in industrial importance in at least a score of towns from Mexico to Argentina. The *obrajes* (textile factories) depended on forced labor of skilled Indian weavers, who were condemned to a twelve- to fourteen-hour workday. Those who failed to produce the day's norm were savagely punished. As with the indentured laborers of plantations and mines, many workers fled into the interior.

By the eighteenth century workshops turned out a range of commodities from soap to gunpowder, leather goods to shipbuilding. With the development of crafts appeared several guilds. Also, as trade expanded, products were shipped not only to Spain but to other points in Latin America by sea or overland. For example, one trail led from Buenos Aires to Lima. Even China played a role in the trade system since the Spanish empire included the Philippines as a point of exchanging products from East Asia.

Two closely related economic institutions had drastic implications for Indian society. One was the *encomienda*, a system with a long history in

Spain. During the period of the reconquest of Spain from Moorish domination, warrior knights were granted title to lands. Similarly, in the New World the *encomendero* had rights to indigenous labor. This type of indenture approached slavery, not radically different from the system of labor in the pre-Columbian empires. However, the evils of the conquerors' expropriation of indigenous land and labor came under widespread criticism from the church. Even as early as the 1520s Charles I (and Pope Paul III in the next decade) issued decrees concerning the intrinsic freedom of the indigenous.

The other institution was the *repartimiento*, whereby temporary groups of laborers were detached from their own land and assigned to the hacienda owner. This system, a counterpart of the *encomienda*, was called the *mita*, if a clear distinction can be made between the two systems—often it was a question of regional differences in terminology. Under the *mita*, Indians were rounded up for work in mines, road building, and *obrajes*. This wanton exploitation seriously decimated the Indian population. Several church leaders complained of the brutal treatment of the natives. As early as 1542 Bartolomé de las Casas wrote his *Brevísima relación de la destrucción de las Indias* (A Brief Report of the Destruction of the Indias). Although the book convinced some members of the court and the Vatican of the need for moderation, others judged his writings as exaggerated, perpetuating the myth of the Amerindians as savages.[6]

Yet, were the Ibero-Americans more unethical in their exploitation of the indigenous than were Anglo-Americans? Even the fate of Indians and Africans in all the grim details of the *obrajes*, *encomienda*, and *mita* would not seem more cruel to many observers than the extermination of the indigenous population of Anglo-America (or the enslavement of Afro-Americans in the U.S. South). Possibly Latin Americans were more consciously inhuman, even sadistic, whereas for North Americans the natives simply happened to be in the way and had to be exterminated. Despite the exploitation of the plantation system, Afro-Brazilians had relatively greater access to manumission. Also, New World economic institutions were hardly more deplorable than conditions in the factories of the Low Countries or England.

Stirrings of Revolt: The Independence Movement

As with Europe, various kinds of oppression plagued the colonies. One notorious example of suppression from the mid-sixteenth century to the late eighteenth century was the Inquisition in its harassment of whites and mestizos. The execution of some two hundred persons and the torture of still others were no worse than in Spain itself. As another reminder of these theocratic practices one has only to recall the persecution of supposed witchcraft in Salem as well as the heavy-handed inroads into religious ideology in northwestern Europe.

As the Enlightenment spread to the Americas repressive institutions became increasingly suspect. Toward the end of the eighteenth century, events in both Europe and the New World were to usher in overt protest to the status quo. The War of Independence in North America (1775–81) and the French Revolution (1789–93) could not go unnoticed in Latin America; yet, local concerns became even more salient.

Slowly but defiantly, Indians reacted against abuses of power. The first attack was in the 1560s when demographic decline and intensified labor demands heightened indigenous consciousness. Concessions had to be made in the operation of the *encomienda* system. *Hacenderos* had to compete with mine owners as new discoveries of gold and silver called for more manpower. For example, in Peru new compromises were reached in the early seventeenth century.[7] By the eighteenth century throughout Hispanic America, Indians resisted economic exploitation—forced labor and high payments for products they were required to buy.[8] Moreover, the crown was uneasy about accommodating to creoles and mestizos. This impasse was combined with a growing consciousness of the general decline in economic and military resources in the home country, now complicated by Bourbon intransigence and the effect of the Napoleonic invasion of the peninsula.

Late in the eighteenth century the French Enlightenment affected portions of the Spanish intelligencia both at home and in the colonies. Young radicals began to voice revolutionary concepts, although they were not always aware of the implications of this new ideology, embracing a wide front of social, political, scientific, and religious overtones, but in the Iberian colonies these radical ideas meant a challenge to royal authority. Among other emerging iconoclasts was the Colombian encyclopedist Antonio Nariño, who translated Thomas Paine's *Rights of Man* (1795) into Spanish and circulated Thomas Jefferson's writings. An ingeniously resourceful Venezuelan adventurer Francisco de Miranda traveled through Europe and the United States, where he was attracted to the ideas of Thomas Paine and Alexander Hamilton. Eventually he was convinced of the need for a full-scale break with Spain. After his return to Europe, his home in London became a meeting place for future leaders of the independence movement, notably Simon Bolívar and Bernardo O'Higgins of today's Venezuela and Chile, respectively.

Among other stirrings afoot were an increasing number of uprisings among Indians and mestizos. Undoubtedly the most dramatic episode of impending revolution was the rebellion led by Tupac Amaru II, a Jesuit-educated descendent of his supposed namesake, a sixteenth-century Inca ruler. In 1780 he instigated an insurrection involving several thousand Indians of southern Peru, Bolivia, and northern Argentina. Frustrated in his hope of removing Spanish injustices, he raised an army, attacked military outposts, but was unable to carry his campaign to success; the whites won out when reinforcements arrived from Spain. Tupac Amaru, his family, and lieutenants were savagely executed. Still, the seeds of rebellion were planted and revolts followed in Colombia, the Plata, and elsewhere.

A definitive break with the past came in Haiti, a French possession. By 1789 Haitians were caught up with slogans of the incipient French Revolution, and in the ensuing years a series of revolts, led by Toussaint

L'Ouverture, dealt a crumbling blow. The French, already deeply involved in their own problems in Europe, suffered a series of military disasters when they tried to dislodge the insurrectionists. By 1804 Haiti had its independence, the first Latin American country to dissolve its ties with the Old World.

All these events produced an instability in the Spanish hegemony. Spain was frequently at war against France, and intermittently against England, adding a further unsettling influence. It was Napoleon's conquests that really aroused hopes for a revolution. Segments of the Spanish population were in revolt against the puppet government in Madrid. In the New World, *peninsulares* were perplexed and divided as creoles began to plot their move for independence. Creole leaders reasoned that if the United States could achieve independence from the mother country, a similar movement could be staged in Hispanic America.

The War of Independence (1810–23) moved intermittently in various directions. Advances occurred in the first years, but staggered with the removal of Napoleon in 1815, and the patriots persisted with increasing numbers until the final defeat of the Spanish. Rival groups appeared, often on regional lines, and disagreements arose about the aims of the revolution. Leaders such as Bolívar, Sucre, and San Martín held different views and strategies about the conflict. It was in large part a war of *criollos* against *gapuchines* (derogatory term for *peninsulares*), with factions emerging in each group. Mestizos generally sided with the creoles but remained in a subordinate role during the independence movement. The nature of the conflict varied by region and time. In Mexico the early phase of the struggle for independence was a social and ethnic revolt, only in the end to become a triumph for conservatism under Iturbide. In South America the revolutionary aims of Bolívar in the north differed from those of San Martín in the south.

The Struggle for National Unity in the Nineteenth Century

The complexities of the battle for independence exhausted Hispanic America, a society already plagued by poverty, an inchoate economic and political system, a caste structure, and inadequate communication. The military campaigns brought only more misery and confusion. The leaders became victims of their own military ideology, and what ideals they had were frustrated by the destruction of life and property. Bitterness between rival factions intensified. Bolívar himself, who dreamed of a unified Latin America, acceded to less idealistic notions held by other leaders. Ultimately, a series of national units developed along the lines of Spanish colonial divisions, that is, the vice royalties of New Spain, New Granada, Peru, and the Plata. Fissions developed as in the departure of Chile from the Plata, and the tripartite division of New Granada into Venezuela, Colombia, and Ecuador. The Central America Federation dissolved into five small republics— Guatemala, El Salvador, Honduras, Nicaragua, and Costa Rica. (Panama was the creation of President Theodore Roosevelt—already contemplating construction of the Panama Canal—when in 1903 he used a rebellion in northwest Colombia as an excuse to recognize the independence of this breakaway province.)

The social structure changed little after Independence. Creoles replaced Spaniards as the rulers, but the mestizos and indigenous remained as pawns.[9] However, the situation was not static. Throughout the nineteenth century, as reflected in the provincial council's records of Tarma, a Peruvian community, economic, legal, and residential norms regarding the indigenous remained confused, almost chaotic. National, regional, and local authorities varied regarding nomenclature and status, particularly in relation to labor and land usage, of the native population. Improvement, however marginal, often through relocation and migration, continued into the twentieth century.[10]

As for political aspirations, Bolívar hoped for a democracy, largely to be shaped according to the goals of wealthy creoles like himself, but found his contemporaries little interested in republican ideals. The creoles, with inadequate concepts of national unity and ideology, could hardly escape chaos. Other liabilities were the limited practice in government and lack of incentive on the part of citizens for self-rule. The breakup of national units into rival regions only served to create new problems. It is little wonder that Bolívar expressed his despair in the often-quoted remark: "Those who have worked for the cause of Latin American freedom have ploughed the sea!"

The nineteenth century proved to be a period of spontaneous and contrived experiments at government. For the most part the only models were feudalistic structures of the past. In the early independence period several leaders tried to incorporate principles of the Enlightenment, but the idea of popular sovereignty gradually gave way to dictatorial rule. Symptomatic of this transition was the change in Bolívar from a liberal before and a cynic after 1815.

Similar ambivalence occurred in all countries; yet, each nation had problems peculiarly its own as the political process took shape through the nineteenth century. A universal phenomenon was *caudillismo*, or the arbitrary rule of a strong man whose success depended on a personalistic following. In certain instances caudillismo descended to the familiar *cartelazo*, or barracks revolt, a pattern that lingered on into the twentieth century.

At a deeper level the conflict was one of creole against mestizo as once it had been between creole and Spanish-born. Certainly, basic problems of the colonial period were not resolved. The schism of rigid social classes remained and any "solution" only added to economic and political problems. Satisfactory political models were lacking in contrast to the newly formed United States, which basically retained Anglo-Saxon representative political institutions, imperfect though they were. In Hispanic America little had changed from colonial times except for new labels and minor shifts in the aristocracy and elite.

Political Schisms

Throughout the nineteenth century, political ideals and processes remained in opposition between centralism and federalism, church and state. At the same time, *caudillismo* persisted as a recurring theme with a deep cleavage between the capital and the provinces. Argentina was a classic example of this conflict in the alternation of presidents favoring a strong central government oriented to *porteños* (residents of Buenos Aires) and their goals countered by federalists who favored the interests of the provinces. Occasionally, gauchos held the balance of power. The dictatorial regime of the federalist Juan Miguel Rosas (1835–52) derived support from this colorful breed of cowboys. In much of Latin America, power shifted between the landed oligarchy and rising merchants of the capital city.

Politically, the struggle in several countries was between conservatives and liberals. Labels might vary, *Blancos* and *Colorados* in Uruguay, *Pelucones* and *Pipiolos* in Chile, but the issues coalesced. Perhaps Colombia most sharply represents a two-party system with conflict for over a century. *Conservadores* and *Liberales* embroiled the country in intermittent civil war from the 1860s to 1900, and again from 1948 to 1957. In nearly all countries, friction emerged from the aspiration of liberals for secular education, separation of church and state, redistribution of the latifundia, free trade, and transfer of control from military to civilian; the conservatives usually voiced the opposite. The differences were often fuzzy with factions developing in each party, as *personalismo* could count more than ideology. The issues at stake, particularly centralism versus federalism, remained roughly the same from Mexico to Argentina and Chile.

With political parties vaguely defined, amorphous, organizational ambiguity, and poor financing, instability persisted into the twentieth century. It is difficult to generalize but two forms of political power arose during the mid- and late-nineteenth century.[11] One was a conservative clique or oligarchy set upon maintaining the status quo. The cone countries generally favored this pattern. Occasionally the traditional oligarchy merged with a newer commercial elite in order to forestall advances of a rising middle class, whose ranks included new immigrants—a trend existing until 1930 in Argentina and Brazil. The second model was the patriarchal, personalistic caudillo as with Porfirio Díaz (1876–1911) of Mexico and others in Mesoamerica. Caudillos might be of parochial orientation, representing traditional values, regional or national, with which the people could identify. Sometimes they sought advice of the positivists (followers of the social philosophy of Auguste Comte, 1798–1857). Their governments might also assume a folk culture.[12] More often than not, these leaders voiced mottoes of Order and Progress. With a populist yet a conservative model, a strong corporatist theme holds, that is, the state represents sacred, collective bonds transcending individual concerns, One not atypical variation is Paraguay, in which a two-party system evolved from a classical caudillo style during the second half of the nineteenth century into the twentieth century. Usually coalitions, each with a relatively specific agenda, remained in power for approximately fifteen years.[13]

Chile offers still another variant of the familiar pattern of clericalism and anticlericalism, dictatorship and constitutionalism, latifundia and minifundia.

The division was principally between two parties, Liberals and Conservatives, but at one time no less than fifteen parties represented the most minute gradations of political ideology. As with other countries, boundary disputes with neighboring nations became another complicating factor. Fundamental issues were fairly specific such as the rights of the church to its autonomous courts and its vast landholdings, primogeniture in the *fundos* (haciendas), and the extension of education to the *rotos* (proletariat) of the city and the *inquilinos* (tenants) on the farms. Personal style varied as one compares the progressive, if sometimes high-handed, methods of Bernardo O'Higgins, the first president, with the reactionary measures of his successor, Diego Portales. Generally in Chilean history *la gente decente* ("nice people"—middle–upper and upper classes) reigned supreme until the Chilean economy collapsed in the 1920s. This conflict of "haves" and "have-nots" remains today for all countries.

Mexico exhibits one of the more intriguing case studies of liberalism versus oligarchy. In the middle of the nineteenth century the country found itself with variations of the same problems confronting most of Latin America. The landscape was composed of *pueblos de indios* (indigenous) and *pueblos de gente decente* (creoles) with different concepts of land ownership, in addition to vast haciendas, the church being the largest landowner, as was true of most of Hispanic America. Tensions between urban and rural led to movements toward universal suffrage, land redistribution, and other issues.[14] In 1857 Benito Juárez successfully challenged the reactionary policies of Santa Anna and his heritage. After Juárez' death in 1872 the pendulum swung again to the right during the last quarter of the century with the *coronelismo* of Porfirio Díaz, who offered carte blanche to capitalists or industrialist entrepreneurs from Britain, France, Germany, and the United States, while federal police and the *rurales* (organized vigilantes) ruthlessly ruled, respectively, the city and the countryside.[15] Today *porfirismo* (the reign of Díaz) appears decadent and profligate, as the lack of monuments to Díaz in Mexico testifies. Still, some compensation may be found in the stability and economic advance of the period, notwithstanding the evils the system brought to peons and Mexican society in general. However apart Juárez and Díaz were in their social and political ideas, leadership in the second half of the nineteenth century partially passed from creoles to mestizos. In comparison to the tumultuous gyrations in Mexico, Brazil's evolution was less violent, as reflected in inset 3b.

INSET 3B   ODD NATION OUT: BRAZIL FROM EMPIRE TO REPUBLIC

As compared to Spanish America, Brazil attained independence in a unique fashion. The country became relatively independent of Portugal largely because of weaknesses in the Portuguese monarchy. Fear of a Napoleonic invasion led to establishment of the Portuguese empire in 1807, Rio de Janeiro becoming its capital and Pedro I its emperor. Already marked by strong regionalism and a rich divergence of ethnic strains, establishment of a relatively autonomous empire further distinguished Brazil from the remainder of Latin America. Pedro's preference for the home country, arbitrary rule, and questionable ethics led to calls for his abdication, which ended the First Empire (1824–31). He was succeeded by his son Pedro II, initiating the Regency (1831–40). Pedro was a child of five and in the relatively free, almost chaotic atmosphere of Brazil—Liberals in conflict with Conservatives—absolutism, republicanism, and other issues were astir. With the Second Empire (1840–89) economic priorities, regionalism, and abolitionism became politic controversies. Like several political figures in Latin America, Pedro II was influenced by the ideas of the Auguste Comte, who looked to science as a means of sociocultural change. Indeed, the motto "Ordem e Progreso" found its way to the Brazilian flag. Pedro's liberalism was in part his undoing as the powerful rural oligarchy resented his antislavery stand and demanded his abdication. Nevertheless, slavery was abolished in 1888 and the Republic was proclaimed the following year—the process of independence being nonviolent as compared to Hispanic America.

The basic pattern of colonial Brazil continued into the nineteenth century; great fortunes were made in sugar and mines. The *casa grande* (mansion house of a plantation) and the *fazenda*, like the Hispanic hacienda, constituted a self-contained world. As in Hispanic America the *fazendeiros* (landowners) dominated a paternalistic and family-oriented elite, which included the higher clergy, bureaucrats, and military chieftains, who had somewhat less power than in neighboring countries.[16] Political and economic processes remained essentially rural as compared to the priority of *municipalidades* in Spanish America. The ethnic and class structures were slightly more fluid, partly because of racial mixture. The Afro-Brazilian was treated as a slave but the chances of mobility and manumission were perceptibly greater than in the U.S. South. Brazil was more secular than its neighbors; the church never had the power or wealth it enjoyed in Hispanic America. Moreover, in the nineteenth century the Brazilian clergy became highly suspect for their questionable sexual behavior. The Masonic orders had even more salience than in the rest of the continent.

Brazilian society was never as highly solidified as in neighboring nations. With its vast territory, geography was inevitably a factor. Also, as a multiracial society the country displayed contradictory trends from colonial times to the twentieth century. The upper class clearly stood for white supremacy, but the fazendeiro indulged himself with female slaves, leading to widespread miscegenation. The North has a huge African population, the South far less. As the fulcrum of economic power shifted to the South, racial strains became more evident from the nineteenth into the twentieth century. Another duality appears in the preference among the elite for a European lifestyle as against the practical realities of a near continent to exploit.

In the late nineteenth century Brazil's economic base of sugar plantations and mines gave way to the emerging coffee culture of the South. The fulcrum shifted further from North to South as a primordial industrialism arose in São Paulo in the 1880s. In addition, the *paulistas* were relentlessly carving out an economic frontier in their West and the area from São Paulo to Rio Grande do Sul attracted large-scale migrations from Europe. During the half-century reign of Pedro II, 1840–89, Brazil increased in population three times, a growth more attributable to immigration than to natural increase.

The Two Americas

For nearly all Latin America, disorganization produced by independence and the slow rate of national development during the nineteenth century led to few solutions to basic problems. The *encomienda* was replaced by the hacienda, most Indians and many mestizos simply remained as serfs or peons on these vast estates. The quasi-caste structure continued.

Again, a comparison may be useful between Latin America and Anglo-America. As pointed out earlier, comparisons can be easily strained. The colonial, feudalistic pattern of the U.S. South parallels somewhat the Latin American socioeconomic structure. To a marked extent the Civil War shattered this feudal system but in the end only accentuated the South's status as an underdeveloped area, whereas in the North industrialism notably accelerated in the 1870s and beyond. The first stirrings of industrialism came to Argentina in the 1880s, but a marginally functional industrial economy did not appear in Latin America until the early twentieth century and with marked regional variations.

Immigration played a major role in this process as it had in North America. Even then, the pattern of immigration differed markedly regionally. Few of the nearly 40 million immigrants to the United States between 1820 and 1950 chose the South. Whether in search of urban employment or as homesteaders in newly opened areas, they preferred the North. Immigration to Latin America also grew (if only a fourth of the ratio of immigrants to the United States) throughout the second half of the nineteenth century, accelerated in the twentieth century, especially to Argentina, Brazil, Chile, and Uruguay. Yet, only a third elected to remain in Latin America as compared to over half who remained in North America.

The Twentieth Century: Challenges and Responses

On the political stage *personalismo, caudillismo,* and the *cuartelazo* continued to haunt Latin America, especially in less advanced nations. Several countries followed the lead of Argentina—notably Chile, Uruguay, and Brazil—into incipient industrialization. Still, all countries exhibited the colonial profile of delivering agricultural and mineral products to Europe and North America as they had done since colonial times. However, intellectual vibrations surfaced. The Spanish–American War and Theodore Roosevelt's "big stick" policy alienated political and literary elites. Among the rousing incursions to the status quo was the invasion of socialist philosophies. Intellectuals such as Francisco Madero and rural activists such as Emiliano Zapata ushered in the Mexican Revolution, which cost nearly one million lives (1910–20, with rumblings until 1929 when political consolidation was reached). Indeed, the disturbance in Mexico, akin to a civil war, was the only upheaval in Iberian America to transform fundamentally the socioeconomic landscape until the Cuban Revolution in the 1960s. Mexico's political-economic development can be viewed in three phases:[17] (1) 1910–40, the dissolution of the *porfirato*, the struggle of various contenders, both in the political and labor sectors with final establishment of a political order;

(2) 1940–70, a period of "stabilizing development" generally more conservative than the first period, with capitalist expansion, increasing state intervention, control of union bosses, and less concern with rehabilitation of the peasants—these phenomena prompting the student revolt of 1968—a sign of the social system's growing complexity; (3) 1970–2000, expansion of industry, consumerism, and social welfare, as set off by the oil bonanza of the 1970s, but the 1982 fiscal crisis catapulted into massive external debt and social upheaval leading to structural economic changes and the neoliberalism of Carlos Salinas (1988–94). The 1990s brought on a new phase of the political process. Rumblings of dissatisfaction with the corruption and clientelism of the ruling, self-perpetuating PRI (Institutional Revolutionary Party) became militant and even violent in Chiapas in 1994. Rival parties intermittently shook up the foundations of PRI leading to its defeat in the 2000 presidential election.

Other countries strived toward basic political and economic reform. Bolivia had its revolution in 1952, which endowed its Indians with a somewhat less feudalistic existence. The Chilean experiment of Salvador Allende (1970–73) was interrupted by a right-wing coup led by General Augusto Pinochet and his lieutenants and supported by the United States. The Sandinista revolution (1979–90) was frustrated by the Contra war, but a few basic changes remain in Nicaragua's socioeconomic fabric.

As will become clear in chapter 4, political processes in the immediate postwar period moved in waves of democratization and authoritarianism. Dictatorships, classical and ad hoc, reigned from Central America to Argentina. By the late 1950s the ride turned, democracy reemerged. In the heady days of the early 1960s arose social movements, related to a more liberal ideology from Europe and North America, as symbolized by John Kennedy's Alliance for Progress, which sadly came only to partial fruition.

These political shifts were often precipitated by the emergence of new political parties led by students and fragments of the urban lower and middle classes. As a consequence, all nations were "democratic" with the exception of Cuba, Haiti, and Paraguay. But by the mid-1960s liberals were demanding more than the oligarchy chose to deliver; hence, dictatorships, whether military or civilian, were back. Even solid democracies such as Chile and Uruguay slid into harsh military, terrorist regimes. All this was to be reversed in the wave of democratization of the mid- and late 1980s, largely as a result of a public outcry over the economic debacle of that decade. As usual, each country had its own pattern of liberation—and lethargy.

This gradual and usually uneven trend to equalitarian governments grew out of extensive social change. The spread of education, an expanding middle class, an increasingly complex economy oriented to industrialization, a variety of services, and mobility, both horizontal and vertical, account for this more sophisticated political process, all in the context of globalization.

Conclusions

This chapter testifies to the enormous debt Latin American society has to indigenous cultures, both classical empires and the many indigenous peoples antedating Europeans and Africans, as well as to the broad features of a culture developing from colonial times to the present. The feudalistic character of the agrarian society, merging of traditional indigenous and Hispanic–Arabic–Christian institutions, and a sharp division of social strata lingered on into the twentieth century. However, we constantly need to remind ourselves that these generalizations cannot be applied with too close a fit to the twenty republics, and even more dangerous would be the failure to recognize regional, ethnic, class, and individual differences within a given nation. Moreover, old values may no longer be functional and will cease to be accepted whatever the rigidities of the total social system.

The twentieth century, particularly in the years after World War II, saw an unprecedented series of changes in Latin America. Rise of the metropolis, solidification of the middle class, changing norms of family life, new models of political behavior, technical breakthroughs in management and communication, all underline this transformation. As with all industrial cultures, social change assumed a geometric rate during the twentieth century as compared to the nineteenth. These changes in the context of an intricate set of cultures constitute the thread of succeeding chapters.

Chapter Four

Society, Economy, and Government

Inherent in any culture are given social categories or subcultures—race or ethnicity, social class, gender, position in the life cycle, among others. Equally basic in the functioning of any society or culture are the social institutions. The most universal or indispensable are how we make a living or the economy and socializing agencies, predominantly the family, and in more advanced societies the educational institution. A means of control is the political institution. Religion and recreation provide for the satisfaction of still other needs. As explained in chapters 1 and 3, Latin American social institutions stem mainly from European and indigenous ingredients, with an African heritage in areas such as Brazil and the Caribbean. As in most of the world, subcultures and institutions alternate between stability and change. In this chapter we analyze ethnicity, social class, the economy, and the political order. Later chapters will explore other institutions.

Race and Ethnicity: An Enduring Heritage

In Europe and North America ethnicity is analyzed in the context of minorities—the variation is noted in table 4.1. However, in several nations (Bolivia, Ecuador, Guatemala, and Peru) indigenous groups are the majority, yet subject to discrimination. Moreover, since the 1980s increasing attention is directed toward minorities, whether based on age, gender, sexual preference, lifestyle—or ethnicity.[1] Still, Latin America has a history of racial exploitation and segregation of indigenous and African populations. Hardly less important are the mixtures forged by this uneven relationship—mestizo, pardo, mulatto, samba, and other labels. In the sixteenth and seventeenth centuries a caste system took shape. Each country has its own scenario in the unfolding of its ethnicity. In addition to the European and African infusions is the Asian heritage. In the late nineteenth and early twentieth centuries thousands of Chinese immigrated to Mesoamerican and Andean countries. Chinese, Japanese, and Koreans, mostly young males, played a conspicuous role in developing fishing, commerce, and other technologies during Mexico's *porfirato* (the tenure of Porfirio Díaz, 1876–1911).

Table 4.1   Ethnic distribution in percentages

	Amerind	European	African/mulatto	Mestizo
Argentina	>1	97	—	2
Bolivia	54	14	—	32
Brazil	<1	50	49	>1
Chile	5	25	—	70
Colombia	1	20	29	50
Costa Rica	0.5	89	3	7.5
Ecuador	40	12	5	43
El Salvador	8	2	—	90
Guatemala	54	4	1	41
Honduras	10	13	<1	76
Mexico	5	9	8	78
Nicaragua	5	12	14	69
Panama	3	19	2	76
Paraguay	3	18	—	79
Peru	46	11	1	42
Uruguay	—	90	2	8
Venezuela	2	19	8	71

Source: Simon Collier, Thomas E. Skidmore, and Harold Blakemore, *The Cambridge Encyclopedia of Latin America and the Caribbean* (New York: Cambridge University Press, 1992), p. 61.

In Mexico, for instance, the status of a given ethnic group varies widely, but the indigenous continues to occupy an inferior status as the 1994 Chiapas revolt and its aftermath document. Still, the struggle toward recognition is far from equal. In the mid-nineteenth century both mestizo and Indian won partial recognition in the election of Juárez to the presidency, but the *porfirato* moved in reverse—mestizos and especially Indians were marginalized, as were also the Chinese and other Asians who arrived on the Pacific Coast, as they did in several Andean countries, working in mining, railroad construction, agriculture, and fishing. Shunned at first, they later had a less ambiguous status.[2] In the aftermath of the Mexican Revolution, the writings of Samuel Ramos and José Vasconcelos lifted the indigenous to a new status in the 1920s. Political forces in the 1930s established a new direction in national identity.[3] After the 1960s this celebration of Indians was given mere lip service in most of Mexico, but *mestizaje* remains an ideological thrust of Mexican political life. Multiculturalism was affirmed in the 1991 article to the Constitution, with the insistence on a "polyethnic nation state."[4]

In Chiapas a different status structure persisted. By 1994 neoliberal economic practices along with oppressive landlords had an even more adverse impact than in other parts of Mexico. Moreover, the plight of the indigenous and their rights to land was long ignored by Mexico City and even the regional capital. An armed revolt was the response of the Zapatista Army of National Liberation (EZLN), in January 1994. In the first days of the rebellion at least 145 *zapatistas* were killed and some 20,000 were forced to leave their homes in order to avoid confrontation by federal forces.[5] The struggle continued more in the hope that *zapatistas* might enter into decision making rather than remain as a quasi-military presence.[6]

For all Latin America, ethnic categorization is a determinant of social mobility. As the universal cliché proclaims, money and schooling "whiten a dark skin." Central America and the Andean nations display a tragic profile of harsh treatment of the indigenous. In Guatemala, Indians were always considered inferior and are only grudgingly accepted once they change their clothing, wear shoes, and learn Spanish. But political and economic demands—specifically the fear of Marxist movements—called for annihilation. Between 1958 and 1988 nearly two hundred thousand Mayan Indians were killed and nearly two hundred thousand sent into exile in a civil war organized by the oligarchy, tacitly supported by the United States, on the pretext that Indians were directly or indirectly Communist agents.[7] Further south, for nearly five centuries the Andean indigenous suffered from the oligarchy's brutalities. These practices continue on remote haciendas in Ecuador and Peru, where sharecroppers still till the soil bound in near-permanent serfdom because of debt to the landlord for land, tools, and seed. Throughout Latin America race is historically linked to slavery.

Ethnicity stems from cultural history more than biology. Race or racism becomes its own ideology; it rationalizes the perceptual world for the European and to a large extent the mestizo population. Although stereotypes are becoming less rigid, a major problem for the indigenous is the inability to form alliances with other Indian groupings. Even more problematic is seeking links with "white" and mestizo organizations. The case of Bolivia is instructive. Through most of the nineteenth century, creoles dominated the economy. By the early twentieth century a bond was forged between creoles (whites or Europeans) and the mestizo landowners, both groups blaming the indigenous for the nation's retrogression.[8] The 1952 revolution raised the status of the Indians, marginal though they remain in the society.

Brazil is often cited as a relatively ideal area of race relations. In reality, the ethnic pattern is complex and varies over the country's vast expanse, between north and south, urban and rural. A heavy concentration of Afro-Brazilians in the northeast and a European population in the south exists along with almost infinite shadings in the mix of indigenous, African, Asian, Mediterranean, and other ingredients. In colonial times a rich vocabulary suggested subtle meanings differently interpreted by the elites and the mass of the population.[9] Afro-Brazilians themselves struggle to resolve their identity, a number of movements arise for equality, yet they are interlaced with their African roots in language, music, dance, and religion. It is an open question whether participation in Western cultural groups, such as evangelical Christians, weakens or strengthens their black identity.[10]

Today, the pattern of relative acceptance of Afro-Brazilians in the north can be compared to a more nuanced discrimination in São Paulo and even greater marginalization in the southeast.[11] In this context it is significant and ironic that in the provinces of Minais Gerais and São Paulo slaves were granted freedom, liberated slaves ("free colored") being involved in a variety of statuses and occupations in the early nineteenth century.[12] Again, race merges with class. For the poverty-stricken, discrimination and segregation can be severe. For instance, Afro-Brazilians are disproportionately subject to harassment and assaults by the police. Democratic principles are conditioned by one's position in the pecking order. Color and class are differentially treated by actual as opposed to formal norms.[13] Still, many Brazilians resolve any cognitive dissonance by saying that "racism simply doesn't exist."[14]

This Latin American ethnic structure continues to the present day, if in a more subtle fashion. Nineteenth-century racist doctrines linger on, elaborated by personal experiences and social realities—sexual exploitation by hacienda overseers, education at a missionary outpost, or insurgence of ethnic groups. A rising indigenous political power, in the Andes for example, is beginning to change the ethnic hierarchy.[15] In the meantime the pattern of race relations differs according to cultural development. Argentines—especially *porteños* (residents of Buenos Aires), unaware of the many mestizos of the country's outer provinces—simply ignore their history of exterminating thousands of their indigenous—similarly to the U.S. heritage—inviting European immigration, and at present boast of the "whitest" population in the hemisphere south of Canada. In a more ethnically mixed society, Venezuelans brush aside any hint that they are other than a color-blind community. In contrast to Anglo-Americans (for whom a drop of African blood traditionally defined the individual as black) Venezuelans like Brazilians distinguish within the Afro-American population the light-skinned mulattos as opposed to the *pardo* (black), but economic discrimination is more relevant than racial discrimination.[16]

The cultural meaning of *mestizaje* has changed. In the nineteenth century the term *lo americano* was used by writers to designate a majority, that is, a mestizo population. By the mid-twentieth century the concept of ethnicity became more contextualized as terms such as Afro-Brazilian or Argentine Jews became current.[17] In a sample of Ecuadorian university students, an identity of *indigena* is preferable to mestizo.[18] Cultural identity is constantly subject to revision. Ethnicity is created and redefined according to the needs of the individual, group, or society. From the beginning of the twenty-first century, a debate is taking place in Latin America as to how social movements can operate in order to protect and advance the needs of indigenous and other groupings. The homogenizing process is acutely slow, infused as it is with questions of gender, class, and politics.[19]

Stratification: Class or Caste?

Inevitably, social class is a complex variable and its definition varies for each culture. Throughout advanced societies no social criterion has an impact on attitudes, values, and lifestyles comparable to social class, whether upper, middle, or lower (or whatever labels fit the given culture). Ethnicity, education, income, residence, consumption patterns, and linguistic styles are basic to one's class position. At the same time, intricate factors enter into the formula by which a class label is assigned to the individual or family. Traditionally in Latin America landed wealth had the most status, but in the second half of the twentieth century entrepreneurial skills altered the equation. Even though new wealth counts for less than old wealth, especially in smaller communities, the source of wealth may be no more significant in Latin America than in Europe or North America (where greed and corruption can bring the individual the trappings of upper class). For instance, operators of the vast gambling industry in Brazil enjoy unprecedented wealth and are accepted among the urban elite.[20]

Class shapes status, power, and a whole range of behavior. Societies vary in the degree to which individuals can change their status. That is, in traditional rural societies one's status tends to be *ascribed* or set by birth with little change of class likely during one's lifetime. In more dynamic communities, opportunities are available for *achieved* status through education, occupational shift, entrepreneurial skills, migration, and other channels. Although Latin America suffers less from occupational and related rigidities than most of the Third World, it offers less intra- and inter-generational mobility than do Europe and Anglo-America. As one Peruvian analyst puts it, for the lower classes in cities and even more in rural areas, barriers to mobility approach the concept of caste.[21] The level or depth of poverty shifts according to economic cycles as in the 1980s. Unemployment and inflation in Mexico during the first years of the twenty-first century intensified poverty, increasing the rate of migration to the United States.[22] However, in a socialist regime such as Sandinsta Nicaragua, class boundaries became somewhat diffuse, and ethnicity a less salient determinant.[23]

The present class structure arose out of the colonial period and internal neocolonialism of the nineteenth century. As outlined in chapter 3, Afro-Americans, Indians, and diverse mixtures made up the lower class, which represented nearly 90 percent of the population in most countries. Possible exceptions include Argentina and Uruguay. With industrialization and the rise, however moderate, of educational opportunity, especially after World War II, new avenues of mobility emerged. Because of cross-national and intranational differences, generalizations are hazardous. Yet, today in most cities the upper class would be 3–5 percent of the population, the middle class 30–40 percent, and the lower class 55–70 percent—allowing for wide differences in the complexion of each class and its meaning for national and regional economic pressures and power structures.

Observers vary in their definition of class according to theoretical viewpoints—Marxist, post-Marxist, structuralist, functionalist. Indeed, there is little agreement as to how class can be measured, whether objectively or subjectively. It would be difficult to exaggerate the significance of time and place—the effect of economic cycles and of city and hinterland. The economic advance from the 1950s to the 1970s along with the expansion of unionization, education, and urban residence permitted a higher standard of living for the middle class and survival potential for the lower class. The Lost Decade of the 1980s brought unemployment, reduced wages, and deeper poverty. Whatever the economic cycle, the income gulf between urban and rural remains staggering. Deprivation in the shantytowns became as acute as in the countryside. Moreover, the many squatter settlements interlacing and surrounding the city document the drive of impoverished peasants to find a niche in the urban world. Poverty continues to haunt migrants, but the possibility of work, education, health facilities means a tenuous foothold on the job ladder, not to mention the temptation of delinquency (inset 4a).

INSET 4A   CLASS, CRIME, AND VIOLENCE

It is well known that Latin America suffers from a high rate of crime and violence, possibly the highest of any region of the planet. As in other parts of the world, economic elites are well known for their corruption or white-collar crimes. However, the economically strangled portion of the population accounts disproportionately for criminality. Although delinquency, including homicide, has a long history in the area the advent of neoliberalism with its ever-growing concentration of income in the upper 1 percent of the population is a factor in the increase of crime. Also, the ups and downs of the economy have had their effect. The severe deprivation during the 1980s magnified by strict monetary demands of the international funding agencies have also had their effects. Except for a few countries, the 1990s showed an increase in homicide and other crimes.

Of course, other factors beyond the economic enter into the equation. The United States, for example, has a high rate of violence, as compared to most Western nations. Similarly, a cult of violence exists among Ibero-American nations, with variations, southern versus northern Brazil. In fact, with the exception of Colombia and Venezuela, the Andean countries along with the cone countries are less prone to violence. Costa Rica also has a favorable profile. Moreover, political and collective violence, as in dictatorships and internecine warfare in Colombia, must be differentiated from individual and gang violence and warfare, which accounts for the recent waves of kidnapping and murder.

Machismo, the drug culture, broken families, educational deficiencies, inadequate housing, and most important, underemployment and unemployment all contribute to crime. On this point, in Chile in 1995, 94 percent of those alleged to be involved in armed robbery were young men, 60 percent between the ages of 15 and 24, three-quarters being jobless or only partially employed. Of those accused of homicide, 46 percent were below the age of 25.

All segments of society are affected by crime; especially vulnerable are urban areas with little police protection. Still, the principal targets are the ultra-wealthy, presumably more for tapping their wealth than expressing rage toward the perpetrators of the economic quagmire. As a result, growth of the quasi-fortresses or "gated" communities is apparent in most cities. The desire for protection from holdups and kidnappings has enormously increased the number of private security services, especially in large cities. In São Paulo one finds three times as many privately employed guards as public policemen. It is misleading to consider the drift toward a free market economy a principal cause of the rising crime rate, yet the frustration of the "have-nots" leads to desperate attempts to better their situation.[24]

Lifestyles, Attitudes, and Survival

Latin Americans continuously struggle to maintain their status. As the personal and family image counts heavily, half of a middle-class family income may go into housing, a fifth into clothing, the remainder for food, with little left if any for recreation and the like. From Mexico City to Santiago the need to preserve a middle-class profile drives the principal breadwinner to hold two or more positions, with supplementary income from employment of other household members. For the lower class the struggle focuses on securing any kind of work. A walk through any city shows the vignettes of informal employment—car watching, selling food, hawking lottery tickets. Since a never-ending search for jobs, fleeting they may be, is the fate of the lower class, the term "unemployment" applies more to the middle class.[25] However, underemployment haunts most workers, whether in the formal or informal sector. For instance, of those working in manufacturing and construction, 27 percent in Mexico, 50 in Brazil, and 84 in Honduras live—or exist—below the poverty line.[26]

Variations *within* each class seem as wide as *between* the three classes. In view of both quantitative and qualitative differences, it is conventional to think of each class as composed of two or three levels—upper–middle, middle–middle, and lower–middle, for instance. One criterion setting off the upper and upper–middle from the lower class is the avoidance of manual labor. A banker, engineer, or professor would seldom be seen working in the garden or washing windows. This ritual applies no less to the wife, who after shopping sends for the maid to carry in her packages when she arrives home in a taxi. Also, as in the United States, being a member of what club or living in what neighborhood can loom large in assigning status. Because of the convoluted status structure in the middle class, a given family might receive eight times the income of another family. Factory workers likely attain lower–middle class status if they earn union wages, which can be well above the income of a clerk or shopkeeper. In some communities class belongingness hinges on life's chances or choices—for one, a church wedding rather than a consensual union. Domestic servants belong to a given niche of the lower class simply as a reflection of the wealth and status of their employer. These subtleties characterize the Western world, but they assume a unique pattern in Latin America.

The configuration of class depends on locality—rural communities offer a different profile from urban ones. With its tradition of regionalism, Colombians speak of Antioquia, which prides itself on a mobile society with a spirit of egalitarianism (if somewhat fictitious), whereas in Cauca, particularly its capital Popyán, it links the individual to a historic aristocracy with his or her *abolengo* (heritage or descent).[27] Moreover, class position and behavior are altered by economic and political forces, as seen in Gurupá, an Amazon community originally studied by Charles Wagley in the 1950s, and reexamined in the late 1990s by Richard Pace.[28] Wagley found a basic divide between an upper stratum (10 percent) of landowners and merchants, as opposed to a broad spectrum of two layers of a working class, as based on steadiness and type of work, housing, and other criteria. By the 1980s financial crises had shrunk the upper class to 5 percent and the class barriers were less salient than a generation earlier. Nevertheless the working class is still subservient to the upper class on whom they depend for their employment. Certain rituals enact this deference, for instance, addressing superiors as "Senhor" or "Dona" and avoiding direct eye contact.

Social class is, of course, reflected in behavior, values, and attitudes. Curiously, the two extremes of the pecking order may coalesce. The upper class is sufficiently powerful to ignore the mores; the marginalized lower class has little at stake when it transgresses social norms. As the middle class has the most to lose it is the most conforming in personal behavior. For example, the middle class is the most religious as measured by church attendance. As compared to lower class, it is also more future-oriented, subscribes to a rationalistic rather than traditional approach to individual and societal choice.[29] As reflected in my research in El Salvador, the lower class, notably in the rural scene, has a more fatalistic attitude, passively accepting *"lo que Díos manda"* ("it's God's will"). These attitudes change with the advent of militant social movements or migration to the city.

Opinion toward political and economic processes is sensitive to the changing social order. Whereas the middle and upper classes lean to conservatism and the status quo, many in the lower class are disenfranchised or simply unable to effect any change. Because of its voting strength, the middle class becomes a fulcrum of political change and stability. As one instance, when militancy from the left becomes unacceptable the middle class tends to accept intervention, even military, from the right, as illustrated in the 1964 coup against the Goulart administration in Brazil. By the late 1970s they were weary of the military regime's bureaucracy and an ineffective economy, paving the way for free elections in 1984. With increasing education and spread of a service-oriented, global economy, the middle class is more accepting of cross-national organizations. For instance, according to an opinion survey of Mexicans in regard to NAFTA, the middle class is more favorable than is the lower class. This attitude seemingly grew out of the urban middle class's increased identification with the United States.[30] The Latin American class structure can hardly escape the effects of globalization and mass communication in a cybernetic universe.

The Urban Milieu

A mark of advanced societies is the rise of cities, as early civilizations of the New World and the Mediterranean document. Significantly, several Latin American cities were built on the sites of pre-Columbian civilizations. Whether early or modern, cities share certain functions in common due to their density, even though the modern metropolis differs sharply from the rigid social structure of the ancient city. Theorists such as Max Weber point to rationality and bureaucratization in the city. Louis Wirth and Ferdinand Toennies stress heterogeneity, impersonality, and segmental character—long before many Third World cities reached the five–ten million population.

A principal feature in Latin American cities is their enormous growth in the second half of the twentieth century. Table 4.2 points to the staggering growth between 1950 and the 1990s. Greater Mexico City grew nearly ten-fold between 1945 and 1995 to an estimated 20 million, and with Tokyo–Yokohama is the largest urban agglomeration in the world. São Paulo, Buenos Aires, and Rio are not far behind. By 1990 five nations (in declining order, Uruguay, Argentina, Chile, Venezuela, and Brazil) overtook the United States in its 75 percent ratio of population living in cities.

The evolution of the city from the colonial period to the present reflects both internal and external forces. The cities were shaped by the Spanish predilection for a central square or *plaza*, surrounded by the cathedral, *cabildo* (meeting house), and *ayumiento* (city hall), and other official buildings. On the adjoining streets resided the elites in homes built around a patio. Further out were the huts and shops of the mestizos. Within a half-century after independence major cities took on a more cosmopolitan shape. Broad, tree-lined avenues radiated from the center of Buenos Aires, Lima, and Santiago. This lavish importation of spatial and landscape designs

Table 4.2 Population of largest cities (in thousands)

Country	Largest city	1950	1993–99*
Argentina	Buenos Aires	5,130	11,298
Bolivia	La Paz	300	1,940
Brazil	São Paulo	3,450	19,528
Chile	Santiago	1,330	4,661
Colombia	Bogotá	680	5,399
Costa Rica	San José	180	1,273
Cuba	Havana	1,081	2,224
Dominican Rep.	Santo Domingo	220	2,135
Ecuador	Guayaquil	259	2,070
El Salvadore	San Salvador	213	408
Guatemala	Guatemala City	337	1,676
Haiti	Port-au-Prince	130	884
Honduras	Tegucigalpa	140	520
Mexico	Mexico City	2,880	15,048**
Nicaragua	Managua	109	821
Panama	Panama City	217	471
Paraguay	Asuncion	200	415
Peru	Lima-Callao	1,010	6,231
Uruguay	Montevideo	1,086	1,303
Venezuela	Caracas	680	1,975

* The figures are for the census of a given year and only for the municipality, not the total metropolitan area.
** The Federal District includes an additional 8.2 million inhabitants.

Source: Latin American Center, *Statistical Abstract of Latin America*, vol. 38 (Los Angeles: University of California, 2002), pp. 67–68.

from Europe were concentrated in the capitals, hardly to the satisfaction of secondary cities.[31] In 1910, a century after independence, Mexico City with its radial *avenidas*, a sense of social order, and economic progress symbolized a new nationalism at the twilight of the *porfirato*.[32] Industrialization in the twentieth century ushered in a tangle of new interstitial areas, waves of migrants, and dilapidated housing in juxtaposition to a fading postcolonial grandeur. No longer could urban problems be hidden, if they ever were. At the beginning of the twenty-first century the megacity evokes images of pollution, traffic jams, hurried commuters, all caught in a global economy, characterized by the Internet, faxes, and cellular phones. With this transformation Mexico City, as an example, became "a fragmented and conceptually indefinable space lacking a center."[33] Demographic, economic, political, architectural, and communication systems transform the definition of the city. The city's role as the focus of various institutional functions—education, religion, recreation, esthetics, to mention a few—continues, but in a more intricate, almost unrecognizable form. These and other factors produce a dynamic in the urban process as seen in figure 4.1.

What perplexes the visitor to a city in a developing country is the vast network of squatter settlements. These shantytowns take many shapes, differing significantly from those of Asia or Africa and vary between and within nations. In Rio *favelas* seem to be everywhere; other cities may set limits, especially under military dictatorships. Each country has its own pattern and jargon: *callampa* or *población* in Chile, *villa miseria* in Argentina, *rancho* in Venezuela, *barriada* in Peru, and so on. No less than 65 percent of Mexico City's inhabitants in 1997 lived in *barrios proletarios*. Some spring up almost spontaneously, others are planned or unplanned invasions. They are as likely to be in interstitial areas both within and on the edge of the city. Dwellings often begin as primitive tar paper and corrugated tin huts to become substantial structures several years later. What the casual observer fails to note is not only the labyrinthine layout of these makeshift communities but also the diversity of the inhabitants, who represent the gamut from marginal low-grade employment to semiprofessionals. On the whole, these settlements are possibly less infested with symptoms of social and personal disorganization than are the *tugurios* (older slum dwellings). Stereotyping of shantytowns by urban middle- and upper-class residents and by foreigners is one more instance of ignorance and prejudice. However, one cannot forget the level of violence permeating city life, with its intra-family conflicts, anomie, miserable housing, drug culture, and not far beyond, the overcrowded prisons of both juvenile and adult inmates.[34]

In other words, cities offer a vast range of functions—economic, political, educational, recreational, esthetic, and religious—which are unavailable in most rural areas. Migrants to the city also find a heterogeneity of lifestyles as well as the option of either anonymity or group orientation, loneliness or new social contacts. Contrasts abide in the vast cultural arena of the megacity, whether Bogotá or Rio, Guadalajara or Montevideo, as opposed to provincial towns. As the service sector of the economy overtakes manufacturing,

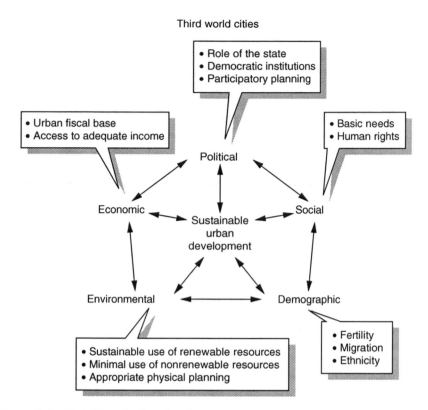

Figure 4.1 Variables of urban development.

Source: David Dukakis, "Third World Cities: Sustainable Urban Development," *Urban Studies*, 1995, 32, 659e–677e.

the urbanite finds a larger range of resources, whether beauticians or psychiatrists, bookstores or health clinics.

Clearly Latin Americans face a crisis in the city at the beginning of the third millennium. Megacities are only marginally livable, smaller cities also suffer from gaps in the infrastructure of schools, transport systems, water supply, and sanitation. Secondary cities, though less polluted, are likewise crowded with migrants and have even greater barriers, at least financial, for dealing with their problems than does the capital city. Political leaders and economic elites are drawn to the capital, which has first call on national resources. Beyond the tenuous relationships between local, regional, and national assets are the ties with the wider world in a growing global economy. As another development are the cross-national urban complexes with a network of maquiladoras, as witness the urban complexes of San Diego–Tiajuana or El Paso–Ciudad Juárez. The political and economic implications of these new facets of urban life demand intensified social planning.

Rural Society

Over a fourth of Latin Americans belong to the rural scene and over a sixth are engaged in agriculture. As noted in chapter 3, the conquest restructured rural society, even though more than traces of the indigenous remain. Fundamentally, Iberian immigrants set up two agricultural systems, the plantation and the hacienda. The *plantation* features tropical crops, sugar, rice, bananas, among the most obvious products. With a factory-like bureaucratic style of management, the process is oriented toward export and profits are mandatory. The *hacienda* has both domestic and export crops, usually patterned after its Iberian antecedent. Products are relatively diverse and *hacendados* treat the estate as their fiefdom, traditionally containing clusters of families, and can be a world unto itself—a store, chapel, prison—but the model is not a fixed one. Continuing from colonial times into the nineteenth century the hacienda actually expanded in certain areas during the early twentieth century as transport systems enlarged the marketing potential.[35]

Still other land tenure and labor patterns are found. One is the *chacra* (parcel) in the minifundia tradition. *Campesinos* still retain tiny acreages, continually parceled by family crises and inheritances. Second is the middle-sized farm, which is more evident in more developed societies such as Argentina and Uruguay. Third is commercial or corporate agriculture, or a more streamlined and mechanized version of the hacienda. No less revealing of the rural scene is the typology of labor. Over three-fourths of *campesinos* are either sharecroppers or wage earners working on the hacienda or other forms of latifundio. Tenancy is another variant as indebtedness for land, seed, tools become a yoke for the *peón*, *colono*, or *yanacono*, as these tenants and sharecroppers are labeled. With the advance of corporate farming, wage or day labor is gradually supplanting the tilling of one's own soil. Capitalism rather than feudalism or paternalism is now the dominant agricultural system, which has complicated the "peasant's survival strategies."[36] Fortunately, some of the peasants' frustration is reduced through ingenuous methods of coping or by new economic development. For instance, in several Andean areas, as in Colombia and Ecuador, new forms of commercial production (for instance fresh flowers for the export market), initiatives from nongovernmental organizations (NGOs) based on community and kinship groupings provide new opportunities. Even so, many *campesinos* move between urban and rural work.[37]

Conflict arising from rural discontent can be a stimulus for radical social change, as will be elaborated in chapter 11. The Cuban revolution began in the countryside in 1953, and led by Fidel Castro, worked its way in rural skirmishes to victory in 1959. In Nicaragua in the 1970s the *sandinistas* (the movement named after General Augusto César Sandino, leader of the guerilla movement against the U.S. occupation of the 1920s) were reacting to the plight of peasants as much as to fighting the Somoza dictatorship. In Peru the violent campaign of the Sendero Luminoso (Shining Path) was directed primarily against exploitation of peasants. Possibly more than any other Latin America revolutionary movement, the Sendero illustrates the ferocity of commitment to a given ideology, charismatic leadership, and the merging of urban intellectuals, largely students, and desperate peasants. As the economic ills of the 1980s worsened, there emerged an ultra-Maoist terrorist group, the Communist Party of Peru, more often known as the Shining Path. Its inspiration stemmed from José Carlos Mariátequi, who dreamt of a better fate for indigenous Peruvians and in 1929 founded a socialist party affiliated with the Communist International. From its origin in a university circle in Ayacuho led by Abimael Gúzman, it recruited over ten thousand adherents with violent attacks over nearly half of Peru. The capture and execution of Gúzman and his lieutenants in 1992 not only enhanced the popularity of President Fujimori, but signaled the twilight of the Sendero. The movement was a model of the potential in its "religious and mystical identity" with which Gúzman inspired his followers.[38] In 2003 a Truth and Reconciliation produced a nine-volume report that the Sendero Luminoso episode accounted for more than sixty-nine thousand deaths between 1980 and 2000—over twice the number previously estimated. Presumably three-fourths of those killed were Quechua-speaking Indians caught in the gunfire between guerrillas and the military.

Each country has its saga of a counterbalance of militancy and government programs. Few situations are as problematic in the clash between peasant mobilization and governmental intractability as that of Colombia. The National Peasant Association of Colombia (ANUC) was founded in 1967 in order to promote delivery of public resources and agrarian reform. As the decades unrolled the organization was caught in rivalries and conflicting policies within and between the two parties, Liberal and Conservative. In addition, ANUC split into Marxist, Trotskyist, and Maoist factions, becoming involved in guerrilla movements and indirectly in narco-trafficking. The shift to large-scale agriculture, migration to the city, and rising rural violence drove peasants off the land. Consequently, peasant participation in agriculture fell from 61 percent in 1960 to 44 percent in 1981. More progressive decision making in the Liberal and Conservative parties resulted in a limited agrarian reform program, despite opposition from the rural oligarchy. About sixty-six thousand families received grants of land during the 1970s. The struggle of the 1980s, the Lost Decade, was less successful.[39]

Throughout Latin America a fundamental cry is for basic agrarian reform in order to convert latifundia and unused land into farms for the *campesino*. Mexico has a history of substantial reform. A principal aim of the 1917 Constitution was to return land to the peasants according to the Aztec *ejido* principle. Under President Lázaro Cárdenas (1934–40) vast haciendas were redistributed. However, the postwar years saw only spasmodic attention to strengthening the ejido system, notably under Adolfo López Mateos (1958–64) and Luis Echeverría (1970–76). During the Salinas regime (1988–94), neoliberalism was in full swing with the adoption of a new agrarian code in 1992, which decreed privatization of communal lands. Yet, according to a 1995 survey comparing an *ejido* with privatized farms, *ejido* residents have a greater sense of equality and community, fewer landlord–tenant difficulties, reduced feeling of proletarianization, and less serious interpersonal tensions. Modernization of the agrarian economy calls for a number of changes, notably raising the value of farm products. Whether ejido or individual farming, income and satisfaction of the farmer hinges on personality, family setting, type of crop (maize, tomatoes, rice) as reported in this Morelos, Mexico, study.[40]

Chile has also extensively experimented with land reform, inevitably reflecting the ebb and flow of sociopolitical forces. The Christian Democrats under Eduardo Frei (1964–70) augmented a system of collective and individual farming arrangements. Salvador Allende (1970–73) enormously expanded the holdings by expropriating the *fundos* (haciendas), followed by Pinochet's terminating nearly all agrarian reform. After the fall of Pinochet, Patricio Aylwin (1989–95) moved to compromise solutions.

All in all, rural more than urban life carries risk and frustration—an almost nonexistent income, lack of social and technical services, exploitive landowners, an indifferent local government, and a remote national apparatus. Brazil offers a disturbing history, especially during the dictatorship (1964–85). First came repression springing from the military's contempt for leftist rural protest as it intervened in rural unions and other worker organizations. Second was the growing preference for commercial or corporate agriculture. Third, an onslaught of measures followed the economic crisis of the 1980s including restriction of credit subsidies.[41] In summary, solution of rural problems awaits a democratic revolution directed to the redistribution of the nation's wealth, Piecemeal answers appear through aid programs, attempts at land reform, educational innovation, and indirectly, demographic shifts—family planning and outmigration. These offer an incomplete response to the economic and health crises that the peasant has endured since colonial times.

Economics as a Driving Force

Societies are organized around basic social systems, usually identified as institutions, which are the focus of several succeeding chapters. Two other institutions are the economic and the political, each the subject of countless volumes. The following discussion is simply to provide a context into which the broader Latin American culture may be placed.

One consistent pattern from colonial times to the present is the persistence of poverty. Whereas the per capita income of the United States in 2002 was roughly $29,080, that of Argentina was $10,880 Uruguay $7,830, but Haiti $1,160 and Nicaragua $2,170 (with extreme stratification, fluctuating exchange rates and buying power being relevant).[42] Domestic variations are even more staggering than international ones. Brazil and Venezuela lead the area in concentration of wealth—over half in the upper 10 percent of the population. (The United States shows an equally shocking maldistribution; however, the lowest 15 percent of the population do not suffer from the same degree of poverty as do their Latin American counterparts.)

A theoretical underpinning of economic processes is in order. At the risk of oversimplification, at least three models or stages were successively in evidence: one, for much of Latin American history, *dependency* theory was the explanation of economic backwardness. According to this form of colonialism, the developed or advanced world takes from developing nations the raw materials it wishes and in return sells their own industrial products, always to their own advantage. In other words, mercantilism reigned from colonial times into the twentieth century. Oversimplified though dependency theory is, it describes the basic economic situation, at least until the 1950s.

Two, in order to combat dependency, *import substituting industrialization* (ISI) appeared in the 1920s as a consequence of the economic isolation during World War I and as a means of making the nation self-sufficient, but became salient in the 1950s. It was supported by the economic doctrine *structuralism* as sponsored by the Argentine economist Raúl Preibisch and the Economic Commission for Latin America (CEPAL) with its headquarters in Santiago, Chile. Structuralists, who were in favor from the 1950s to the 1970s, assert that an economy can only flourish with the supply of private and especially public investment in industries as a means of counteracting the rigidities and bottlenecks of a traditional economy. They perceive different segments of the economy developing at uneven rates; for example, as urban areas expand, food demands may outstrip supply, requiring adjustments that call for state intervention. Moreover, the primary economic sector, notably agriculture and mining, could never sufficiently absorb the labor force; consequently, industrialization is mandatory.

Three, the *monetarists* criticized structuralism as being fraught with the dangers of fiscal irresponsibility, price controls, subsidies, trade imbalances, and budget deficits. Also, for monetarists a market-oriented economy assures relatively full employment. According to monetarists, as the government pumps money into the economy, inflation is an inevitable byproduct. Indeed, hyperinflation reached staggering levels, particularly from the 1960s through the 1970s, in most countries, disastrously so in Mexico, Brazil, Peru, and Argentina. Despite its weaknesses, especially the fate of being an unequal partner in the world economy, structuralism held sway into the 1980s.

The Dilemma of Neoliberalism

By the 1980s the new economic philosophy of *neoliberalism* became current. The Pinochet military regime led the way in adopting the advice of the economists of the Catholic University in Santiago and the "Chicago boys" (graduates of Milton Friedman of the University of Chicago) in establishing a free market economy.[43] The response to the 1980s economic crisis was to slow down state welfare and other types of governmental intervention in the economy. This return to a "free" economy brought results in increasing production and profits, but at an enormous cost to workers when wages are cut and welfare suspended. Even though labeled as a "laissez-faire" system, neoliberalism involves considerable state intervention and coalition building. Clearly this arrangement favors the export and financial sectors.[44] Argentina, Brazil, Uruguay, and other relatively developed nations turned to the neoliberal path with varying results. The poorer countries followed suit.

In the 1970s most Latin American countries borrowed heavily from abroad. When the Organization of Petroleum Exporting Countries (OPEC) raised nearly tenfold the price of oil, fuel costs skyrocketed in almost the entire region. In 1982 several governments were forced to default on their international debts. Eventually, foreign banks rescheduled payments but at stringent conditions set by the International Monetary Fund (IMF) and the World Bank. Unemployment, wage reduction, welfare cuts, lower subsidies, and other stabilization efforts meant a reduced standard of living for the middle class and extreme poverty for a third to half of the population during the remainder of the 1980s—the Lost Decade. Fortunately the 1990s saw a turnaround as debt reduction and lower interest rates prevailed. New investments paid off. Neoliberalism was now in vogue in nearly every country, if unevenly as dictated by resources, leadership, and the sociopolitical climate. As societies become more pluralistic, neoliberalism calls for compromises between corporate, commercial, and labor sectors. Yet, a major aim is reducing trade barriers and expanding capital flow.[45] Wealth became ever more concentrated in the economic elite: the number of multimillionaires and billionaires in U.S. dollars rose by over 500 percent between the mid-1980s and mid-1990s.

There is little question that neoliberalism ushered Latin America into a new era. Foreign investment more than doubled between 1991 and 1995. International trade trebled in the same period. However, these financial profits had little effect on the lower class as the gap widened between the two extremes of income levels. Inflation continues, if to a reduced degree. Most Latin Americans realize that they are caught between an old oligarchy, new corporate structures, and governmental bureaucrats, civil or military. Neoliberalism brings seemingly no more prosperity to the masses than did structuralism, that is, privatization does not reduce insecurity for a majority of the population.[46] In a survey of fifteen hundred of Brazil's political and economic leaders, almost 90 percent held that the nation had achieved economic success but at the cost of a severe social displacement. The issue of waste is staggering, for instance, of the some one hundred million persons of working age, only seventy million Brazilians represent the economically active population. The question of resource management extends to every aspect of the economy, from highway construction and nuclear energy to social services and education.[47] Centralization of the economy and government is an acute problem, but decentralization has its drawbacks— corruption, cost overruns, and accountability.

The most serious charges against neoliberalism and the global market have come from the left. For one, *maquiladoras* or other sweatshops condemn workers to the lowest wages with few if any fringe benefits, in other words, the perils of the informal economy. Other criticisms include the inability of exports to maintain a balance with imports, unpredictability of profits, dominance of large over small enterprises, and the uncertainties of a world market.[48] In 1994, Canada, Mexico, and the United States signed NAFTA with favorable—and unfavorable—results, particularly for Mexico. Not least of the problems is the vulnerability to economic swings in the powerful nation to the North.[49] Complications also arise from cross-national technical and cultural differences in standards. For instance, exportation of U.S. corn into Mexico leads to potential conflicts regarding cross-hybridization in addition to intruding on cultivation practices that Indians in Oaxaca have used for generations.[50] Finally, both worker and entrepreneur are at the whim of national decision makers. These questions impact on a central issue of social security (inset 4b). Again, the economy hinges on the political institution.

INSET 4B SOCIAL SECURITY IN BRAZIL: REALITY OR ILLUSION?

Throughout Latin America the search for economic sustainability is a source of anxiety for most individuals. Brazil represents this problem on a mammoth scale. A 1996 commission revealed the poverty and desperation of the nation's elderly, including those in government-supported homes such as the Saanta Genoveva Clinic in Rio. Reform is broached but seldom realized. As with the rest of the world, today's payroll taxes on workers underwrite the income of present-day retirees. Yet, much of these proceeds go into meeting other priorities—the federal government relying on inflation to reduce pensions and other allotments for the next generation. The 1988 constitution expanded the pool by adding nearly five million rural workers, irrespective of their contributions to the system. Because of demographics, the sheer weight of the burden is itself frightening. In 1940 there were thirty workers for every retiree, in 2004 the ratio is 2.3 active workers for every retiree.

Among the system's injustices is the fate of the informal economy—over half of the workers receive no benefits. Also, entitlements begin at differing ages, depending on the system of retirement, contributory or universal. Another disparity is between types of employment. Employees in the private sector receive a fraction of that of civil servants, who retire with pay equal to their terminal salary.

Further, as with other countries, men retire after 35 years, women after 25 years. Depending on the system, retirement age may be as early as 53 or as late as 65. Moreover, the security system appears to perpetuate social inequalities. The 1990 resolution to peg pensions to the minimal wage—a resolution only partially maintained— hardly solves the problem of poverty among the aged.[51] In 2003, because of budget deficits the government met almost violent opposition when it tried to raise the retirement age for given classes of workers.

Beyond the inequities of pensions is the question of basic financing. A number of Latin American nations are wrestling with the inadequacy of their social security structures. Especially in the 1990s motions to privatize social security gained momentum. Neoliberalism has its impact, along with urging of international funding agencies, to which most nations are in perpetual debt. Fiscal crises, inflation, corruption, and political confrontation and polarization are involved in the debacle.

Political Structures and Processes

Does the political process control the economy, or the reverse? Is democracy viable for Latin America? To what degree can various sectors in the society share power? Do individual citizens feel they are intrinsic to political structures? How is the political integrated with other social institutions? These questions perplex political actors, whether lay or professional.

As noted in chapter 3, the painful unfolding of feudalism inherited from colonial Spanish domination led to the fumbling and arbitrary rule of nineteenth-century *criollos*. In most republics political parties were established to represent competing ideologies—federal or provincial, state or church, free or controlled trade. More often than not, the *caudillo* (leader) embodies political power, with or without party affiliation. This tradition lingered on into the twentieth century, but as societies became pluralistic and interest groups more articulate, political styles were carefully crafted, at least among those countries undergoing industrialization. However, there is no fixed standard. For instance, by 1910 Mexico was in revolution with a continuing conflict among *caudillos* but arriving at partial stabilization in the late 1920s. Later, charismatic leaders such as Getúlio Vargas in Brazil (1930–45) and Juan Perón in Argentina (1946–55), both later returned briefly to power, offered a new populism appealing to special segments of the population, primarily urban workers. This model of populism with a *caudillo* flair appears occasionally, as with Peru's Fujimori (1990–2000) and with Venezuela's Hugo Chávez (1999–).

In several countries after World War II, democratic regimes were solidified but still fragile vis-à-vis both internal and external pressures. Voter turnout increased in the second half of the twentieth century, varying according to the political and social climate of the particular era. For instance, after the 1944 revolt against Guatemalan dictator Jorge Ubico, a socialist Juan José Arévalo was elected president and succeeded by fellow Marxist Jacobo Arbenz. A discontented rural oligarchy abetted by the United Fruit Company and the White House set the stage in 1954 for a CIA operation that returned the presidency to the old order.

As another turnaround, in the 1940s Colombia moved to the left, in electing Liberals, only to be replaced by harsh Conservative regimes— Ospina Pérez and Laurerano Gómez—as violence flared across much of the rural landscape. In desperation the nation in 1953 turned to General Gustavo Rojas Pinilla, whose virtual dictatorship was only partially successful in quelling the violence. University students were the first to protest in a general demand for a return to democracy, ushering in a novel experiment, that is, the National Front, which set up a sixteen-year plan (1958–74), by which Liberals and Conservatives would alternate in the presidency of four-year terms. Although other parties are on the ballot, the two main parties continue to assure a relatively democratic process.

In another series of reversals Venezuela in 1958 saw a return to democratic rule. After the despotic rule of Juan Vicente Gómez (1908–35), less repressive military officers assumed the presidency. A 1945 military coup ushered in the Democratic Action Party's candidate Rómulo Betancourt. This excursion into democracy was short lived; a military takeover installed Maarcos Pérez Jiménez (1948–58). Happily, Betancourt returned to the presidency, to be followed by quasi-democratic regimes ever since.

Similarly, other nations adopted varieties of democratic regimes. Consequently in 1961 when John F. Kennedy introduced his Alliance for Progress every country except Cuba, Haiti, and Paraguay were nominally in the democratic camp. This wave was more of a promise than a reality. From the mid-1960s to early 1970s every government except Colombia, Costa Rica, and Venezuela fell into dictatorships—Mexico being a seemingly perpetual monopoly or as the Peruvian novelist Mario Vargas Llosa once described it, a the "perfect dictatorship."

South America experienced tragic reversals in the 1960s. Brazil was the most dramatic case of political reversal. In 1964 the economic elites regarded the João Goulart regime as too leftist and financially irresponsible; predictably the military staged a coup followed by a severe repression of civil liberties.[52] Argentina presents a special case. Free elections were firmly established in 1916, with the victory of Radical Party's Hipólito Irigoyen, but economic disaster set the stage for the 1930 intervention of General José Uriburo. Most defining of Argentina's authoritarianism was Juan Perón (inset 4c), followed by three more years of quasi-military rule. After a shaky democratic regime (1958–66), dictatorship returned to continue until 1983 (with a brief hiatus of a façade democracy, 1973–76).

Inset 4C   A Meteoric Rise and Fall

Few political careers in Latin America during the twentieth century rival the bluster and charisma of the rise, fall, and return of Colonel Juan Domingo Perón (1895–1974), Part of his military training was in Italy, where Benito Mussolini's fascism favorably impressed him. Appointed in 1943 as minister of war and head of the Secretariat of Labor and Social Welfare he built an enormous following among labor, increasing union membership from 10 percent to over 60 percent of the workforce. His wife Eva was no less appealing than her husband in creating an enthusiastic following. His drive toward power worried both civilian and military leaders and in October 1945 he was arrested and sent to prison on Martín García Island on the river La Plata. Within two days some thirty thousand workers, led by Eva Perón, demonstrated and Perón was released.

Under the banner of Justicalism, Perón voiced an authoritarian political philosophy based on his vision of social justice, with a penchant for nationalism, militarism, and fascism as expressed in Italy and Spain—hardly alien to Argentina in view of its Italian and Spanish ties.[53] Perón was elected president in 1946. Once in power, paying his debt to his followers, he increased wages, social benefits, and indirectly the political power of workers. He nationalized a number of foreign-owned companies. His most successful years were between 1946 and 1949. Nonetheless, Perón was reelected for a second term in 1951. In the 1950s the economic situation worsened, with severe inflation. Seemingly, he had more talent for political maneuvers than for solving economic problems. Although his support for labor unions irritated the military, it was conflict with the church that spelled his doom. After Eva's premature death from cancer in 1952, he was angry that the church frowned upon his and his cohorts' clamoring for the canonization of his late wife. Perón was also troubled by some of the clergy's espousal of Christian socialism. In response he decreed abolishing the laws against divorce and prostitution, closed a number of church schools, and advocated a complete separation of church and state. By 1955 the end was at hand, he boarded a gun boat to Paraguay, ultimately finding refuge in Spain. With the breakdown of the government in 1973, Perón was again the savior, but his tenure was brief; he died in July 1974.

Especially tragic was the fall of two vibrant democracies—Chile and Uruguay. In 1973 Pinochet violently drove out the democratically elected socialist Allende experiment, albeit the fact that Allende had the support of little more than a third of the electorate. The well-intentioned but flawed social reforms of Allende were decried by the middle class. A later survey found that over 70 percent of Chileans regarded Pinochet's coup as "saving democracy."[54] However, the doom surrounding the Allende regime was in part contrived by the oligarchy with support from abroad, notably the U.S. State Department and CIA.[55] In Uruguay, the strongest democracy south of the Rio Grande, economic disaster precipitated dictatorial rule, largely as a reaction against the Tupamaros (Movimiento de Liberación Nacional), who represented urban guerrilla warfare and other terrorist protests, precipitating a military takeover in 1973.[56]

The tide turned once again. By the mid-1980s most electorates realized that authoritarian regimes had not delivered. In the case of Argentina the military floundered badly in the 1982 Malvinas/Falklands War; consequently, civilian leaders had little difficulty reestablishing democracy in 1983. In Brazil an *abertura* (opening) led to the end of military rule and new elections in 1985 against Great Britain. Finally, by 1993 all countries except Cuba were nominally "democratic," even though most suffered from symptoms of stress and arbitrary rule, especially Haiti, Paraguay, and Peru. The transition from authoritarian to democratic regimes is seldom a simple process, Among the problems are establishing legitimacy, stability, accommodation to a pluralist system, and confidence in the future.[57]

Peru also illustrates the clash of political and economic forces and the inherent conflict between civil and military forces. As with other countries, an oligarchy was challenged in the 1960s by a number of progressive civilian leaders. However, a military government led by Juan Velasco Alvarado (1968–75) attempted to resolve indebtedness, inflation, and other problems of his predecessors, but was unable to find a coherent plan for his economic reforms. Velasco was followed by even less effective military leadership. In 1978 civilian rule returned. The economic debacle of the 1980s along with ravages of the Sendero Luminoso movement brought additional chaos. Into this maelstrom stepped Alberto Fujimori in 1990. As a personalist, neoliberal *caudillo* with considerable mass support he centralized power—legislative, judicial, and other governmental operations suffered—and imposed technocratic decision making. His 1992 *autogolpe* increased the military grip on Peru, further weakening democratic institutions. As no meaningful alternative appeared Fujimori was reelected in 1995, even though many elites, including the military, were concerned about the future of democratic institutions.[58] His third electoral victory in 2000 set a near-record for duplicity, which led to his resignation later that year. Aspects of this scenario surface from China to Mexico, Singapore to Venezuela, yet few are as politically defiant as the Fujimori model—or as economically successful as the Asian ones.

Political Parties: Structure and Process

Any classification belies the complexity and subtleties of political parties. As compared to Europe or North America, political organizations tend to be diffuse, lacking in grassroots support, and/or are underfunded. Also, political parties are subject to wide ideological swings.[59] By Western standards, parties are ephemeral and personalist. Some parties have a lengthy history, others are short-lived, often the creation of a single charismatic leader or an opportunistic clique. Further, in view of relatively weak legislative and judicial branches, the executive has considerable power, and party politics is of varying relevance. Most nations have a tradition, however wavering, of leaning to a single party, two major parties, or multiple parties. For example, in the 1993 election in Paraguay nine parties competed for the presidency, three for Congress; in Peru in 1995 the ratio was reversed. In Bazil's 1994 presidential election three parties supported Fernando Henique Cardoso, each of five other parties gathered around a single candidate. Despite the arbitrariness of party systems, several orientations stand out:

1. *Traditional* or nineteenth-century parties, as noted in chapter 3, include the Conservative and Liberal of Colombia, the *Colorado* (Liberal) and *Blanco* (National) of Uruguay. The Radical parties in Argentina and Chile have an enduring history with more than occasional shifts in ideology, often toward conservatism (opposition to the Peronists in Argentina) through the decades. Inevitably new ideas, new leaders, new challenges, and a changing sociopolitical landscape call for innovation.

2. *Action-oriented* parties stem from Radical and other parties and are of two fundamental orientations. One is the Democratic Action (AD) of Venezuela, which was founded in 1941 and won the presidency in 1945, but

was ousted by the military in 1948. With the return to democracy in 1958, AD under the leadership of Rómulo Betancourt initiated various social programs but was forced to struggle against disaffected leftists on one side and the traditional oligarchy on the other. In the 1970s and 1980s the party floundered particularly under the presidency of Carlos Andrés Pérez. Other nations have similar parties—notably the Brazilian People's Democratic Movement (PMDB) or the Popular Action (AP) in Peru.

The other major genre of broad-centered social action is the Christian Democratic (DC) party as found visibly in Chile and Venezuela. Its tenets belong to the applied ethics enunciated in papal edicts dating from Leo XIII, Pius XI, and most recently in John XXIII (1958–63). Like Acción Democrática in Venezuela, the movement began in Chilean university enclaves before World War II under the leadership of Eduardo Frei, who became its first president (1964–70). Frei tried to combine his Christian humanitarianism with practical economics. The achievements of DC were not sufficient to forestall the Marxist thrust of Allende, whose actions drove the DC to grudgingly accept the Pinochet takeover. Horrified by the terrorism of the dictatorship the DC moved again to the left-center. After the removal of Pinochet, Chile elected in 1989 Christian Democrat Patricio Aylwin, who retained with some modification Pinochet's neoliberalism. The stratified society remained intact, if less painfully as compared to the past. As most voters seemed happy with a centrist position, Frei's son assumed the presidency in 1995, followed by Ricardo Lagos in 2001.

The other qualified success story of Christian Democracy belongs to Venezuela's Comité de Organización Política Electoral Independiente (COPEI). Its leader in the 1960s was Rafael Caldera, who worked in tandem with AD leaders. As the AD party split into two factions in 1968, Caldera was elected president and like Frei in Chile was committed to social justice and welfare. COPEI was out of power following Caldera'a term, but returned in 1979 with Luis Herrera Campins as president, who proved to be less than competent. Discontent with the drift of both COPEI and AD, Caldera formed a new movement *Convergencia* and was returned to the presidency in 1993. By the late 1990s the political structure was in shambles. Consequently, an impetuous reform-oriented military officer Hugo Chávez appeared on the scene. In 2000 he was elected president followed by the adoption of a new constitution. In a far different social setting a less successful DC party appeared in El Salvador, led by an idealistic José Napoleón Duarte, who won the 1980 election, but failed to rein in the military and their death squads and the landed oligarchy, who blocked his agrarian reform program.

Progressive politics attracts, if to varying degree, urban labor, peasants, and the indigenous. Charismatic, even authoritarian leadership helps, but the message, usually antiestablishment, is essential. A classic example is APRA (American People's Revolutionary Alliance) in Peru, founded in 1924 by Haya de la Torre, who urged a national program for all Peruvians, that is, a recognition of the indigenous. The party finally came to power with Alán García (1985–90). By entering into coalitions on both the left and the right, APRA had a significant effect on politics in Peru and beyond its borders. Hardly surprisingly, a minority party is rarely able to overturn the status quo.

3. Through most of the twentieth century *national "revolutionary"* parties embodied popular movements based on the hope of a widespread awakening, often dedicated to uplifting a given sector. Getúlio Vargas's Estado Novo, proclaimed in 1937, belongs to the national popular movement tradition; Juan Perón's *Juaticialismo* appealed to workers and in the 1990s Carlos Menem operated in the shadow of Peronism; Costa Rica's National Liberation Party (PLN) as personified by charismatic José Figueres, who framed a new democratic sociopolitical order in 1948. Other examples are Panama's Democratic Revolutionary Party (PRD) and Bolivia's National Revolutionary Movement (MNR), which sparked the revolution of 1952.

The most enduring model of popular revolutionary party is Mexico's Institutional Revolutionary Party (PRI). Spawned by the critical years 1910–20 of the revolution it did not become a political party, Partido Nacional Revolucionario, until 1929. The party found its mission most sharply defined by Lázaro Cárdenas (1934–40)—and a new label, Partido de la Revolución Mexicana—in a massive program of agrarian reform and national self-assertiveness as in the government's expropriation of foreign oil companies in 1938. When Miguel Alemán assumed the presidency in 1946 the party was redefined as the Party of the Institutionalized Revolution or PRI, and Cárdenas's commitment to the indigenous was never again fully honored—although both Adolfo López Mateos and Luís Echeverría were conscious of the "other Mexico." In successive years corruption, bureaucracy, and inability to deliver on promises increased malaise on the left and the right. Finally, in 2000 the center-right Partido de Acción Nacional (PAN) Vicente Fox won the presidency.

4. *Marxist* or communist parties appeared in Latin America in the 1920s reflecting the advent of the Soviet Union. In several nations they are legal, in others they operate underground or as guerrilla organizations with specific goals, methods, and venues. They may exist as a quasi-parlimentary function or different paramilitary operations as in Colombia and Peru. Increasingly since the 1950s communist parties have been joining forces with other left-center parties. The intricate relationship of Marxism to politics and reform movements can be illustrated by a few case studies: in El Salvador, Farabundo Martí, who as a youth was shocked at the exploitation of *campesinos*, organized the Communist Party (CP) in 1931, planning a massive revolt in 1932. As word spread, the oligarchy ordered the murder of approximately thirty thousand peasants—their thumbs tied behind their backs—a trademark repeated a half-century later by the death squads. After decades of feudalism, peasant movements emerged in the 1960s followed by sporadic violence and in 1980 appeared the Farabundo Martí Front for National Liberation (FMLN), which worked in concert with other guerrilla groups as well as the Communist Party.[60] Similarly, in Nicaragua sundry militant organizations, notably the Marxist–Leninist Popular Action

Movement (MAP-ML), had connections with the CP; however, the CP received only 1.4 percent of the votes in the 1990 election.[61] In Guatemala the CP endorsed the 1952 candidacy of labor's Jacobo Arbenz, who also had wide middle-class support. In other words, a communist party serves as a protest against the status quo. Throughout the hemisphere, leftist like rightist political movements frequently form alliances with centrist parties.

Leftist parties may also emerge from ethnic tensions as racial and indigenous groups form social and economic suppression. Mestizos or ladinos may combine with the indigenous in joint grievances in collective defiance, especially in Bolivia, Ecuador, Guatemala, and Peru.[62] This opposition may begin in a social movement but often is incorporated into a political party, as in the Guatemalan National Revolutionary Union (URNG). The enduring case of socialism is, of course, Cuba (inset 4d).

5. *Military* parties dominated the political landscape from the 1960s to the 1980s. Highly authoritarian, they come to power because of a vacuum in the political status quo or they simply oust the incumbent, as in Pinochet's 1973 takeover under the auspices of the National Party. In Brazil the generals devised their own party National Recovery Alliance (ARENA) in order to give legitimacy to their assumption of power. Another ARENA party (National Republican Alliance) was formed in El Salvador led by the notorious Roberto D'Aubuisson (who received support from the Reagan administration). Actions by death squads prompted Washington to call for a less savage political apparatus; consequently, ARENA rule was temporarily replaced by the Christian Democrats.

6. *New wave* or antiestablishment parties appear when an outsider offers a strikingly novel approach. An outstanding example is Cambio '90 of Alberto Fujimori, whose opponent was the distinguished writer and intellectual Mario Vargas Lloso. Both protested the inept Alan García presidency, but in the 1990 election *limeños* and other Peruvians saw "El Chinito" closer to the poor and their needs. (Latin Americans often identify most East Asians as "*chino*"—*Chinese*—just as those from the eastern Mediterranean, Lebanese among others, are somehow "*turcos*"—Turks!) Strongly authoritarian, Fujimori took on the Sendero Luminoso, reduced inflation, and won a second term despite his violation of civil rights for which he was dubbed "Chinochet." The epithet became even more appropriate in 2000 when fraudulently winning the presidency for an unprecedented and ill-fated third term, he fled to Japan. By 2005 with the economy floundering some Peruvians were nostalgic for Fujimori's ruthless will.

Even in strongly democratic countries, political processes are subject to questions of legitimacy, sustainability, and fragmentation. Voter participation has gained, but with variations, since 1950. For instance, Costa Rica had a voter turnout of 80 percent, especially in the late 1950s, but dropped to below 50 percent by 1998. On the other hand, Brazil shows a reverse trend over the decades (figure 4.2). Several factors are involved, not least the lack of charismatic candidates.[63] As another aspect of political phenomena, coalitions between parties, is a recurrent trend in Latin American politics, as

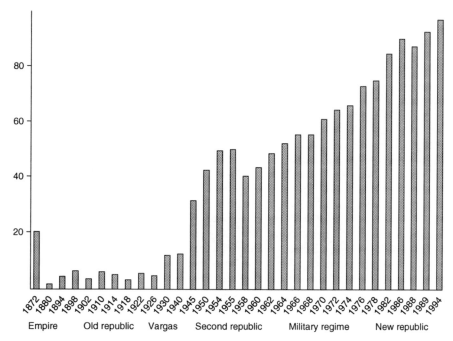

Figure 4.2   Electorate as percentage of voting age, 1872–1994.

Source: Lincoln Gordon, *Brazil's Second Chance: En Route Toward the First World* (Washington, DC: Brookings Institution, 2001), p. 144.

illustrated in the uncertain years of Chilean "cupola politics" or *política de acuerdos* (politics of mutual agreement), which characterized relationships between leaders. However, during the 1990s polarization weakened the possibility of accommodation. For instance, in a 1995 poll 40 percent felt that further reconciliation between the left and the middle was unlikely.[64]

INSET 4D   CUBA: A DIFFERENT PATH

All too often, North Americans deny the existence of Cuba, or simply reduce the island to empty stereotypes. Historically the United States has since the Monroe Doctrine (1823) assumed a domineering attitude toward Latin America, particularly Cuba as a legacy of the 1898 Spanish American War. The present impasse dates from the arrival of Fidel Castro to power in 1959 and his drift toward Marxism in the following year. For over four decades Cuba was transformed into a socialist state with a focus on social welfare if not always accompanied by civil rights. The country's economic problems were magnified by the demise of the Soviet Union, cutting off resources from Eastern Europe. At the end of the Cold War in 1989 hostility and intransigence of the United States did not abate and its economic embargo embarrasses all those who are seriously concerned with hemispheric unity and human rights. Havana's rigidity is as pathetic as Washington's; even Castro admits that the present U.S.–Cuba relationship is counterproductive.[65]

Cuba's political philosophy is based on a principle of equal justice for all, including women—rather unique for Latin America. Its constitution even guarantees freedom of speech and press, as long *as it is* "in keeping with the objectives of a socialist society."[66] The idealistic goals of the revolution are only partially realized. The lack of civil liberties in Cuba and its economic deprivation are well documented. Still, the improvement in education and health standards in rural areas as compared to pre-Castro Cuba is seldom mentioned in the United States; nor is the malnutrition of children because of the U.S. embargo. As another point, critics of Cuba fail to acknowledge the degree of flexibility and innovation in political and economic arrangements in the 1990s: reforms, albeit modest, in jurisprudence, direct election of the National Assembly, acceptance of a dollar economy, and joint economic ventures, as Canada and Western Europe moved into the vacuum left by the United States and the Soviet Union. No less than other Latin American leaders Castro is forced to compromise with capitalistic themes such as neoliberalism. Cuba continues in a bleak and uncertain future as its people and leadership confront a "no-win" economic and political stalemate. Considerable uncertainty also exists because of Castro's advancing age. Cuba's fate rests to an appreciable extent on ideological shifts in the United States.[67] The United States, yielding on one side to given business interests, anxious to recover their losses of over four decades earlier, and on the other side ingratiating the Miami Cuban Americans, remains in an ideological straightjacket with little empathy for the Cuban people.

Ideology, Legitimacy, and Political Goals

As will be explained in chapter 11, political processes are in more rapid change than perhaps any other time in Latin American history. A growing chorus of voices is demanding a redirection of the economy and an increased democratization, or more precisely, mobilizing societal segments to implement economic reforms in welfare and social security.[68] As the citizenry is ever more cynical about the possibility of strategic change, voters are failing to go to the polls on election day and reluctant to participate in public opinion polls.[69] The history of voting turnout sows a growth over the decades (figure 4.2). Brazil stands as a primary example of a nation with a history of populist leaders who depend on flamboyant public relations campaigns, only to be ousted from office on corruption charges, for instance Collar de Mello in 1992. As one Brazilian critic points out, democracy is not a product to be marketed but must be inherently institutionalized.[70] The 1994 election of Cardoso gave the country a progressive leader, who struggled against political and economic blockages in his attempts to help the poor. Similarly, he tried to persuade congress to adapt austerity measures in order to remove indebtedness. Unfortunately, congressional and economic leaders operate according to their special interests and clients. His successor Luis Silva is also caught between the demands of the left and the right in his attempt to promote a more just society.

The boundary between authoritarianism and democracy is generally precarious. Populism more often becomes rigid rather than permissive. Whatever its history, populism appears in special periods of crisis or impasse. Predominantly an urban phenomenon it emerged as a challenge to a rural elite, however differently expressed by Cárdenas, Vargas, and Perón from the 1930s to the1950s. Yet, a return to populism emerges periodically, as in the 1980s and 1990s with Salinas and Fujimori, each in his particular blend of personality. Both pushed economic neoliberalism while also launching into populist approaches. Venezuela's Hugo Chávez in his appeal to the proletariat demonstrates a more radical form of populism.

The term "neopopulism" suggests a merging of economic and political populism (see inset 4e). This balancing act of selling neoliberalism to the masses is not without risk. The Salinas financial debacle in Mexico, largely due to his own corruption, left his successor Ernesto Zedillo unable to revive the party apparatus nor to resolve economic disparities. Yet, all three parties, PRI, PAN, and the more recent MDR (Revolutionary Democratic Movement), still turn to populist appeals. The Peruvian scenario is no less complex and Menem of Argentina could never resolve the debate of how much neoliberalism could be combined with populism.[71]

INSET 4E   POPULISM, OLD AND NEW: ECUADORIAN STYLE

Populism flourishes throughout the western hemisphere—but in different styles. Probably no populist quite equals Ecuador's Abdalah Bucarem in his 1988 appeal to varied publics. Ecuador has is own unique history of traditionalism and authoritarianism, but populism crept in during the twentieth century. The most enduring leader of the mid-century was Velasco Ibarra, whose personalist approach captured the electorate's emotions in his attack on the oligarchy. After his departure from office in 1972, militarists followed, but from 1979 to the present no military intervention occurred. The 1970s and 1980s were periods of modernization with urbanization and limited industrialism moving forward.

Labeling himself as the "father of the poor," Bucarem's style was one of machismo, street jargon, and promises of a better future. He assured the hungry and the homeless—over 65 percent of Ecuadorians live in poverty—that they would be rescued from the tentacles of the wealthy and the powerful. Whether campaigning on the plaza or on the TV screen he offered a number of crude theatrical techniques, not least singing baddy ditties, seemingly inspired by guitarist Menem and saxophonist Clinton. Also, he arrived on the scene when television became a compelling medium. Moreover, the Lebanese colony and other allies contributed handsomely to his campaign. To reinforce his supermacho image he indulged in sexual innuendo in his speeches—graphically portraying the oligarchy's exploitation of the people as equal to the rape of an innocent woman. He also chose to dance in public and on television with his vice presidential candidate, a immaculately dressed and socially proper woman.

Of Lebanese heritage, mayor of Guayaquil, Bucaram was voted into the presidency in 1996. Once in office he borrowed heavily in order to fund social programs. The result was staggering inflation, higher taxes, and an economic depression, including massive unemployment, Six months later, over a million people massed on the plaza and adjoining streets to demand his ouster.

It is instructive to compare the versions of populism. Bucarem's form stands in contrast to the emotional but tasteful exhortations of Velasco Ibarra. Beyond his disastrous economic policy, Bucarem was simply too crude and banal for the Ecuadorian public.

More than earlier populists Bucarem had the TV screen at his disposal. He was aware of the impact of the photogenic Collar de Mello of Brazil and the unrelenting Fujimori of Peru. Perhaps even more than the earlier populism, telepopulism had to seek approval from a variety of publics. Bucarem focused on the mestizo, both middle and lower class. Little attention was directed to the Indians, as few qualified on literacy level to vote. Whether old or new, populism usually seeks a wavering course between authoritarianism and democracy. The ousting of Bucarem shows that the norms of respectability cannot be totally transgressed.[72]

Personality needs of political actors and the contour of networks and organizations determine the pattern of political processes. An interesting scenario in this constellation is seen in the comparison of the elections of Nicaragua in 1990 and El Salvador in 1994. Both focused on the transition from one political and socioeconomic system to another. In El Salvador the players were more willing to subdue their expression of power in order to enhance interpersonal trust. Outside forces also have their role. In El Salvador the UN remained continuously in an advisory capacity before and after the election. In addition, an Ad Hoc Truth Commission, sponsored by domestic and international agencies, served as a conciliatory agency. In Nicaragua the monitoring was more complex. Not only were the UN, OAS, and the commission led by Jimmy Carter's staff seemingly more concerned with procedure than substance, but a severe deterrent was the presence of still-armed *contras*.[73] Monitoring of the political process also appears in Colombia, where President Ernesto Samper (1994–98) was widely attacked in the press for his apparent acceptance of drug money for his electoral campaign. Removal from office was a threat but the charges were never absolutely proven.[74]

With the new century, the Latin American political agenda finds itself in conflict, alternating between personalism and bureaucracy, civil and military governments. Economic imperatives are all important. An observer of Costa Rica asks why the "social democratic model," a dominant theme of the PLN from 1948 to 1982, became secondary to neoliberalism.[75] The future will determine whether neoliberalism can deliver a just economic return to the majority of the population. Governmental decision making will depend on an uneasy equilibrium between powerful elites and the voting public.

The question remains whether liberal democracy will continue to grow in Latin America. Throughout the region's history, competing economic and political elites shaped the course of government. Is a truly pluralistic society possible? Part of the answer lies with the nonelites, who come into play when instabilities arise. Institutionalization of political parties and the relevance of their values are highly significant. The "deepening of democracy" depends, of course, on social and economic factors.[76] Of growing promise are grassroots groups, organizations, and movements, in which the Internet plays a growing role.

This chapter has focused primarily on the twentieth century in its exploration of Latin American society, notably its economic and political institutions. In view of the region's entry into a global market place and world system, sociopolitical change is even more critical than it was in the past. The complexity of this transition will appear in later chapters, particularly in chapter 11.

Chapter Five

Identity, Modernity, and the Arts

What are the basic values of Latin American society? How do myths and rituals express culture and identity? What are the identities Latin Americans actually feel? How does the region perceive modernism and the postmodern? How do these two concepts differ from those of the Western world in general? These questions are analyzed in a conceptual framework in this chapter. In addition, we shall explore the sociocultural implications of literature and the arts.

The Cultural Matrix

In approaching the question of core values, myths, and, most important, identity—both national and personal—the Latin American landscape reflects contrasting images and realities, as documented in chapters 1 and 3. Brazil has likely the most diverse heritage with its Portuguese, Dutch, German, Polish, Italian, especially African and indigenous roots. The plurality of cultures, African cults, medieval Catholicism, native medicine, with their folklore, mythology, and other symbolic content provide a richness and intense atmosphere, as a study of Recife on the Northeast coast of Brazil suggests.[1] Indeed, all countries of the region have their ethnic and class distinctions. Several nations, such as Bolivia and Paraguay, appear to be relatively isolated. Others are more infused by influences beyond their borders. More than other nations, Mexico has intimate, if ambivalent, ties with its northern neighbor. Argentina and Uruguay pride themselves on an enduring bond with the Hispanic and Italian, from which most of their inhabitants stem. Each country has its unique profile. Latin America often seeks unity in its culture, yet its heterogeneity appears to be salient.[2]

Values and the Cultural Matrix

Whatever the diversity of Latin American culture, several value clusters apply in varying degrees to all nations. Values are seldom given quantitative documentation and are based on the observation of both Latin Americans and outsiders. Nor would a profile of Anglo-American value orientations be less jagged in view of the society's diverse subcultures. As human beings and societies are seldom known for their rationality or consistency, values are often contradictory. Consequently, the order of the following values reflects neither their relative importance nor their validity:

Personalism or the *self as an entity*. Universally, human beings are self-oriented, however, Latin Americans display a unique pattern of self-involvement or saving face.[3] *La dignidad de la persona* or self-dignity is critical, especially for the male as in the trait *hombriado* (maleness)—somewhat different from machismo, which has more of an overtly sexual connotation. The consciousness of dignity is positively correlated with social status—more visible in the middle than lower classes. This need is a means of compensating for the sense of inferiority that many Latin Americans feel when they compare themselves, implicitly or explicitly, with Europeans and North Americans. Also, most of the population, especially indigenous and Afro-Americans, are sensitive to the feeling of superiority or even hostility by the upper and middle classes. Moreover, an insecurity appears in the most European of Latin American nations. As one example, Argentines are reported to be the most concerned with personal appearance and inner security, as shown by a high rate of plastic surgery and psychiatric care.

Self-concern appears in a variety of behaviors. One index of the need to achieve a meaningful identity is the large proportion of the personal budget going to clothing, particularly for women. The personalist aspect of politics is another extension of self or personality as a major preoccupation. Self-serving techniques are universal, but for Latin Americans in their struggle for survival, the search for personal connections is fundamental. Brazilians are highly adept at the use of the *jeito* (literally knack or dexterity) as a means of making the right connection in advancing their mobility.[4] Throughout Latin America *clientelismo* becomes almost an art form involving kin, friends, even casual acquaintances. Whether in job seeking, housing needs, or entrepreneurial expansion the individual turns to a power broker at the local or national level.[5] Further, through most of Latin America the self is protected by verbal devices. On asking street directions rarely have I encountered a *salvodoreño, bogotano,* or *limeño* who would say *"no sé"* ("I don't know"), but offers some response, the information being at least as often wrong as right. (My practice is to ask four or five people and take the modal response as a hopeful route to take.) Expressing ignorance is seldom an option for preserving the self-image—a stance not unique to Latin America.

Familism is not unrelated to personalism. The individual is deeply involved with both the nuclear and the extended family, as will be explicit in chapters 6 and 7. This desire for family support is not surprising for one who lives in what seems to be a forbidding world. This strong family tradition cuts across the society. (Latin Americans who visit or move to the United States comment on how loose the ties are for both the nuclear and the extended family.) For instance, an urban bus driver on Sunday may take along his wife, often with an infant in her arms, to help collect fares or simply as an escape from their one-room dwelling. As another example of familism, an unmarried adult rarely moves out of the family setting—tradition counts as much as the financial burden of renting a room or an apartment. Even narcotrafficking mafiosi adhere to familial norms—the notorious Colombian drug czar Pablo Escobar provided an opulent castle for his extended family. Finally, familism expresses the orientation of Latin Americans to the particularistic rather than the universalistic, as illustrated in the tendency toward nepotism—the *jefe* (manager) is expected to find a job for a close relative somewhere in the office or plant.

Supremacy of being over doing, or a kind of transcendentalism, points to the preference for religious and esthetic values. Although this orientation is less visible in Latin America than in parts of Asia, it stands in opposition to sociologist Max Weber's concept of the Protestant ethic. Avoidance of manual labor by all who can avoid it expresses this detachment, which has social class overtones. Ascribed, or inherited, status traditionally counts more than achieved status.[6] Systematic concern with applied science, technology, and engineering, like industrialization, came to Latin America considerably later than to Europe and Anglo-America.

Fatalism, which suggests a variant of mysticism, is evident in the fascination with magic, mystery, and even death. Beyond the folklore of the Iberian Peninsula are belief systems and rituals of the New World as practiced among the indigenous from Mexico to Patagonia. This fatalistic attitude is seen in the love of bullfights, helmet-less motorcyclists, pedestrians (including myself when I'm in Latin America) madly crossing an intersection with buses and trucks coming from all four directions, or preference for folk remedies over scientific medicine. As an aspect of fatalism, the ceremonies surrounding death stem from both the Ibero-Arab past and customs of indigenous peoples. Moreover, one is struck by the extent of *el luto* (mourning). Widowhood traditionally means six months to a year of black clothing, followed by *medio luto*, of a black band and gray clothing, in addition is the *veloria* and *novena* as a means of paying tribute to the recently departed. In today's urban scene these rituals are noticeably abbreviated. Still, on *El Dia de los Muertos* (Day of the Dead or All Saint's Day, November 1 and 2) not only can the family enjoy a meal at the grave of a loved one, but a sort of fiesta reigns in which death is mocked. Particularly in Mexico, toys, cookies, and the like appear in the shape of skeletons, skulls, and coffins—a means of detraumatizing the threat of death.[7]

An *affect component* or emotionalism characterizes many Latin Americans. Preference for feeling over reason is reflected in the love of poetry, art, and music. A politician or a business executive can write poetry and still retain his masculinity. Religious expression—Roman Catholicism or Pentecostalism, different as they are—points to the strong role of emotion. As another instance, in El Salvador I recall watching moviegoers' reactions to events on the screen. Women more than men graphically emote with sighs and shrieks at the protagonist's questionable fate.

A *sense of hierarchy* refers to the pattern of subordination that Latin Americans accept as given, notwithstanding the ripples of revolt that periodically appear. One example is the class—or caste—system traditionally found in countries marked by the acute separation between Europeans and minorities. The occupational structure or pecking order of an office staff suggests a greater rigidity than found in most Western nations. When in more than one capital at the Fulbright office I asked for the list of housing possibilities, the clerk referred to her superior (who was out of the office), even though she had access to the file. On arriving at the university the guard (who earned about a sixth of a professor's salary) would seldom fail to say "*Buenos Días, Doctor*." (In several countries "Doctor" is addressed to almost anyone who has a postsecondary education; in Chile apparently only a medical doctor is given that title.)

The *realm of the undefined* can be thought of as a blurring between the subjective and objective. This value obviously overlaps with personalism and fatalism. It surfaces in the verbal magic by which the individual convinces himself or herself of a nonexistent reality (a trait that also persists well north of the Rio Grande). Occasionally an innovational or charismatic leader may infuse the public, if only temporarily, with a new version of reality, as seemingly is presently occurring in Argentina (inset 5a). This attitude toward an ambivalent reality possibly stems from the mix of Roman Catholicism with indigenous traditions and beliefs.[8] Another facet of this sense of the undefined is the perception of time, as noted in visitors' remarks about the vagueness of appointments as compared to Western norms. Much has been made of the "mañana" spirit or a "laid back" approach, but this attitude varies according to the urban–rural dimension, temperature, and other factors. For instance, a survey of samples from thirty-one nations found the pace (e.g. walking speed) in three countries (Brazil, El Salvador, and Mexico) to be relatively slower, but Costa Rica was comparable to the United States—a reminder of the arbitrariness of judgments regarding cultural traits.[9]

INSET 5A   A SHIFT IN PRESIDENTIAL VALUES

Only days before the 2003 presidential election in Argentina, Carlos Menem dropped out of the running, making Néstor Kirchner, a relative unknown, the winner by default. In the first run-off with multiple candidates he won only 22 percent of the vote. Governor of a remote province in Patagonia, the fifty-three-year-old Kirchner, nominally a Peronist, was a mystery to most Argentines. In less than two months in office he took on the Establishment, embedded in years of corruption, notably in the Menem presidency, 1993–2003. In assaulting one institution after another, he lost little time in purging the armed forces and national police of officers associated with human rights violations. In his expressed desire for an independent judiciary he turned his reformist eagerness to ousting what he considered incompetent or unscrupulous judges. His relentless pursuit of public figures prompted several critics to accuse him of a perpetual witch-hunt.[10] Possibly the most critical remaking of the past is his insistence, with the support of the majority of Argentines, on abrogating the amnesty awarded in the late 1980s to the military officers responsible for the torture, disappearance, and murders of an estimated thirty thousand persons during the "dirty war" (1976–83). As another attack on the Establishment he reopened a far-reaching investigation of the 1994 bomb attack on Associación Mutual Israelita Argentina, which caused the death of eighty-six people—an inquiry that Menem had allowed to languish.

Two important cultural dimensions are *time* and *space*. The heritage of classical New World civilizations—Aztec, Maya, and Inca, for example—and the perceptions and attitudes of more recent indigenous sources have left their stamp. Space and time reflect ancient forms of perception and cognition. As time was circular or cyclic for the Mayans and the Inca, *campesinos* view the life cycle as reliving the fate of their ancestors. There is not the linear concept of ever higher goals (unrealized though they may be) permeating Western nations.

It is unclear whether space represents for Latin Americans more of the division between secular and sacred than it does for North Americans. Certain streets and plazas have special significance for Latin Americans as they are involved in the "geography of protest," whether the competition for urban space among the poor in Rio or on the Plaza de Mayo in Buenos Aires (during the 1976–83 dictatorship) with the silent mothers grieving for their lost children.[11] A Brazilian military parade assumes special significance as it takes place in commemorative governmental areas in Rio, whereas Carnival occurs in the streets.[12] The barrios of Buenos Aires—Belgrano, Monserrat, and others—evoke specific images, drawn from history, life experiences, and the writings of its authors, Jorge Luís Borges especially.

Lest the reader consider the previously mentioned value orientations to be interpreted as unfavorable to Latin Americans, she or he might compose a list for North Americans. It might well include materialism, hedonism, self-satisfaction, gullibility, generosity, gregariousness—in other words a value composite that might be tested for validity, as might any catalogue of value orientations of Latin Americans or Europeans, with a wide range of cultural and subcultural variations—nationality, ethnicity, regionalism, and so on.

The Search for Identity

Because of diverse cultural traditions Latin Americans experience a difficult task in establishing their meaning and identity—nationality, region, family, and personality. Cross-currents compete against and between foreign powers, desperate caudillos, restless minorities, and contending institutions. These pressures account for an insecurity in most citizens, particularly those whose destiny is least favored by economics and politics. Clearly, the struggle for identity draws on folklore, the arts, linguistics, sociopolitical movements, and international relations, to mention several sources. Sometimes identity expresses what one is not. A citizen of the Dominican Republic draws some of her or his identity from being anti-Haitian or anti-American.[13]

Myth and Ritual

Meaning and identity in any culture strongly involve myth and ritual, along with cultural stereotyping. It is desirable to place the question of myth and ritual in a conceptual and historical framework. For half a millennium Latin Americans have lived from crisis to crisis with the occasional vision of a new social order, as articulated by a philosopher or perhaps a new-wave politician. The dream is hardly novel. Even the Inca state is likened to Thomas More's Utopia.[14] The utopian dream appears among depressed segments of Latin America's population. It also finds expression among thinkers. The abrupt superimposition of Spanish feudalism on a number of New World cultures was further complicated by a lack of sociocultural direction after the Wars of Independence, in contrast to the relative continuity in the United States following its break with England. Since the 1960s the Latin American hope of achieving a truly functioning democracy appears at a latent or marginally conscious level and overtly in social movements, as in Chiapas in 1999. The wrath aimed at the governments in Venezuela, Ecuador, and Peru in 2004 appears to signal the onset of a violent social movement. Moreover, throughout Latin America the drift toward pluralism hides itself behind "an egalitarian mask."[15] That is, even when a nation democratizes, political leaders and other elites tend to gloss over ethnic, economic, religious, and other cultural barriers.

Aspiration toward an equalitarian order assumed differing expression over the centuries. Bartolomé de las Casas and Pedro Claver pled for humane treatment of Indians and Afro-Americans, respectively. They were followed in the seventeenth century by the Jesuit Antonio de la Vieira. New voices were heard in the following two centuries, but in the early 1900s the quest resurged for elevating the indigenous to their rightful position. In his book *La Raza Cósmica* (The Cosmic Race) the Mexican philosopher José Vasconcelos proclaimed the Indian Hispanic blend as a dynamic contribution to world culture.[16] Carlos Maríategui of Peru focused on the pre-Columbian cultures as being the ideal of perfection possible in a New World Marxism. In contrast to most of his compatriots, Argentina's Leopoldo Zea looked to Latin American unity based on *mestizaje*, which he felt would eventually bring the region to a cultural status that might compete with the image of the United States. A more complete exploration of these concepts of identity is found in inset 5b.

INSET 5B   CHANGING VISTAS IN COPING WITH IDENTITY

The opening of the twentieth century brought a cry among Latin Americans from the ideologies of the nineteenth century. Writers and philosophers moved to assert an identity of their own—a need to liberate themselves from the cultural dominance of Western Europe and the aggressive U.S. materialism—even though they were to borrow from European and a few North American thinkers, but at least different ones that they had followed before. This struggle for cultural liberation or "Latin American modernism" arose among university staff and students in 1904.

Particularly vulnerable was Auguste Comte's positivism (progress as based on scientific advance), a major theme in Brazil. Antipositivism emerged in Mexico, Peru, and Argentina, spreading to Uruguay, Chile, and Bolivia. The philosopher William James' pragmatism crept into the salons and universities. Antonio Caso and José Vasconcelos of Mexico and Alejandro Korn of Argentina were among the more prolific in the movement to find a Latin American version of pragmatism. The movement against positivism was centered in Henri Bergson's intuitionism, as a contrast to the scientism of Comte. Probably most influential in this breakaway from the past was the eclectic philosophy of Spain's José Ortega y Gasset, whose *La Rebelión de las Masas* intrigued readers on both sides of the Atlantic.

As another direction of Latin American *modernismo* a host of writers turned to attacking sociopolitical anarchy, *caudillismo* (classical strongman leadership), clerical power, despotism, injustice, and inadequate educational systems. This plea for reducing social inequality was articulated by Justo Sierra, who transformed public instruction in Mexico, especially his reorganization of the national university in 1910.

In a different vein is Uruguayan José Ennrique Rodó's *Ariel* and *Liberalismo y Jacobinismo*. These books had an enthusiastic readership in the theme of superiority of the intellectual minority over the masses as well as Rodó's preference for the spirituality of Latins as compared to Anglo-Saxon utilitarianism—a thesis particularly acute to many Latin Americans' identity.[17]

These intellectual, mostly humanistic, movements had a marked influence among intellectual and occasionally political elites, as Latin Americans were trying to find an equilibrium between democracy and authoritarianism. By the 1930s new waves of dissent directed to economic issues and the needs of the indigenous, notably in Mexico and Peru. In addition, nationalism as voiced by populists became increasingly visible on the political stage. Still, the intellectual writings of the first two decades of the century lingered on in academic circles.[18]

In several concepts Argentina contributed its share of image making and aspirations toward a meaningful world. The nineteenth century had its philosophical rivalries rooted in governance. Following the ruthless Rosas dictatorship (1835–52) was the relatively enlightened presidency of Domingo Faustino Sarmiento, who in *Civilization and Barbarism* (1845), often referred to by the title *Facundo*, outlined the opposition between the urban and the rural. For Sarmiento the rural represented not only the heritage of Rosas in his rustic primitiveness, but also the image of the gaucho, who roamed the pampa in his struggle to survive on beef and *maté*. The legend of the gaucho was set in José Hernández' poem *Martín Fierro* (1872). Inspite of the hero's drunken brawls and other behavior patterns reminiscent of the North American cowboy, the gaucho had a code of ethics based on comradeship, to which not all gauchos conformed.[19] Hernández himself softened the portrait in *The Return of Martín Fierro* (1879). Railroads, settlement, and public education were to bring the twilight of the gaucho culture by the end of the nineteenth century. Still, the gaucho remains as a historic icon and must be defined in the context of time and place.[20] No less than in Argentina the cult lives on in southern Brazil in the Gaúcho

Traditonalist Movement with its two million members, largely urban yet glorifying a rural tradition. Myth and ritual fuse as specialty shops, restaurants, radio stations, and other avenues of popular culture celebrate this bucalic image.[21]

The extent of myth making continues in the search for meaning. Societies and individuals arrange their thought processes in order to establish a coherent self-image. For instance, Mexicans have a myriad of images of their cultural heritage, but two stand out: (1) consciousness of being a nation long before the conquest, (2) the presence of God delivered to them in the image of the Virgin Mary at Guadalupe.[22] Further, the façade of democracy extends well beyond Latin America as does the theme of racial equality. On this point chapter 4 pointed to Argentina, Brazil, and Venezuela as examples of hypocrisy, notwithstanding a marginal improvement of ethnic relations over time. Another example of myth making is Cuba, where the late-nineteenth-century poet and reformer José Martí urged justice for the Afro-Cuban—a goal only slightly realized in the early twentieth century.[23] Reports from contemporary Cuba reveal that the problem is not yet fully solved.[24]

The mass media are a vehicle for assigning identity, sometimes intended but more often unintended. From the 1920s to the 1950s the actor Cantinflas communicated something of Mexican charm in his warm, humorous, and occasional mischief. In a different mode Carmen Miranda conveyed to the United States in several films between 1935 and 1955 an image of a Brazil that was more myth than reality. Brazilians resented the Hollywood culture industry's portrait of gender and national identity.[25]

The Need for Ritual

For all societies, ritual is a primary means of achieving meaning, continuity, and authenticity. Andean cultures are particularly rich in incorporating ancient civilizations in today's poetry, music, dance, art, and religious rites—for instance, pre-Inca designs in today's textiles. Similarly, the Aztecs, Maya, and Quimbaya still penetrate cultural institutions in rituals from Mexico to the Andes.

Many of Brazil's rituals are derived from African and indigenous groups fused with Christian themes from Portugal. Most striking of Brazilian rituals is Carnival, which takes place in the three days preceding Lent. First of all, its dazzling, sumptuous display points to a psychological escape from life's drabness. As Roberto DaMatta points out, carnival removes the individual from the home to the street, from the sacred to the profane. Its deeper meaning lies in the transformation and displacement that enables Cariocas (Rio residents) to survive *a dura realidade da vida* (the harsh reality of life).[26] Even more than soccer and *telenovelas*—two other national obsessions— Carnival expresses both the outer and inner selves—the conscious and unconscious. Carnival also delivers an ethnic message as it involves African themes with a stress on "black power" and "soul music," combined with poetry and music of the Candomblé and other religious cults.[27] The inversion or reversal from the ordinary self to a transfixed state has a remarkable psychological significance. Seemingly, this annual ritual has an emotional impact beyond what most North Americans and many Latin Americans fail to experience. For some observers, it is questionable whether the fantasies of carnival are not more real than the daily routine of everyday life.[28] To a slight degree, carnival brings the extremes of society together; however, the upper classes have their own celebration in private clubs and costume balls. The carnival moved gradually from the cultural hegemony of upper classes to lower-class Afro-Brazilians, who fostered the rise of the samba, which became a dominant symbol by the 1930s.[29] Moreover, in recent decades tourism has infiltrated this pre-Lenten ceremony.

Parameters of Identity

Latin Americans probably have greater identity problems with their societal climate than do most Westerners. Harassment by powerful states of Europe and North America began in the colonial period with Spain as the principal enemy. However, through much of the nineteenth century a referential feeling toward Europe, especially France, persisted.[30] By the early twentieth century it was the heavy hand of the United States and a questionable relationship continues to the present. Latin Americans ask themselves: "Who am I—European, Indian, African?" According to observers, many Central Americans wonder, "are we superior or inferior to those about us?" "What is our heritage?" Economic, ethnic, and linguistic discrimination is a factor. Mayans in Guatemala and Afro-Caribbeans on Costa Rica's east coast find their identity complicated by a subordinate relationship to the dominant Spanish speakers.[31] Often the question is posed as a duality: Christian or non-Christian, lower or middle class, American or European?[32] Other contradictions complicate the formula—the sacred versus the secular, individualism versus equalitarianism, official versus "real" history. With these alternatives the search for models is seldom resolved. Because of isolation and cultural controls, each group holds fast to its own culture. For instance, central to the identity of many Mexicans is the survival of indigenous roots, which first became lost in a national culture, later in a global one.[33]

Language is a significant variable in determining one's identity. For the vast majority of Latin Americans, Spanish or Portuguese is the linguistic orientation, both languages being interwoven with indigenous and African borrowings. In the Andes and Central America millions of speakers feel isolated from the national culture because of their language identity, even when they adopt various cultural traits (occupation, clothing, technology) from the national culture. On that point, over a million-and-half Guatemalans speak one of twenty indigenous languages. Also, as in Honduras, Nicaragua, Costa Rica, and Panama, Creole and Caribbean speakers confront each other.[34] In these and other countries, degrees of bilingualism pose divided loyalties between the majority and minority language. The individual's preference shifts in the choice of "code" or language. Life experience, formal and informal socialization, socioeconomic forces, and definition of the situation are among the inputs influencing language choice and consequently cultural or national identity.[35]

Another factor determining the individual's place in history is the changing style of political and economic development. For Latin Americans identity cannot be separated from changing political and economic development. As one Mexican describes his country's shifts: (1) 1940–70, rise of capitalism, (2) 1970–82 populism, (3) 1982 to a shaky present, neoliberalism.[36] Moreover, in the search for identity, individuals struggle against a tide of economic frustration and its psychological stress. In the 1996 World Human Values Study of forty-three nations of the world, a positive correlation appears between the feeling of well being and GNP per capita, including four Latin American nations (Argentina, Brazil, Chile, and Mexico), which lagged Western Europe, although they surpassed Eastern Europe.[37] More to the point, the study also shows the struggle for human survival looms larger than in the world's more developed areas. Beyond statistics we are almost constantly aware of this societal crisis in Central America and even more acutely the Andean nations, as the political convulsions in Venezuela since 2000. Also, Ecuador suffered through the chaotic presidency of José Velasco Ibarra intermittingly from 1934 to 1972 followed by military and civil governments, both regressive and progressive, reaching a crescendo of disarray by the end of the century.[38] Worst still was the impact of the Shining Path on the Peruvian consciousness during the 1980s (inset 5c). In the same vein, Colombians may see themselves as the victims of the ultimate tragedy in an unending cycle of violence, narcoterrorism, and guerrilla warfare. However, parallel to this physical and psychic anguish are cross-cultural comparisons and the feeling of being trapped in the past. As one social critic puts it, "we Colombians still cling to nineteenth-century thinking and traditions." For many Colombians, Argentina and Mexico (also Venezuela until the year 2000) are ahead culturally.[39] Still, inferiority feelings on the part of Latin Americans are less evident today than a generation ago.

INSET 5C   NATIONAL TRAUMA: IDENTITY IN REVERSE

Identity, both personal and societal, can assume many forms. For a society in disequilibrium, a succession of traumas, a leader, and a movement can bring on a momentous change in attachments and identities. A notable instance is the Sendero Luminoso (Shining Path, SP) that shattered Peru beginning in 1980.

This movement resulted in the death of over twenty-five thousand people, largely in small towns and rural areas, but later the SP moved into the capital Lima. The movement's leader, Abimael Guzmán, an overweight, bespectacled middle-aged philosophy professor, stands in vivid contrast to other charismatic revolutionaries such as Che Guevara or Fidel Castro nearly a generation earlier. Guzmán was official leader of Peru's Communist Party, and usually called by his followers as "President Goncalo," inspired by Mao Ze Tung of China. In the Guzmán dogma, no compromise can be made with the authorities. Strikes are called, houses blown up, all resistance met with violence. Men, women, and children are to be executed before the eyes of family members.

Because of the revolutionary fervor in university circles, the movement appealed as much to middle- and upper-class youth as to landless, indigenous peasants. The explanation stems from the almost steady rise of poverty from the 1970s to 1990s. Intransigent and inadequate national leadership brought many Peruvians to question their loyalties and identity, leading to a consequent shift to a radical allegiance.

With the Fujimori presidency, the struggle by the military, 139,000 strong, increasingly pressed its campaign against the Sendero. On September 12, 1992 Guzmán was captured in an upstairs dance studio—the abode of his devoted young wife. He remained his theatrical self, recalling the martyrs before him, saying "it's now my turn." In prison, he attempted to continue his leadership, but the movement was finally limited to occasional demonstrations and strikes.[40]

At the base of establishing identity is the *sociocultural* setting. Mass media, notably television, and a growing use of cyberspace are principal instruments in bringing about this awareness of a larger world. Even the ubiquitous bus takes the *campesino* beyond the *município*. In a world of multi-textured stimuli (linguistic, religious, and other symbolic systems) imagery becomes a mark of identity. Argentines have the tango, Brazilians the carnival, Mexicans may identify with bullfights and mariachis, also recently reinforcing their uniqueness through a specific cuisine.[41] Soccer is a national passion nearly everywhere. Nonetheless, beyond the more salient signs of identity are the intellectual contributions. Octavio Paz represents Mexico to the larger world, not only in his poetry but in his *Labyrinth of Solitude* (1950). Similarly, Gilberto Freyre in his *Casa Grande e Senzala* (The Masters and the Slaves) (1933) offers a penetrating vista of national character, bringing the tragedy of the parched, poverty-stricken Northeast to the Brazilian public.[42] For many Colombians, literature, notably García Márquez, is vital to their cultural pride.[43]

Still, Latin America with all its cross-currents appears to lack sufficient homogeneity to foster a secure identity. As with many cultures, contrasts or polarizations occur in the juxtaposition of young and old, tumult and tranquility, independence and dependence, autonomy and control, poverty and affluence, wisdom and ignorance.[44] Through the centuries, elites contrived a false identity.[45] Traditionally, loyalties are fixated on the military, the hacienda, or the church. Democracy is not yet sufficiently effective to permit the *clase popular* (working class) or even the *clase media* (middle class) to feel comfortable with the state. Social stratification remains a principal obstacle to a meaningful national identity.

Space, Politics, and National Identity

Universally, *territoriality* is a factor in identity. As in other parts of the globe, rural inhabitants look to their local heritage, whereas nations gravitate to a wider assortment of moorings. For instance, Ecuadorians are uncertain about national identity as they are aware of three disparate entities—*costa, sierra, oriente* (coastal plain, Andean mountains, and eastern basin), each with its distinct cultures and economies.[46] Still, Ecuadorians are surrounded by landscape, architecture, and other images, including the equatorial monument just north of Quito reminding them that they are literally at the center of the planet. History counts too. By the 1880s a monetary system, the Quito–Guayaquil railroad, secularization of schools were established; other nationalizing events were soon to follow.[47] Even so, national unity is seldom complete. Ecuadorians assign Africans to the coast, Spanish and mestizo to the *sierra*, and indigenous to the *oriente*. Also, cultural traditions and historical events can produce local variants (inset 5d).

INSERT 5D   AN ANDEAN COMMUNITY
IN SEARCH OF IDENTITY

Possibly more than other Andean nations, Ecuador represents a patchwork of local cultural variations, not least its periodic festivals. In the parish of Salasaca the indigenous are demanding a more legitimate means of establishing their religious festivals. They desire closer attention to the sacred images in their history and landscape. This search for identity intersects with ethnicity, politics, religion, and the natural landscape.

A principal aspect of the festival is the choice of sponsors. Festival sponsorship has its roots in the early colonial exploitation of Indian landholdings, in addition to the accretion of other traditions such as communal work projects, for instance, upkeep of the cemetery and cleaning irrigation canals. The system of establishing sponsors by means of *alcaldes* (mayors or, in this context, community elders) is by an election process supervised by a committee of nuns of a given order. Essentially it is a competition among the parish's leading citizens, largely based on the propriety of their religious practices, including advising one's peers to maintain Catholic morals such as marital fidelity. Among other requirements, the candidate must visit every parish household.

Moreover, because of a rising indigenous consciousness, the degree of a sponsor's knowledge of shrines or sacred places has become salient. For example, the top of a mountain may be where a shaman once said a prayer or made an offering, or where a saint appeared. For the Salasacans, identity is attached to a mystical image of pre-Columbian, colonial, and even modern images and rituals.[48]

National unity and identity is difficult in small countries, but hardly less so in large, highly diverse ones. For example, Brazil is not only divided by North and South, but also the West, notably the Amazon basin is ever a source of anxiety as well as pride.[49] Chile extends twenty-eight hundred miles from north to south and many provincials refer to Santiago as a corrupt, smog-laden city. As another instance, Colombia was not really unified until the air age. The people of Medellín seldom referred to themselves as *colombianos* until the mid-twentieth century. Even today many Colombians think of themselves as, for example, *antioqueños* or *tolimenses* (residents of Antioquia and Tolima, respectively). As another touch of regionalism, each

Colombian province until recently imposed its own taxes on their alcoholic products. It would be difficult to find a country without regional preference (the prejudice of a Milanese toward a Sicilian or a Bostonian toward a Texan or vice versa). Urban–rural and regional conflict is universal but in Latin America it has a special historic significance, notably regional leaders as the history of Brazil illustrates—São Paulo, Rio Grande do Sul, and other states have their traditions and favorite *politicos* (politicians).

Inevitably a fundamental hurdle to regional and national identity is the realization that one's destiny is often determined beyond the national border. This insecurity was most articulately voiced at the beginning of the twentieth century. The Uruguayan José Enrique Rodó in 1900 wrote in his *Ariel* of two types of civilization—Calibán, a mass-driven materialism exhibited by the United States versus Ariel, a spiritual–rational order. Although Rodó admired to a limted degree the utilitarianism of the North American colossus, he felt that it contradicted the basic tenets of civilization. Appearing on the eve of Theodore Roosevelt's seizure of Panama, Rodó's book had no small effect in Latin America. The theme of establishing national and regional identity continued in different flavors throughout the century.

Modernity and Postmodernity

In view of rapid social change, intellectuals are raising the question of what constitutes modernism and postmodernism. This inquiry, of course, begins with the definition of the term "modern." Although little agreement exists about the meaning of modernity, analysts assert that it is about change— flexibility in the ebb and flow of events.[50] For many historians modern begins with a cluster of sociocultural events, the advent of science and the Protestant Reformation beginning in the sixteenth and seventeenth centuries, accelerating into the Enlightenment of the eighteenth century. For others, the term refers to the rapid industrialism and urbanization of the nineteenth century. Then there are those who refer to the culmination of technical and ideological innovations in the twentieth century. Spain and notably Latin America lagged much of Europe in these changes, but over the last several decades the region moved rapidly forward to fill in the vacuum. Throughout urban society a vibrant dialogue took place in industrializing societies even in the first decades of the twentieth century. In the political sphere, modernity trends made possible Vargas's Estado Novo (1937–45) in Brazil.

For Latin American observers, modernity assumes a variety of meanings. One asks: are modernity and globalization the end of colonialism, or simply a new form of dependency?[51] Modernity appears predominantly in the metropolis, notably highly urbane ones such as Buenos Aires, Mexico City, Santiago, or São Paulo, where intellectuals discuss the latest scientific theory, varying lifestyles flourish, and the arts reflect movements such as expressionism or surrealism drifting in from abroad, especially in the 1980s and 1990s.[52] Although modernism may first appear in the upper and upper–middle classes it gains momentum in the popular culture of the middle class, as depicted by sociologist Gabriel Careaga.[53] Another Mexican writes of modernity as the Europeanization and North Americanization along with personalization of his culture.[54] Still another observer questions whether Latin America ever attained modernity as the region never realized the rationality implied by the concept of modernity—or the region failed to resolve its economic, social, and political grievances.[55]

Theories of Modernity and Postmodernity

Most theories focus on modernity in the context of advancements in the twentieth century. *Postmodernity* develops in varying directions, with questions about these advances. Fundamentally, the movement rejects the rationalist, bureaucratic, and scientific–technical aspects of our society in preference to a reformulation of feeling and even the spiritual component of human behavior. Postmodernity refers to questions surrounding the foundations of belief. The sociology of knowledge, hermeneutics (analysis of meaning), and various other viewpoints, such as neo-Marxism and feminism, offer a skepticism about the validity of science, irrationality of political and economic systems, and the tenets underlying social institutions, education, and religion, among others.[56]

In exploring the meaning of modernity and postmodernity in Latin America it is useful to introduce the philosophical movements underpinning these two concepts. For one, Karl Marx (1818–83) proclaimed that relativity of knowledge as bourgeois capitalism deluded us into attaching market value to all phenomena—a theory accepted by many Latin Americans through the twentieth century. Not only was Marx the inspiration for the Cuban and Nicaraguan revolutions, but his ideas became the rallying cry of communist parties in other countries. Also, Marx's theory, particularly his sociology of knowledge, found its way into social thinking and literature, in addition to his unorthodox analysis of social sciences, which had appeal to Latin American intellectuals.

In the context of Latin American culture, a number of scholars are examining the meaning beyond conventional interpretations of history. For instance, the legend of the Virgin of Guadalupe affects various levels or segments of the national consciousness, poor and rich, powerful and powerless, as visitors to Mexico realize as they see the image in almost every bus and taxi assuring the driver of security. When the Virgin appeared to the peasant boy Juan in 1531, the church hierarchy interpreted the vision as proclaiming the land as sacred, thus belonging to God and therefore to be maintained by state and church. In the contemporary context Latin Americans debate the meaning of neoliberalism or the challenge of democracy, with little agreement on how these concepts are to be examined or analyzed.

Has, or has not, Latin America reached a crisis in the confrontation in the extremes of modernity? As one critic maintains, Latin America has no more, perhaps even less, of the schizoid symptoms—from hyperbolic mass media to technocratic corporations—than does the rest of the world. In other words, society may be less out of control than in major Western nations.[57] Other observers are uncertain. In reference to Mexico, the betrayal of the revolution and the unsavory features of modernity—economic practices, a frozen political process, synthetic mass media—are real.[58] As with the United States, gadgetry, consumerism, televised political campaigns have invaded ever-growing urban centers.[59]

The Meaning of Modernity in Latin America

The signs of modernity are as imprecise for Latin America as they are for the rest of the world. Still, a number of changes took place in the twentieth century that brought much of the region out of feudalism, including a political system based on *caudillismo* ("strong man" leadership):

1. Economic development. Industrialization encouraged urbanization, educational expansion, and at least a marginal recognition of labor's rights. By the beginning of the twenty-first century an increasingly sophisticated service economy involved larger numbers of professionals and semiprofessionals.
2. A remarkable demographic shift, including a decreased birthrate, a greater life expectancy, and moderate rather than high growth rate.
3. Secular values as opposed to traditional religious values. Pluralism began to threaten a monochromatic culture. A bipolar culture—rich and poor, past-oriented and future-oriented, religious and secular— gave way (however imperfectly) to a multipolar society.[60]
4. Emergence of a new identity. If far from complete, this revised image of the society at least found a partial compromise between a European and an indigenous heritage and a modus operandi with the outside world, notably Europe and North America.
5. Questioning of myths and rituals. Urbanization, education, and the mass media introduce relatively objective approaches to reality.

Obviously these indices of modernity are only partially visible and affect a minority of the population in most countries. In viewing modernization in a cross-national framework, ideology, politics, economy, and religion shape differing patterns of cultural change. Yet, despite secularization, traditional values persist. For instance, in the comparison of value scales for eighty-five societies around the world, overlappings prevail. Nonetheless, Latin America forms a cluster somewhat different from other national groupings (Catholic Europe, South Asia, among others), that is, more traditionalist than secular–rational, but intermediate on the survival versus self-expression dimension (in other words, the contrast between socioeconomically marginal societies and the urbane consumer cultures).[61] Moreover, modernity does not guarantee democracy, nor have the impoverished significantly improved their situation, but a middle class is now more visible and vocal in contrast to the pre-1960 Latin American scene. International events and processes can alter a nation's search for both modernity and identity (inset 5e).

INSET 5E   CRISIS AND IDENTITY

The conflict that many Latin Americans feel in their search for identity characterizes Panama. This country arose in 1903 when a popular uprising conveniently permitted President Theodore Roosevelt to recognize Panama's independence in order to realize his plan of building a trans-isthmus canal. In succeeding years the country had to search for its meaning in a culture with conflicting values. Its elites looked to the United States in their urge for progress and the concept of liberalism, but realized that their cultural ties were European, notably Hispanic. The increase of Afro-Caribbean workers during the building of the canal only intensified ethnic strains. This sense of national identity has dominated the political scene for over a generation, especially during the presidency of Arnulfo Arias Madrid, who assumed the presidency in 1940, declaring an exclusionary version of nationalism. By the 1960s a new national identity emerged as ethnic groups—Hindus, Arabs, Afro-Caribbeans, and Hispanics—gained political force. General Omar Torrijos became president in 1968, initiating reforms in housing, labor code, public health, and education. Consequently, national identity focused on the needs of marginalized segments of the population.[62]

Clearly the definition and parameters of modernity and postmodernity are slippery, that is, less than objective, even though a number of scales attempt to measure variables such as traditional and rational values, future-orientation, achievement motivation, among others. According to Ronald Inglehart, the distinctions between materialism and postmaterialism, modernity and postmodernity are arbitrary. He also warns of the danger of using Western dimensions or attaching moral superiority to a given definition. Modernity represents the overthrow of tradition in favor of rationality, science, and technology, whereas postmodernity means a reevaluation of traditional and contemporary societal trends and the advent of new lifestyles, cultural diversity, and individual choice.[63] Postmodernity may also connote the "commodification of culture"—sale of body parts, prospects of cloning, cultural transformation by tourists, to mention a few instances.[64]

For many Latin Americans, modernization signifies accepting the norms of other Western nations. In the 1920s, modernism in the arts in Argentina, Brazil, Chile, and Uruguay, reflected a European linkage. At the same time, counterarguments in Mexico and Peru were urging a meeting ground with the indigenous. Modernity also suggests an escape from colonialism. In several countries the keynote of modernity was a frozen progress clinging to an outworn positivism—"Baroque modernism" in the words of Laraín Ibañez. For him, modernization never brought together the individual and the institution, the private and the public.[65] However, anomie or the loss of personal identity emerging from the growth of the industrial machine from the 1950s to the present digital, global universe is visible proof for many Latin Americans that they are not adjusted to this new rationalized order. Fundamentally, as one Latin American critic points out, modernity—or supermodernity—calls out a rigid examination of belief and behavioral systems or *desideologización* (de-ideologizing)—in other words, a new ideology based on a global democracy.[66] By the 1980s cyberspace signified a new meaning of modernity, even beyond the Western world. By the 1990s this new model of communication and processing data together with a globalized neoliberalism engulfed Latin America.

The Questionable Meaning of Postmodernity

According to many Latin Americans, their society is moving into a postmodern abyss similar to what they perceive in Europe and especially North America. These changes fall into several categories. One is the obscenities of the economy—inflation, national debt, starvation, and so on. The glaring maldistribution of wealth is anything but new, but appears even more depraved in view of a glossy mass-media-driven society exacerbated by hyperinflation. Mexican sociologist Néstor García Canclini regards rampant consumerism as a leading symptom of postmodernism. As handcrafts give way to manufactured goods, production and consumption are almost totally mechanized. Two, the mass media reach a new crescendo with their "hyper reality" and a shattering of traditional norms. For instance, by the late 1980s North American style rock supplanted *La Onda* on Mexican sound waves.[67] Three, with urbanization and over a dozen cities reaching beyond five million population, lifestyles in Latin America move from traditional modes to new levels of personal disorganization, from anomie to delinquency. On that point, one Venezuelan asks: if the blaring car radios on the streets of Caracas are not due to a fear of quietude?[68]

Again, to what degree can one designate the parameters of modernity and postmodernity? The definitions remain arbitrary. As the unsavory was with us well before the postmodern or modern, it is difficult to assert that the obscene constitutes postmodern. What is clear is the geometric rate of change in contemporary society. Latin America is no exception.

Literature and Modernity

As we approach Latin American literature it is worth mentioning that the term "*modernismo*" first surfaced toward the end of the nineteenth century with the Cuban José Martí and the Nicaraguan poet Ruben Darío. Later came a number of transformations, especially Argentina in the 1920s, pointing to literary themes emerging in Europe. Modernity also referred to the abandonment of older styles such as romanticism, naturalism, and realism, or what new authors referred to as pseudo-realism. Novel expressions in literary and artistic production arrived in successive waves, notably after World War II—the 1970s and 1980s being a high point in a near flood of creativity, in no small part due to the explosive societal events of that era. Moreover, in the twentieth century the world was exposed to cataclysmic breakthroughs in philosophy, science, technology, and lifestyles. The works of Darwin, Marx, Freud, and Einstein reverberated throughout society. The effect of two world wars, the space age, electronics, and cyberspace have no small impact on the way that individuals view their universe. Writers and artists can hardly fail to respond to this transformed world.

In addition, Latin American literature and art reflect the conflict of ethnic and social class divisions, revolutionary social movements, struggles for democracy and equality, and the search for identity, both societal and personal. As Mexican writer Octavio Paz suggests, a binary principle expresses itself in literature and the arts: new themes emerge between "you and me," "them and us." Confrontation becomes the dialogue of history.[69] Identity and power are recurrent themes in Latin American literature.

The term "modernity" usually refers to a mélange of societal changes, "postmodernity" more often is used in the contexts of the arts; yet both terms are relative. Still, as implied earlier in this chapter, the distinction between modernity and postmodernity remains murky, with no agreement about heir meanings.[70] Literary historians may use postmodernity as they are confused about what modernity means. Others think of postmodernists as simply neo-modernists. As postmodernity usually refers to a period of rapid social change, in the context of literature and the arts the term can refer to a startling innovation of themes and styles.

Whatever the evolution of modernism, it represents both experimentation in subject matter or technique or both. Throughout the twentieth century, literature tried to come to terms with its European antecedents, its indigenous roots, creating a product reflecting a regional or national setting. A common theme was a reaction to the materialist and mechanistic aspects of European and Anglo-American culture. However, the outpouring of creativity assumed a number of directions as fashioned by nation, region, individual experiences, and social forces. Surrealism, Cubism, and other avant-garde movements swept the intellectual elites, especially in the cone countries. Yet, Latin American writers struggled to find their own equilibrium and identity. The outstanding figure through much of the twentieth century was Borges (inset 5f).

INSET 5F   BORGES: A LITERARY GIANT

A far-reaching break with the past were the writings of Jorge Luís Borges (1899–1986). Although a *porteño* (resident of Buenos Aires) Borges was steeped in

European culture. In 1921 he founded the *ultraísmo* movement as a revolt against earlier *modernismo* advocates. His early poetry has an almost mystical tone. In the 1930s he turned more to fiction, particularly short stories. Still later he became primarily an essayist.

Throughout his writings is a sense of irony, his antiheroes become heroes or the reverse. His writings reflect the irony, skepticism, and intensity reminiscent of a number of writers—Thomas Swift, Edgar Allen Poe, G.K. Chesterton, and Franz Kafka. His essays and more than occasionally his fiction range between ethics, esthetics, metaphysics, and religion, questioning the meaning of civilization. His dualist treatment of the material and idea, particularly in his poetry, reflects the idealism of the eighteenth-century philosophers George Berkeley and Immanuel Kant. As a writer in the formal structuralist tradition, his poetry, fiction, and essays mirror the chaos of political and international events of the twentieth century, including the implications of the Holocaust. He was also aware of fascist tendencies in his own country from Peronism to the Falkland/Malvinas War. However, observers today ask whether Borges, with his prestige and clout, sufficiently attacked the military dictatorships (1966–73, 1976–83)—a possible explanation of why he failed to win the Nobel Prize for literature. His seemingly apolitical stance still confounds critics.

Still, Borges' highly intellectual approach continued in various Argentine circles. One of the more influential was the journal *Contorno* from 1953 to 1959 and its attempt to facilitate the end of *peronismo*. After Perón's fall in 1955, *Contorno*'s articles focused on the means of realizing a democratic government and the end of economic injustices they found in a society of privilege. At the same time, its writers were not happy with the violations of individual freedom they saw in classical Marxism. Although several writers, including Borges, turned to French literary figures such as Sartre, de Beauvoir, and Camus, they strove to maintain an Argentine orientation.[71]

Some early expressions of modernism tinged with social protest came from poets. Possibly the first was the Cuban Afro-American Nicolás Guillén. However, two outstanding figures were César Vallejo (1892–1938) of Peru and Pablo Neruda (1904–73) of Chile. Vallejo was greatly influenced by Mariátequi and his Marxist ideas. Unlike many avant-garde poets, Vallejo's work did not choose an escape but rather the search for understanding the nature of personal and social disintegration. Neruda established a critical outlook—both of despair and a sense of humanity. Neruda, the second Chilean to receive the Nobel Prize for literature (preceded by Gabriela Mistral, who in 1945 became the first Latin American recipient) was the son of a laborer in southern Chile. Like many Latin American intellectuals, Neruda was disillusioned not only by the plight of the greater part of the masses, but also with the collapse of civilization that World War I exemplified. His pessimism deepened with the world depression and as a leftist he went to Spain to give support to the Loyalists in the Civil War. His most important work, *Canto General* (1970), drew on the history of the Americas from the pre-Incas of Macchu Picchu to the liberation struggles of the twentieth century. Along with other Latin American poets Neruda felt akin to Walt Whitman, who he said taught him to be an "American."

Fiction and "Magical Realism"

Beyond the flow of poetry in defining identity and the meaning of modernity is the even greater impact of fiction. A number of writers worked in both areas as well as in essays and drama. Again, questions of conflict and power, Old and New World, individual and society are at the core of the authors' search. As in Argentina, a modernist movement took shape in Brazil, especially in São Paulo's attempt to assert its cultural primacy over Rio, with the Pan-Brazil Manifesto (1924) urging a "Brazilianizing" in order to combat foreign literary influence. Perhaps the most daring writer of that genre was Mario de Andrade (1893–1945) in his *Macunaíma* (1928) with its exploration of Brazilian folklore and the birth of "magical realism."[72] This outflow of mythology, social reality, ethnicity, and regionalism was fist awakened by Euclydes da Cunha's *Os Sertões* (Rebellion in the Backlands) (1902), an examination of the drought-ridden, poverty-stricken northeast. In this connection, Brazilian fiction is to a marked degree influenced by the epic style of *Os Sertões*—a merging of sociology and fiction in its strong and moving imagery of a forsaken landscape.[73] In the same vein, Gilberto Freyre's *Casa Grande e Senzala*, vividly describes the stratified plantation system.

Likewise, the Cuban Alejo Carpentier (1904–80), an activist in the 1923 revolt against the Gerardo Machada dictatorship, wrote *El Reino de este Mindo* (The Kingdom of this World) (1949) celebrating Afro-Caribbean cultural clashes. In somewhat the same genre is Guatemala's Nobel Prize winner Miguel Angel Asturias (1899–1974), who, inspired by *Macumaíma*, gives an anthropological touch in his *Leyendas de Guatemala* (Legends of Guatemala). His most famous work is *El Señor Pressidente* (1946), a somewhat surrealist satire of a dictatorship patterned after Porfiro Díaz and his own country's Estrado Cabrera, written in the 1930s but could not be published until his country was restored to democracy in 1945. His *Hombres de Maíz* (Men of Corn) (1949) probes indigenous mythology, notably the magic of the Mayas and its effect on the peasants' "subconscience."[74] The Mexican Revolution also serves as a setting for his magical approach to reality. Marino Azuelo's *Los de Abajo* (The Underdogs) (1915) offers a biting naturalism not known before. More than a generation later, in 1955 Juan Rulfo published *Pedro Páramo*, a study of corruption and betrayal of the revolution.

The "Boom" of fiction came well after World War II, especially in the 1960s and 1970s. Perhaps the first sign of this creative outburst is Mexico's Carlos Fuentes (1928–), whose *La Región Más Transparente* (Where the Air is Clear) (1958), a portrait of Mexico City, created a sensation. However, his masterpiece *La Muerte de Artemio Cruz* (The Death of Artemio Cruz) (1963) ushered in a novel treatment of identity. Influenced by John Dos Passos and William Faulkner, Fuentes presents his protagonist Cruz in shifting images, convoluting space and time, and significantly contributing to Mexico's cultural revolution.[75] Like Octavio Paz he offers a critical analysis of his country's ills and the universe beyond. Another startling entry into the Boom is Argentine Julio Cortázar (1914–84) with his many short stories. However, he is best known for his novel *Rayuela* (Hopscotch) (1963), an intricate contest for identity and freedom as it moves between Buenos Aires and Paris. In a Joycian and surrealist mood, akin to his fellow countryman Borges, Cortázar pierces the surface of bourgeois life. Besides his criticism of the artificiality of Western society, Cortázar brings to his readers an insight in to the dilemmas they encounter in a culture based on logic and repression.[76] The third outstanding novelist of the Boom is Mario Vargas Llosa (1936–), who lost his bid for the presidency of Peru to Alberto Fujimori in 1990. His first novel, *La Ciudad de los Perros* (translated as *Time of the Hero*, 1962), written at the age of twenty-six, reveals his Flaubertian naturalism and dispassion. In *La Casa Verde* (*The Green House*, 1966), Vargas's deconstructiveness toward class culture and his contempt for mass culture places him in the postmodernist fold. He also clearly views gender and ethnicity as power relationships.

Few would question that the most significant breakthrough of the later phase of the Boom is *Cien Años de Soledad* (One Hundred Years of Solitude, 1967) of Nobel Prize winner Gabriel García Márquez (1928–). This tale vividly recounts the imaginary town Macondo, the Buendía family, and its patriarch José Arcadio over the better part of a century. This hypercritical look at corruption, licentiousness, and incest is not only a critique of Colombia and Latin America, but of human nature itself. His next book *El Otoño del Patriarca* (The Autumn of the Patriarch) (1975) expresses the many-voiced responses to a dictator. García Márquez has lived through Colombia's bloody events of the twentieth century—from the banana strikes of the 1920s, *la violencia* (the rural civil war) 1948–57, to the guerrillas and narcotraffickers dominating the 1980s and beyond.[77] Perhaps no other Latin American writer probes society and its actors in as deep a Faulknerian style as García Márquez.[78]

The Boom's strength arises from its break with the literary past. It brings a mystification of reality, end of romanticism, and an assault on taboos—notably of sexuality and religion, along with exploration of our inner or secret life, abandon of linearity, questioning of chronology, elusiveness of time and space, role complexity, symbolism, and ambiguity. The movement introduces a new structuralism with its symbolism of natural phenomena and novel patterns of semiotic relationships. In *Cien Años de Soledad*, for example, incessant movement in the narrative appears as a reworking of reality and fiction.[79] Not only García Márquez but other Boom and post-Boom writers established a new *postcolonialism*, Spanish American fiction declaring its autonomy from European writers.[80]

Despite the impression that literature is almost exclusively a male product, a host of women writers are making their impact in Latin America, notably since the 1970s. Among the outstanding women authors is Elena Poniatowska, whose *Hasta No Verte, Jesús Mio* (*Until We Meet Again*) (1968) is a tale of women who survived the crises of the Mexican

Revolution. Another Mexican writer who merges fiction with documentation is Silvia Molina, with her treatment of various episodes in history, especially Mayan, in both fiction and drama. Isabel Allende of Chile offers in *La Casa de los Espíritus* (The House of the Spirits) (1982) a saga of her country over the twentieth century through four generations and of men's violence against women culminating in the Pinochet reign of terror. Another two of her best sellers are *Eva Luna* (1987) and *Aphrodite* (1997), which explores the world of sensuality. Among the Brazilian authors is Clarice Lispector, whose *A Hora da Estrela* (The Hour of the Star) (1977) recounts a girl's transition from the impoverished northeast to the materialism of São Paulo.

Beyond the contribution of women, a lengthy tradition of indigenous literature should not be ignored. For instance, both Borges and Cortázar turned to ancient works of a worldview and cosmology such as *Popul Vuh*, a Mayan chronicle, and the Nahuatl *Inomaca Tonatiuh*. Also, the Inca heritage is still found in contemporary Quechua poetry.[81]

Identity and the Performing Arts

Drama and Social Consciousness

If we turn to drama and theater we find new directions of modernity and social protest. In a culture where more than half the population has only recently become literate, theater in all its forms becomes a major means of transmitting knowledge. Even the Aztecs relied on a ritualized drama to enact their sacred legends. Church and state continue their dialogue to the people by theatrical means. For instance, the Cultural Missions organized by José Vasconcelos in the 1920s used one-act plays to illustrate the slogans of the Mexican Revolution, "Tierra y Libertad" (bread and liberty) and "Tierra y Escuelas" (land and schools) to peasants and villagers in the hope of changing their society. In the mid-1920s urban scene a zeal for cultural nationalism brought on a renaissance in the theater. This period saw more stage performances than had occurred in the previous quarter of a century.

The dramatic outpouring of social distress through the theater sprang forth on a major scale in the 1960s, but the process was captive to the vagaries of political restrictions. Confrontational dramas suffered severely in Argentina, Brazil, and Chile—countries traditionally oriented to a vibrant theater. The advent of authoritarian regimes and the negative reaction of the hemisphere to the Cuban revolution, led by the United States, had devastating consequences on the theater.[82]

Still, Brechtian and similar themes persist in disguised satires. Among the dramatists who battled the status quo were Griselda Gambato of Cuba, Enrique Buenaventura of Colombia, and Egon Wolff of Chile. Their dramas engulf the saga of Latin America and the outside world, in a gamut of daring images from the terror of Auschwitz in Gambato to the theater of the absurd in Wolff.[83] An outstanding example is the "memory theater" inspired by the appalling encounter at the Plaza de Tletelolco in Mexico City on October 2, 1968 when government troops fired into a crowd of thousands, killing an estimated three hundred protestors and bystanders. The Institutional Revolutionary Party (PRI) chose to downplay the massacre, but over the next few decades at least a dozen plays depicted various aspects of the tragedy. In their enactment of personal reminiscences, rituals surrounding grief, satire of the government, weaving of past and present the public is exhorted to relive what many Mexicans, notably intellectuals, consider the most tragic episode since the violence of the revolution.[84] As will be documented in Chapter 8, this spirit reverberated in Liberation Theology as the message of Pope John XXIII inspired a revolt of many New World priests against a hierarchical society.

Music, Dance, and Regionalism

Even more than in drama, the indigenous, mestizo, and, to some extent, Europeans find their essential meaning in the most emotionally evocative of all the arts. The compositions of Latin America's greatest composers— Mexico's Carlos Chávez, Brazil's Heitor Villa-Lobos, and Argentina's Alberto Ginastera—are in the standard repertory of the world's symphony orchestras. Also, the *mariachis* and trembling love songs of Mexico (see inset 5g), the rumba of Cuba, the samba of Brazil, the tango of Argentina have their enthusiasts throughout Western culture. However, our concern in this limited discussion is how music expresses symbolism, identity, modernity, cultural change, and social reaction.

INSET 5G   TROUBADOURS AND MARIACHIS FROM *AMOR* TO
DROGAS

Mexico's rich tradition of musical escapes as expressed by *mariachis* focus on love as in "*Cielito Lindo*" or in the many *rancheras*. The themes fundamentally derive from colonial times. However, new touches were added by twentieth-century *mariachi* singers, notably José Jiménez. Usually the performers are a group of guitarists with a violinist and a trumpet player, all in traditional costume.

Despite the fantasy of love and escape, *mariachis* are by no means detached from the contemporary scene. As an example, the haunting sounds from the guitar via microphone now convey provocative meanings of the drug culture. The *canción* (song) communicates the smuggler's devotion to his rooster, parrot, and goat, the lingo for marihuana, cocaine, and heroine, respectively. After all, these are *the* sources of livelihood. Another ditty, "Most Wanted Men," evokes the voice of a leading trafficker who glories in his control of politicians. These songs are a counterpart of the rock music found in Europe and the United States with their vistas of a narcotic culture. The public enjoys these new *corridas* as they rationalize a culture representing a different type of corruption from that of religious, governmental, and economic elites perpetuated for generations.

Although the Ministry of the Interior has banned this new version of the *mariachi* over the radio, the public looks upon the ballads as depicting a reality, not least because they are accompanied by beepers, a machine gun barrage, and police sirens. At the same time, the band prides itself on its own moral code, derived from its rural origins in the state of Sinaloa in northwest Mexico. In contrast to the traditional *mariachis*, the band members imbibe water, rather than alcohol or drugs during their performance, even though they may have a gun in their halter. As one observer comments, the traffickers themselves personify several masculine values such as bravery, cunning, and ferocity. As men of humble background, they are simply defying the power structure.[85]

A striking aspect of the musical scene is its regionalist and ethnic blend—whether Caribbean, Central American, Andean, or indigenous, African, and Western. Hardly less important is the composer's response to the mechanical and electronic innovations of the twentieth century. Nearly every country has its particular musical idiom, often associated with dance.[86] For instance, in the Dominican Republic the *bachata* with its acoustic guitars and sentimental lyrics in vernacular spread from rural areas to the shantytowns of Santo Domingo. Another case of grassroots musical innovation is the incorporation of three cultural traditions—Hispanic, African, and Amerind—in Evitar, a village in the coastal area of Colombia. In this folk music the principal agent is a European and indigenous instrumental heritage, strings, wind, and percussion. Moreover, in the sixteenth-century campaigns of Cortés, Pizarro, Alvarado, and Valdivia, soldiers brought along their instruments, mostly variations of the guitar.[87] Similarly, romantic songs from the Renaissance remained relatively pure in Colombia and Chile, with some restructuring in Argentina and Ecuador, yet drastically altered in the ballad or *corrida* of the Mexican Revolution.[88]

As Latin America urbanized, musical expressions became mass phenomena and reflected the social needs of a class-conscious society. The most salient example is the Argentine tango. It apparently emerged in the proletarian bordellos of Buenos Aires toward the end of the nineteenth century. After 1900 it migrated into the middle and upper classes and spread to Europe and North America, reaching its height of popularity in 1913, continuing to be favored through the 1920s and 1930s. Although somewhat ossified it is far from forgotten, symbolizing Buenos Aires, both for Argentines and the world beyond in the mass media as well as on the dance floor. Besides exploring the dynamics of the sexual relationship and demarcation of form and space, the tango in its ritualized yet varied movements defines the specificity of gender.[89]

Art, Architecture, and Modernity

Although art belongs to every Latin American nation, Mexico's visual artists receive the most attention. From 1920 to 1960 the frescos and murals of Rivera, Orozco, and Siquieros remain as both artistic genius and social commentary. The catalyst for these muralists was Gerardo Murillo, better known as Dr. Atl (he chose this name from "water" in Nahuatl, the Aztec language). His revolutionary fervor took several directions; for one, he was leader of the World-Wide Workers' Organization. In 1910 he denounced the "bourgeois" program of the Academia de San Carlos (which was later named the National School of Fine Arts). In 1913 he persuaded Rivera, Siquieros, and other young artists to join the Revolution, specifically the forces of Carranza.[90]

Mexico's three "greats"—Rivera, Orozco, and Siquieros—drew on pre-Columbian and Mexican history, studied at San Carlos and abroad (particularly Paris), and painted in the United States as well as in Mexico. Still, they differed in their political ideology: Siquieros the most militant, the *enfant terrible* Rivera (inset 5h), and Orozco, though socially conscious, fearful of mixing art with what he defined as propaganda.

INSET 5H   RIVERA AS ICONOCLAST

Most widely known is the work of Diego Rivera (1886–1957), who as a youth visited Aztec and Maya sites, fusing these images with Cubism, which he later abandoned as he felt its abstractions would not be understood by the masses he hoped to reach. Like other artists he was influenced by Giotto's frescos in Padua, Michelangelo's Sistine Chapel, and similar works of the Italian Renaissance.[91] His murals depict the revolution, especially peasants and their leader Emilio Zapata. At the same time, his focus is on Western history in the rising industrialization and the fate of workers. His works adorn the Ministry of Education, the National Palace, the Cortés palace in Cuernavaca in addition to his Detroit creations on machines and workers. His mural on the history of capitalism for Rockefeller Center in the 1930s stirred sufficient controversy (especially as the mural included the face of Lenin) to get it destroyed. The United States was—and probably still is—not ready for such a provocative public representation of its economic history. On the other hand, Mexico welcomed Rivera's quasi-Marxist outlook as it gave the PRI a legitimacy that it hardly deserved.[92]

José Clemente Orozco (1883–1949) left a rich heritage in murals ranging from Guadalajara to Mexico City, from Pomona College in California to Dartmouth College in New Hampshire. His epical frescos extend from the banal, unknown, and tragic to despotism, science, religion, and human deprivation. His apocalyptic visions are woven into the history of civilization, particularly Mexico, including national myths such as the first and second visitations of the Aztec deity Quetzalcoatl.[93]

Of the three, David Alvaro Siquieros (1896–1974) was the most revolutionary as he pled for workers' and peasants' rights—a commitment he came by naturally as his grandfather was a *guerrillero* in Juárez' troops. After his career as drummer boy in the revolutionary army Siquieros joined Rivera in Paris as both military attaché and student artist. Returning to Mexico in 1922 he was active in unionizing miners as well as fellow artists. He experimented with various techniques on large-scale projects—at one point with a spray gun—feeling that the artist should incorporate tools iden-tified with the worker. At the front in the Spanish Civil War and later imprisoned by PRI authorities, 1960–64, he never doubted his commitment to a better life for the masses. His greatest work may well be his *March of Humanity* (in Cuernavaca), a mural of nearly fifty thousand square feet involving both abstraction and strident expressionism.

Beyond these three muralists are many other Latin American artists. To mention a few, Rufino Tamayo, son of Zapotec parents, passed through a sociopolitical phase into Picasso abstractionism. Similarly, the Cuban Wilfredo Lam was a disciple of Picasso, whom he met in Paris after Lam's participation in the defense of Madrid in the Spanish Civil War. He is known for his surrealist paintings of primitive cults along with an array of other Cuban images. Among those best known to the international world is Frida Kahlo, whose life was marked by intense pain due to an accident. Also, her marriage to Diego Rivera could be described as turbulent—hardly surprising in view of the personality of both partners. She brought into her paintings exciting contradictions—atheist and religious symbols, despair and wit, alternating dualistic visions of the personal and political, tradi-tional and avant-garde.[94]

Architecture and Social Statement

For much of the world, the breakthrough in building styles in Latin America is a wonder. Besides the contributions of Zapotecs, Aztecs, Mayans, and Incas are those of the colonial builders of churches and governmental palaces. Both of these phases are staggering architectural contributions. However, it is the twentieth-century accomplishments that are relevant to our understanding of contemporary society.

Brazil led the way. As early as the 1920s the young architect Lucia Costa was infused with the Bauhaus movement in Germany and later with Le Corbusier. Within a decade Costa designed the startling Ministry of Education building. His disciple Oscar Niemeyer further innovated in a series of structures ranging from the 1939 New York World Exposition to public buildings in his native land. His most daring design was in collabo-ration with Costa for the new capital Brasilia. The use of curvilinear forms and the merging of visual arts with building design appear in the congress hall, cathedral, and other structures in Brasilia. This vast complex with its use of glass and concrete parabolas is visually exciting, but because of its radical features, maintenance—including temperature control—poses problems.[95]

In the early twentieth century Mexico ventured into new forms as in the Palacio de Bellas Artes, planned by Dr. Atl on the eve of the revolution—the palace remains today as a tourist monument. In the 1930s Juan O'Connor and Luis Barragán innovated in both public and domestic architecture in the novel use of glass. Daring designs became a means of renouncing colonial architecture as it was a reminder of the colonial past and its exploitation of the indigenous. Most striking of all is the Ciudad Universitaria of UNAM (National Autonomous University of Mexico) built 1950–53, designed by Mario Pani and Enrique del Moral. Dozens of architects, engineers, and artists created a set of buildings inspired by pre-Columbian and other native themes, especially in murals covering entire walls. This dazzling achievement led other countries to innovate, Venezuela in particular created a number of private and public complexes during its years of oil prosperity. Also, despite budgetary limits, several countries including diverse economies, as for example Cuba and Argentina, are experimenting with new forms of public housing. At the same time, architects and city planners are caught in the debate of how funds are to be prioritized in view of limited resources. In other words, can monumental structures be created for the elites when the bulk of the population is marginalized?[96]

The Arts, Identity, and Modernism

One may ask whether many Latin Americans could survive their inequities were it not for the escape into literature, visual arts, and particularly music. Also, these contributions offer inspiration and strength to social movements— even though the outcome of these stirrings seldom realize the artists' dreams. For instance, Rivera, Orozco, and Siquieros authenticated the Mexican Revolution expressing aspirations beyond what became an ossified PRI. In another direction, part of the anguish for Nicaraguans after the fall of the Sandinista regime was the removal of revolutionary murals from the streets.

Popular music transmits identity. Despite inroads from rock-n-roll, *corridas* and *rancheras* evoke for traditional Mexicans a distinct style and sentiment with their *portamento*—a kind of lagging drawl.[97] In addition, music becomes a powerful force for social change. An example is the significance of popular music in El Salvador's national liberation movement between 1975 and 1992. Besides recordings, no less than fifty musical groups were active in a country roughly the size of New Jersey, singing *Vamos ganando la paz, Llegó la hora,* and *Por eso luchamos* ("We are winning the peace," "There came that moment," and "That is what we're struggling for"), which gave strength to the general population as well as to those in clandestine guerrilla warfare.[98]

Moreover, artists and writers bring forth a meaning to themselves and others. García Márquez in *Cien Años de Soledad* extends a national identity to Colombians, whether the intricate relationship between European, mestizo, and indigenous, the succession of civil wars, or colonialism imposed by the United Fruit Company.[99] From Sarmiento to Borges, Argentines experience their culture, their values, and their sense of being. Yet, the tumultuous nature of Argentine history—the alternation of dictatorship and democracy from 1930 to 1983—may account for the pessimism of its literature.[100] In specific reference to ethnic identity, writers provide a kind of bonding, for instance, the Cuban Nicolás Guillén and Haitian Aimé Césaire, both in their poetry and political activity document the anguish and aspirations of Afro-Caribbeans.[101] In a different context, a number of Jewish woman writers in several Latin American nations find belongingness in their writings, principally memoirs.[102] In another direction, gay and lesbian writers are exploring a variety of themes—a phenomenon unthinkable during the dictatorships, when exile (or suicide) was an all too common fate.[103]

Another aspect of the relationship between society and the arts is the cultural climate at a given time interval. One instance is the effect of political and economic situation. An Ecuadorian writes of the tenuous state of the arts during the economic crisis of the 1980s—the difficulty of securing funds for staging a play or concert or publishing a book. Yet it was the escape into the theater and concert hall that lessened somewhat the pain of the financial crunch.[104] During that same decade, after years of rigid censorship in Havana the Cultural Renaissance of the 1980s opened the doors, if only marginally, to new expression in drama, music, printmaking, photography, and film. Also, changes in the 1990s provided a fervor in creativity. For instance, in Mexico the Zapatista movement, the debates about globalization and NAFTA gave rise to novel expression in the arts.[105]

Finally, one facet of the outpouring of fiction is the struggle over modernity. More than their Western colleagues, Latin American writers merge past and present. García Márquez sees a continuing state of dependency for his country and its neighbors.[106] In his writings Vargas Llosa urges a temporary halt to modernity because of its surrender to rationalism. Rather, he suggests that one should take time off and "stare hypnotically into the mystery of a lost presence."[107]

Chapter Six

Gender, Socialization, and Women's Roles

A tantalizing phenomenon in Western culture over the last one hundred years is the shift in sex roles and the search for gender identity. Beginning with the 1960s several marginal groups were demanding a redefinition of their status in society, first in Europe and North America, with Latin America nearly a generation behind. Not the least vocal were women. Accompanying the continuing debate over gender are still others over ethnicity, age groups, differing disabilities as with physical handicap, and sexual orientation. This chapter analyzes the meanings and consequences of gender. Especially important is the changing role of women in the home, workplace, and the political scene.

The discussion of differences between female and male in Western and non-Western societies fills volumes, with implications not only for social relations but also for the economy, politics, education, and other institutions. The debate over masculinity and femininity draws from biology and culture. Behavior develops out of genetic differences, life experiences, and definition of the specific situation the individual encounters. That is, gender-oriented actions represent a complex process of social and individual interpretation of models and symbols of other people's attitudes and behavior. We all differ in our choice of options in gender behavior, and in the salience we place on gender in our search for self-identity. Our culture abounds with given directives: the male is rational, active, and powerful; the female emotional, passive, and nurturant—these stereotypes are continuously subject to review. Given characterizations of gender draw from the viewer's philosophical stance. For some observers, gender is purely subjective as it is defined by language differences, unconscious urges, and the feminist struggle in a phallocentric universe.[1] Neo-Marxists turn to the differential participation of the two sexes in the production process and how money, power, and politics operate in the "culture industry."[2] Contemporary feminist theory asks some fundamental questions: (1) How are women's status, roles, and experiences different from men's in given settings? (2) What are the roots of gender inequality? (3) How does gender inequality and oppression relate to structural oppression—the societal barriers to equality?[3]

The conflict in gender relations calls for an examination of the psychological bases of sex roles in Latin American society. Again, several questions emerge: How is the separateness of roles shaped by culture? How did machismo become entrenched, how is it perpetuated, and is it changing? What role does the "absent father" play in this pattern? How can gender values and psychological adjustment become articulated? How does sex identity compare between Latin American and Western cultures? Later in the chapter we examine the struggle of women to establish their rights in the home, employment, and political involvement.

Gender Roles, Society, and the Value Complex

One cannot understand the relationship between the two sexes without considering the totality of Latin American culture. First of all, we recall that the social structure originated in a conquest based on slavery and serfdom. According to the Mexican writer Octavio Paz, the conquistador represents masculine dominance subduing or raping an indigenous society by physical force.[4] Consequently, males, predominantly European and mestizo, today reenact the *conquista* as in expressing their machismo through the conquest of women. Relatively few indigenous cultures lean toward machismo.

A second factor is the strong masculine bias in Western and notably Spanish culture, as influenced by Moorish male dominance. What status women gained in the Renaissance, Enlightenment, and French Revolution had little visible effect in Spain and its colonies. Nor did the humanitarian movement of the nineteenth century with its marginal extension into women's rights have a perceptible impact on Latin America. Women have full civil rights in only a few countries, but these rights are hardly significant outside urban and middle-class subcultures.

Third, certain features, if somewhat hypothetical, demarcate the sexes. The sense of hierarchy, dichotomy of the mystical and the material, acceptance of forces beyond the control of the individual, all mold a culture in which men and women are conditioned to develop separate feelings and behaviors. This set of cultural values is only partially responsible for creating the psychological differences between the sexes but lends its support.

Fourth, in view of the dualism deeply implanted in Western and Eastern thought, it is assumed as given that the two genders represent almost uniquely separate biological and psychological entities. As a charismatic editor of *La Mujer*, a Colombian periodical, wrote in 1878, "men are external and visible, women are hidden and interior."[5] This gender division continues into the present, blended by class and ethnic shadings.

In Latin America the man–woman relationship became almost uniquely a sexual one except for the bond of the family institution as marriage provided for various societal needs. Affection and empathy appear to be negligible in marriage for the hard-pressed lower class and to some extent at other social levels. For the general population, romantic love is, after all, a late development in Western history—the seventeenth and eighteenth centuries. It had only marginal impact in most Latin American cultures until the second half of the twentieth century. More than most Western cultures, Latin America tends to exaggerate traditional sex norms. Hence, we ask: are these roles satisfying to the individual? To what extent are these modal responses changing? These questions involve the range of individual personality as well as the differences imposed by the particular subculture.

Gender Options: Maleness and Machismo

Gender is socially sanctioned and shaped by acquiring appropriate behaviors at given points in the life course. With the intricacies of contemporary urban society, gender more than ever is a complex pattern of traits shifting with one's total life experience. Hence, masculinity and femininity are never far from inner conflict. The contradictions are more evident in men than in women. In Western society men appear to struggle against two possible dangers: one of giving in to whatever feminine impulses they have, the other of displaying their masculine identity through hyperaggression. In Latin America gender roles tend to be explicitly defined.

A prevalent but far from universal option for the male is machismo. The model assumes many forms. In Mexico an ultimate ideal is the matador (bullfighter), who embodies the image of bravery and recklessness. New sports such as soccer convey an aggressive expression of the male ideal. (It is relevant that Latin America like Europe chooses soccer, whereas the United States clings to football as a more violent and macho game!)

Another variation of machismo is the gaucho. As mentioned in chapter 5, the culture of the cowboy as portrayed in Martin Fierro or Don Segundo Sombra—two examples in Argentine literature of legendary proportion—embodies a solitary and primitive existence. Few women were available on the nineteenth-century pampa and whatever contacts existed with them were largely physical. As Argentina drifted toward modernization in the twentieth century, the gaucho was idealized for his rugged independence in a hierarchical, paternalistic society.[6] Today, urban, middle-class life in Buenos Aires, Montevideo, or Santiago represents almost the antithesis of the macho principle, as contemporary drama, literature, and mass media reflect. Machismo enacts diverse traits—from courage, stoicism, warmth, generosity to chauvinism and violence—an extreme being the drug culture in Colombia and elsewhere (inset 6a). Wife beating is another outlet. Machismo is often dependent on the eye of the beholder. Sometimes machismo is defined as what it is not—a *mandelón* or "hen-pecked" husband.[7] As one Brazilian expresses: "To be a man, a real man you have to be a *machão*. A machão in bed and in the street."[8] Machismo is a multilayered concept in its assignment of statuses, roles, and other behaviors. Besides the power relations between men and women, it is a set of stereotyped signs and gestures in the male–male relationship, proving a masculine identity to one's self as well as to others.[9] Nor is the macho shield absent from the United States.

INSET 6A   SOCIALIZATION, MACHISMO, AND
NARCOTRAFFICKING

Machismo assumes a variety of expressions. Indeed, devotion to guns and the culture of violence in the United States may have psychosexual dimensions. In another context, in Medellín, Colombia, dozens of *bandas* (gangs) emerged during the 1980s and early 1990s oriented to the Escobar empire or what is often called the Medellín cartel, later to be largely replaced with Luís Ochoa's Cali cartel. In this novel entrepreneurial universe, police, politicians, and judges can be corrupt—or honest with painful consequences.

The new culture of murders, kidnappings, and gang warfare not only changes the national community but also leads to a new language, with words such as *basuca* for pasta *base de cocaine*, *pistaloco* (gun crazy), *paniquar* (to panic), and any number of terms combined with narco. Moreover, institutionalizing the drug trade is interpreted as a "democratization of wealth," as money finds its way into the lower classes. The drug culture is also a "democratization of pain" as when the public sees the photograph of former president Cezár Turbay mourning over the casket of his daughter, a victim of kidnapping and murder.

This wave of violence persists into the twenty-first century. Many of these gangs are *gamines* (homeless boys on the street), still others are from fatherless homes. With little or no employment beyond marginal ones of car watching, lottery ticket sales, and shoe polishing, the lure was big money in the narcotics trade. More than occasionally, the gangs are not so much directed to drugs as allegiance to the neighborhood or *comuna*. In many instances, teenage boys simply learned to kill in order to establish their male identity and their personal worth. Generally they are surrounded by narcotraffiicking culture with a new vocabulary. Words such as *violentólogos* (students of violence) and *cocaína* are in the air. The wave of violence also drifts in from revolutionaries and counterrevolutionaries in rural areas. Most of the boys who join this cult of violence come from broken homes or no home, roaming the streets, with neither school nor work. With chieftains as their models, killing becomes an expression of their ego, the gun conveys power. Drugs are another catalyst. Money becomes a reality, if there is a family, emergence from poverty can be the outcome.

However, it is unwise to stereotype those who live by the gun. An interesting case is that of twenty-seven-year-old "Larry," who grew up in a home with an alcoholic father. He was ushered into begging at age twelve. By age eighteen the military conscripted him in a round up of the neighborhood. After his release followed a series of unrewarding jobs. Continually frustrated in his attempt to become a *persona decente* he joined one of the *comuna* (municipality) guerreilla groups, which had encounters with military and paramilitary. After a heavy attack by the army the guerrillas retreated. Into this vacuum moved Larry, who organized *a grupo de autodefensa* or death squad. This development was spawned by witnessing his brother Tony's near death in trying to defend a helpless old man against a *banda*'s holdup. After years of humiliation Larry now found a personal vindication in forming a self-defense group. In an interview with journalist Alma Guillermoprieto he described how his group, dressed in black with a hood, operated against what they considered "undesirables," accounting for some thirty executions. Larry asserted that the task gradually grew easier after the first killings. His consolation is that the *barrio* is now calm and clean. When asked about the police's alleged murder of given gangs, Larry responded: "The cops are murderers. They massacre everybody. I'm a Christian. I only take a human life when it's absolutely necessary."[10]

Whatever its variations, machismo is inherent in the socialization process. With the approach of puberty, boys are encouraged toward overt sexual behavior. As a cultural ideal the male should prove that he is virile and desired by women. In colonial and nineteenth-century Brazil as soon as a boy reached puberty he was to have his first sexual experience with a *preto* (black or Afro-Brazilian), that is, a slave, usually a servant in the home.[11] Whether in the casa grande of Brazil or in the Colombian lowlands mestizo village Aritama, the boy who is not "muy macho" is subject to ridicule. Today, if less than in the past, in conventional courtship a young man is assumed to have more than one *novia* (girlfriend) at a time. Besides these casual flirtations he may turn to a prostitute for a physical relationship, or to a girl from the lower class, for whom the mores are relatively permissive. Or more accurately, lower-class women are even more subject to exploitation than those from the middle or upper classes. In the working classes, the social order demands that male sexuality should be active and dominant. Ideally, women are to be submissive and monogamous.

Occupational activities, especially in the rural scene, affect the male's preferential status. In Huecorio, a Mexican highland village, women carry the heavier load and go barefoot, whereas men wear shoes. When the *campesino* is asked why men have this status, he explains it as the inevitability of nature. In the Colombian village Aritama male superiority is underlined in the responses of Aritama women, who nearly all wished they had been born male.[12] As the urban milieu, particularly television, is creeping into the countryside, the gender profile is changing.

The woman's role as *abnegada* (self-denial) and the male's superior status and its reliance on machismo are now threatened with urbanization, womens' employment, contraception, and other mobility norms of the middle class. Yet, machismo with its implied inferior status of women dies hard as it is rooted in the value structure as well as in the socialization process. Machismo also meets the unconscious need system of many men, at least those who feel insecure.

The Parent–Child Relationship

In theory the father rules the home, but as he is often absent, the mother becomes the primary socializing agent on a day-to-day basis, possibly more in the lower classes than in the middle classes. As the lower class "wife" may have a succession of "husbands," identification with a "father," if he is present, is problematic. From the report of women living in a popular *bzrrio* (working-class) in Cuernavaca, Mexico, the absence of or abuse by the father produces acute frustration and anxiety among children.[13] Moreover, lack of an adult male model with whom the boy can identify is regarded as one cause of machismo. According to many psychiatrists, the urge for male dominance is the conscious motive, but fundamentally the boy's obsession with maleness is a "reaction formation" against the anxiety brought on by his feminized environment. The inferiority complex sometimes attributed to the Latin American male is part of this process. Machismo appears to be in part a compensation for inner weakness.[14]

Although the father's absence is a widespread phenomenon, the child or adolescent can identify with other male figures—an uncle, older brother, or *padrino* (godfather). Still, other situations cause psychological problems. The relatively large number of children, natural and adopted, is one factor, at least until the birthrate began to drop in the 1970s. Servants in middle- and upper-class homes pose another barrier to parent–child identification. As lower-class mothers are increasingly employed outside the home, neglect or an intermittent parental substitute is the outcome. Mothers also abandon their children. It is significant that in my Santiago survey of schizophrenic, delinquent, and "normal" samples, over a third of delinquents were without father or mother for much of their childhood in contrast to the relatively intact homes of the other two samples.[15] These inadequacies in the home may, or may not, "explain" delinquent behavior, which might relate to an excess—or a lack—of machismo. In this connection, it must be emphasized that machismo has to be contextualized. It assumes many forms, and for many Latin American males is practically nonexistent.

It would be misleading to create a stereotype about parent–child relationships. Although most men generally leave the parenting task to the spouse, others provide nurturance. In Santo Domingo (a Mexico City *barrio*) residents feel that fathers are occasionally more affectionate than mothers, possibly because they leave the disciplining as well as the care to the mother. However, there is little doubt about the differential in the social-ization of the two genders. In Managua, Roger Lancaster noted several practices prescribed for boys but denied to girls: taunting by older siblings in order to provoke anger, laughing at and tolerating profanity, encouraging autonomy by assigning tasks such as fetching water, permitting teenager boys to roam freely through the neighborhood with their peers. In addition, boys are allowed to express a certain level of independence from the mother. They are constantly reminded of the need to demonstrate their machismo with strong hints such as "No sea cochón!" (Don't be a faggot!).[16]

An index of the father's relationship to the child is found in the responses of students in Guadalajara and Michigan to a semantic differential scale.[17] The Mexican father is perceived to be more critical, forbidding, irritable, and at the same time more perceptive and persuading. The North American father is judged as more relaxed and permissive, if less certain of himself. In other words, the Mexican father is less consistent in his behavior, but likely more conspicuous in the child's consciousness. According to another Mexican survey, adult subjects say that their deepest tie was with their mother rather than the father; 51 percent see their mother as "very understanding" as compared to 42 percent in regard to their fathers— hardly different from most cultures.[18] In extreme cases, children are known to protect the mother from the father's violence, as revealed in a study of a Guadalajara *barrio*.[19]

For any sample involving complex human phenomena, overlappings occur and hasty conclusions should be avoided, not least because questionnaire and interview data have less than ideal validity. Yet, the typical Latin American father is around the house less hours than the North American father—a dubious generalization in view of the life of an upper-mobile, managerial, commuting suburbanite—but a Mexican or Colombian father is highly visible when he is at home and his influence operates even when he is away. Still, in the more European culture of Chile, children are six times as likely to turn to their mother than to their father for help in their homework.[20]

Understandably, the mother may have a masochistic attitude toward her status. If she has been reared to think of her role in life as *abnegada* (denied), she may or may not feel the urge to pass on this model to her daughters. Her attitude toward children is one of affection and at the same time strictness. Traditionally, her focus of love is mainly toward the youngest child on whom she devotes maximum attention at the expense of older ones. The rejection presumably produces latent hostility and anxiety in the child, or at least a degree of sibling rivalry, as happens in many Western cultures. The remoteness or mutual mistrust toward the other person, occasionally reported for Latin Americans, may be a result of their being treated with indifference by their father and decreasing attention by the mother as they grow up. These generalizations must be tempered by the decreasing family size since the 1970s.

According to a study of Lima female-headed homes, mothers permit their sons some nonmasculine behavior, such as crying—distinctly in opposition to *criollo* standards.[21] However, relationships between parents and children are not necessarily polarized. Certainly they vary by nationality, ethnicity, and social class. This variation is well illustrated in the survey of *colonia* Santo Domingo in Mexico City, where approximately a third of the men engage in intimate contact, such as holding the child (a virtual taboo among the more macho) or take total care in the mother's absence. As working-class families are usually without servants they may have more parent–child interaction—both mother and father—than in the middle and upper classes.[22]

The responses to my questionnaire in Santiago among late adolescents and adults in the control or normal sample did not reveal a situation essentially different from other Western cultures. Forty-one percent thought of the father as dominant in decision making; for 18 percent the mother was dominant; the remainder considered the relationship as an equalitarian one or could make no judgment because of a lengthy absence of one or both parents. Only a tenth reported strong feelings of rejection by the father, even less by the mother. Most respondents described a relatively consistent emotional climate in the home. However, no less than 83 percent among the delinquents felt they had been rejected by their parents, as reflected in remarks such as "I never saw my father except once or twice a year, and when I did a beating took place," or "my mother was always at work and most of the time we kids were alone." On the whole, a less strong preference for one parent over the other characterized the normal sample. Most revealing in the context of the North American parent–child regime is the finding that in about four-tenths of all three samples (delinquent, schizophrenic, and normal) the father was reported as never having played games with his children. In summary, although most Santiago "normals" perceived no conscious rejection by the father, the father's position might be described as remote.

These questions impinge on punishment norms. Parental control is affected by social class. As in the United States, middle-class parents turn more to manipulative means such as denying the child a material reward or withholding affection, whereas the lower class relies more on direct punishment. In the middle- and lower-class Santiago control (normal) sample, approximately half the parents habitually resort to punishment of one type or another. Less than a third of these parents use physical punishment (as compared to over half among the delinquents, who came disproportionately from lower-class homes). The mother may use punishment as a form of scapegoating, victimized as she is by her husband's aggression.

Even more serious is family breakdown in many cities. One symptom is the presence of homeless boys, *los gamines*, as found in Bogotá. A number of factors—delinquent parents, absence of the father, harsh punishment—complicate the situation.[23] Even for relatively intact families, various constraints limit what parents can do for their children, or what children can do for themselves. Substitute parents, particularly godparents, are occasionally available. But networks in the lower class are not as extensive as in the middle class; the family moves, or other resources disappear. Networks in the community and neighborhood are an option only in certain settings.

Inevitably, a harsh economy affects family relationships and the ability to care for children. The 1982 Mexican fiscal crisis brought intense poverty to many families as women were forced as never before to find work outside the home with deleterious effects on children. In this crisis social welfare, including health services, became ever tighter.[24] A survey of Santa Ursula, a shack town immediately south of Mexico City, found widespread malnutrition among children and adolescents.[25]

Since the 1970s counseling and social service agencies are available for family adjustment, especially parent–child relationships. Unfortunately this service improves the situation more for the middle than for the working class. However, this focus on interpersonal relations permits individuals and families to perceive their roles and behavior more clearly.[26] On the whole, imagery about women is changing with the mass media's processing new norms of social behavior.[27] Feminism and Liberation Theology have redefined the woman's role with direct and indirect effects on the lives of men. In recent decades governments have changed the civil and penal codes to restrict patriarchalism for both fathers and mothers. However, these deterrents have been directed more toward the lower than the middle class.[28]

Relations with Siblings and Peers

More often than not, the bond between siblings is an affectionate one. For nearly all Latin Americans the family represents a sacred value, partly as a refuge from frustrations outside the home. The older daughter assumes care of younger children as her mother turns to her newest sibling. Children train each other in household chores. Despite the separateness of gender roles, brothers and sisters share in many activities. In that context, games are a source of mutual interaction. Children in the same family play together within their dwelling. Even if the play is in a group setting as in the patio of a *mesón* (a multiple slum dwelling built around a patio, mostly in Central America) a sort of family solidarity persists. This in-group feeling and isolation from neighbors possibly influence the individual to be somewhat reserved toward his or her peers, especially strangers, in later life. Particularly in the lower class a degree of tension exists among siblings. In Aritama, communities living on the margin of existence, the older child may take food from the younger sibling. Competition for food is keen, and the male, especially the older one, has the better bargain.[29] To what degree these events produce sibling rivalry is not known, but sibling relationships are hardly free of ambivalence—if indeed they ever are for any two intimately associated human beings in an uncertain universe.

As in most cultures, friendships follow gender lines. For example, in Andean communities *cuerdas* are composed of males, who in childhood form a friendship group, often lasting until early adulthood. Meeting evenings, the *cuerda* has a distinctive name or whistle known only to the in-group. The members act for each other as go-betweens in arranging meetings with a girl. With marriage or migration from the village the group tends to break apart. A less formalized group is the *cantina* relationship or a drinking club around the bar, where men discuss the affairs of the day, whether politics, fights, cattle rustling, or sexual adventures. Significantly, women seldom have a social group comparable to the *cuerda*.[30]

Conflict, Machismo, and Male Anxiety

Latin America seems to be a man's world, but it is at a certain cost. Traditionally a male may experience a relationship with at least three female images: (1) a mother who he cherishes as a reflection of the Virgin Mary cult, (2) a wife who is a homemaker and mother of his own children, and (3) his *querida* or a prostitute (depending on social class) to whom he turns for sexual satisfaction. In rural areas males cannot always resolve the dilemma of the pure and fallen woman; in many villages they attempt to maintain their wife in the former category by refusing to arouse her sexual impulses by affection and foreplay—lest she be tempted to wander outside the marital bond. Whether the formally chaperoned courtship of yesteryear or a casual relationship of today, neither offers opportunity to develop adequate communication with or knowledge of the other person. This incomplete bonding can continue into marriage—the two spouses often fail to enter into meaningful communication or a deep psychological relationship with each other. Again, these norms are shifting in the urban scene.

Machismo is, after all, a model trait and not a universal one. One variation is homosexuality. Although no solid data exist, it can be assumed that about as many Latin Americans are homosexually inclined as elsewhere in the world. In Aritama a dozen or so of each sex were believed to be alternating between heterosexuality and homosexuality. Likewise, homosexuality along with other unconventional sex practices takes place in towns such as Minas Velhas.[31] Homoeroticism assumes various forms in most cultures of the world, but in Latin America it has even more diverse nuances, with a marked polarization between a "stigmatized passivity and unstigmatized activity" in homosexual roles.[32] In a multiyear study of homosexuals in Guadalajara (which almost uniquely among Mexican cities tolerates gay meeting places such as bars and baths), a preference for the active or insertive role versus the passive role in intercourse was observed. In addition to these two roles is the *internacional*, which reflects the European or North American preference for the oral approach. However, many among this sample are bisexual married males in this "limited gay subculture," marked by secrecy, including a strong desire to prevent their family, especially their parents, from learning about their sexual behavior. Further, a number of married males indulged in homosexuality.[33] Although the causes for homosexuality and its overt expression are complex, a Mexican study finds secularization, urbanization, changing mass media, and pornography to be relevant.[34]

While a number of unfavorable epithets are attached to homosexual and other "deviant" behaviors, the individual is not as severely condemned as in puritanical cultures. My observations of colleagues in Colombia and especially El Salvador point to a degree of tolerance toward personal deviations. While ridicule and jokes are commonplace, ostracism or legal duress is rare. These deviations from the norm imply that machismo is often more an ideal than a coercion.

As personal maladjustment exists in any society, machismo is not necessarily more a source of maladjustment than is any other basic pattern of gender relationships. Still, machismo has more ill effects on women as it constrains their opportunities, whether in public appearances, employment, or other activities. Moreover, the prevalence of self-destructive outlets is relevant. Alcohol consumption, a sign of machismo, is relatively high in many countries. In El Salvador a large portion of lower-class males are known to spend over a fourth of their income on alcohol. In view of the inadequate health standards (low caloric and protein intake, for one), intoxication is inevitable.[35] I remember on watching Saturday afternoons, while sitting in the plaza in San Salvador, how a man whose diet consisted of rice and tortillas with its limited quantity of calories, protein, and vitamins could become drunk on a bottle of beer. Still, it would be presumptuous to place the cause in sex conflicts predominantly in machismo, in view of socioeconomic frustrations. For many Latin American communities drinking is no more extensive than in other areas of the world.

This lack of psychological resources can trigger a certain degree of violence—albeit a level no higher than that in the United States! Homicide becomes a means of striking out against the world, proving one's manliness, or simply the expression of a subcultural trait, as in Colombia with thirty thousand homicides annually, in large part a result of the drug war. In traditional rural areas a double murder is, or was, culturally prescribed if the husband finds his wife in the arms of a lover. He would be subject to derision if he did not protect the family honor. Until recently the law was in his favor. Nevertheless, the tide is turning. For instance, in Argentina— scarcely a haven of machismo—the imagery in media and the arts (theater, TV, and cinema) displays aggressiveness and violence.

The Status of Women

Gradually, a new environment is unfolding for women. The first stages of women's liberation in Latin America drew attention to the long-neglected area of women's statuses, roles, rights, and contributions to society. Males have dominated societal institutions including the family in most cultures throughout human history, but exceptions are evident. Certainly, for most of the world, women have a double work role with few economic and psychological rewards. This gender stratification has both micro and macro effects in nearly all societies. As men control the economy, women have little bargaining power. This reality shifted slightly in Western society with the romanticism of the Victorian age as well as by urbanization, the franchise, and the spread of new psychological doctrines in the first half of the twentieth century. Along with other marginal groups, women began their search for liberation in the 1960s—a struggle continuing to the present.

This shift of women's status must be placed in a broad theoretical and historical context. Although a number of indigenous societies confer on women various property rights as with matrilineal, matrilocal, or even matriarchal patterns, these societies are in a minority. Even when women have these rights, men occupy a compensatory status in government, religion, or other domains. One case is that of the Hopi, who are matrifocal, but the male plays the critical role in the important *kiva* ceremony. In Western society from the classical civilizations to the modern age the male has been dominant. Spain, which shaped Hispanic America, was steeped in Roman and Arabic traditions. From colonial times through the nineteenth century, men in Western nations regarded women as chattel, In addition, according to one Mexican writer, women are traditionally thought of as an abstraction, indeed, they are only powerful in the abstract—the least abstract is the prostitute, who represents the fallen versus the pure.[36]

Nonetheless, women have been asking questions. Even as early as 1879 in *La Mujer*, a women's periodical in Colombia, the editor asks: "Why cannot women enter industrial employment or the university as they do in Europe?" In a 1926 edition of *El Hogar* (a twentieth-century successor to *La Mujer*) a significant question was whether a woman should request her husband's permission to have her hair bobbed![37] Even in liberal countries such as Chile, where in the early 1900s women could enter the professions, they had only secondary rights in child custody and until recently could sign no legal documents without the husband's approval. With the exception of the Pinochet dictatorship, Chilean women—unlike their sisters in neighboring countries—probably today have as open access to the professions as almost anywhere in the Western world.[38]

Theories and Issues

As the women's movement for identity and equality surfaced in the 1960s it struggled against an enormous antifeminist bias. In the Western world nearly all social philosophies have been heavily oriented either directly or indirectly toward male supremacy. As a striking instance, Marx built almost his entire theory on the ownership of the production process but failed to note that women were and are even further removed from ownership than are men. Except for a few reformers, most social analysts took gender inequality for granted or rationalized this discontinuity within a larger system. An example is functionalism, the leading sociological theory of the mid-twentieth century. According to its leader Talcott Parsons, integration of society is based on instrumental roles for the male and expressive roles for the female. In other words, the mother is to provide the nutrient role in the home, leaving to the male his position of power in the larger world.[39] As gender roles have shifted over the last half-century, this bifurcation has been revised.

With the redefinition of gender statuses and roles in the 1960s and beyond, several theories emerged to clarify the underpinnings of gender inequality. These derive from several disciplines ranging from biology, psychiatry, psychology to anthropology and sociology, with several theorists representing cross-disciplines. For some observers of preliterate societies the subordinate status of female stems from "menstrual defilement." As hinted earlier in the chapter, psychoanalysis, at least the Freudian version, places gender development within the parental relationship. However, these interpretations are under increasing criticism by behavioral scientists. Moreover, feminists reject orthodox Freudianism, which stresses the interplay of personality, particularly the asymmetry in parental and sibling relations, that is, the mother assumes a passive relationship, and the son asserts the autonomy he acquires from his father.

Beyond the discussion of biological and psychiatric approaches is a fundamental debate about the feminine role. Feminist theory is divided into various directions, ranging from "radical" to "liberal." It is an oversimplification to refer to the radicals as viewing women as free agents, whereas liberals lean to the woman's role as rooted in the family, but would prefer a widening of her options. The distinction also hinges on differing stances on methodology as well as ideology, interwoven with Marxist, neo-Marxist, and anti-Marxist themes along with analyses by a host of theorists. As elsewhere, the debate continues in Latin America. As one Brazilian Marxist explains, capitalism turns to gender in order to limit the number of individuals who may enter the competitive labor market. Marginalization of women arises from the incapacity of capitalism to use all available workers. The search for a final theoretical framework has been less intense in Latin America than in Europe and North America, at least until the 1990s. Still, in most countries women now question the pressure points they find in social institutions—economy, government, church, and education—that shape their own self-definition, as noted by several eminent Chilean feminists.[40]

More critical to women, especially in the Third World, is the question of social meanings, both implicit and explicit, derived from socioeconomic institutions, including the family, often itself an economic operation. In this clarification of meaning, women operate as change agents. In most societies, change assumes many forms—end of patriarchy, liberation from sexist stereotypes, and occupational equality with access to professions. This transition conveys involvement in various social movements. For the mothers of the disappeared in El Salvador's civil war, change signifies new identities, among others, a sense of human rights and entry into feminist ideologies and organizations.[41] In a Maya community in Chiapas, Mexico, the "dynamics of gender" involves negotiation of power and status in the male–female relationship, or how to redefine boundaries in a society ruled by males from colonial times to the twilight of the twentieth century.[42] One sign of the change in feminine roles is the growing acceptance of contraception, which varies considerably between nations (table 6.1). As a Brazilian observer notes, feminism derives from the destruction of myths, whether biological ones such as weakness and disability presumably imposed by the menstrual cycle, and stereotypes interlaced with ethnic or other subcultural biases. Particularly, Afro-Brazilian women are singled out for stereotyping and discrimination.[43]

In all societies women live in a condition of risk. In Latin America the pitfalls are particularly severe: exploitation—whether in the workplace, the home, or on the street—miserable pay, rape, abandonment are only the more grotesque hazards. The whole social fabric is tilted toward male prerogatives. Most intriguing are the effects of conflict, violence, and revolution on gender roles. In the Zapatista struggles of the 1990s for the redress of social grievances the indigenous population in Oaxaca and Chiapas were called upon to assume behaviors different from their traditional ones. Women took to combat and men were deprived of their masculinity as they were forced to submit to federal troops.[44]

Table 6.1   Fertility and contraception for Latin America

Country	Births per woman		Women aged 15–49 using contraception
	1980	*2002*	
Argentina	3.3	2.4	—
Bolivia	5.5	3.8	49
Brazil	7.9	2.1	77
Chile	2.8	2.2	—
Colombia	3.9	2.5	77
Costa Rica	3.6	3.3	—
Cuba	2.8	2.3	—
Dominican Rep.	4.2	2.6	66
Ecuador	5.0	2.8	66
El Salvador	4.9	2.9	60
Guatemala	6..3	4.3	38
Haiti	5.0	4.2	28
Honduras	6.5	4.3	62
Mexico	4.7	2.4	65
Nicaragua	6.3	3.4	60
Panama	3.7	2.4	—
Paraguay	5.2	3.8	57
Peru	4.5	2.6	58
United States	1.8	2.1	64
Uruguay	2.7	2.2	—
Venezuela	4.3	2.7	—

Source: World Development Indicators, United Nations, 2004.

Women's Work Roles: Factors in Discrimination

In other words, the advent of women's liberation in Latin America drew attention to the long-neglected area of women's statuses, roles, rights, and contributions to society. A number of variables interplay with the entrapment of and the escape for women. In Latin America, and to a lesser extent in Anglo-America, the "feminine mystique," with the childbearing role, continues to remove women from the men's world, inhibiting orientation to social change, including family planning. We return to the question of *marianismo*, or the sacred cult of the woman (based on the figure of the Virgin Mary), which draws on the classical civilizations as well as Near Eastern doctrines—a viewpoint often accepted by women who are dedicated to humility and self-sacrifice enacting the stereotype the male-oriented culture decrees. The church proclaims a cult of purity and a ritualized suffering when the husband philanders.[45] Consequently, being culturally forced to accept a special category, women implicitly support a sharp division of labor between the sexes and create a climate in which neither competitiveness nor equality is obtainable. As an example in feudal Vicos, Peru, before agrarian reform, women far more than men assumed a variety of tasks including working the fields. Accordingly, the inferior role is assigned to women by men. Yet, when in the late 1950s the Vicosinos attained ownership of the fields and crops were commercialized, men assumed the fieldwork, as it was to their profit.[46] In reality, women work in agriculture throughout Latin America. Globalization calls upon both genders for its "factories in the field," often on a migratory basis. Not surprisingly, Brazil has a lengthy history of women producing fruits and vegetables for the world's market.[47]

Participation of women in higher education and the professions is growing. The more Europeanized countries took the lead. Chile permitted women to attend university as early as the 1870s. São Paulo was one of the first cities to grant women the right to work in white-collar occupations (hiring a woman for the post office in 1886) and matriculation to the law school in 1898. For over a half-century Latin American women have entered the professions of law, medicine, and academics. They also distinguish themselves in the arts and election to public office. In several countries women become mayors, still others reach the rank of ambassador. Also, in 1990 Violetta Chamorro defeated the sandinista Daniel Ortega for the presidency of Nicaragua. But for the most part, their work roles are in the traditional occupations of servant, bench work in the *maquiladoras*, and clerical tasks. In several nations approximately 90 percent of the elementary school teachers are women, further femininizing the stimulus world of children.[48]

In several respects, Latin American women are at a disadvantage in the economy. As in nearly all cultures, women are considered a poor risk because of a supposed general incapacity and their periodic interruption for childbearing. Ironically, further, modern capitalism places women at an even greater disadvantage than they knew in preindustrial times. In the handcraft period women used the skills they learned at home—ceramics and weaving instead of routine technical tasks on the assembly line.[49] Moreover, women's limited access to a career and to power is reflected in the educational attainment they are allotted. In most countries, males are permitted lengthier schooling. As a result, of Mexico's recorded illiterates in 1985, 2.5 percent were male, 3.9 percent were females.[50] Although this ratio improved over the next two decades, limited access to education has far-reaching effects. As one statistic, for women with no or little schooling the infant mortality rate is at least three times higher than for women with secondary or university education.[51]

Political and economic change is relevant. In Mexico neoliberalism has led to further abandonment of the *ejido* system. As inheritance and other legal norms favor men (many of whom migrate to the city or to the United States for work), women struggle against the government in order to hold on to their plot or receive proper compensation. By organizing they are able to pressure local authorities, who concede, at least in some communities, to their demands for retaining their plot or receiving compensation.[52] Another variable is the technological revolution, including cyberspace, which increasingly affects both genders. As suggested by a Chilean study, women in the 1990s advanced significantly in the service sector with greater job access, training, and promotion. However, less mobility occurred in the industrial sector.[53]

An inescapable determinant is social class. In colonial times classes operated as castes, but this barrier did not prevent the upper white's sexual exploitation of a servant. One contributor to social class is education. Social class overrides the minority status of the woman, but only within limits. A middle- and upper-class woman can give orders to a male underling, but she is seldom encouraged to enter into decision making in the political or economic sphere. In the contemporary world, gender and class have their eccentricities—middle-class women tend toward conservatism, however, their daughters are exposed to liberal and radical ideology of teachers and fellow students at the university. They often join social movements, even violent ones, for instance the Shining Path in Peru and the Tupamaros in Uruguay.

The class factor is well illustrated in domestic service, traditionally the largest outlet for urban single women. In 1979 prior to the Sandinista revolution a quarter of all Nicaraguan women working outside the home were servants, whether live-in or daytime, that is, women with children constitute the "dailies."[54] A similar scenario is found in most of Latin America. Approximately three-quarters of Salvadorian domestics are reported to work sixty hours a week and subjected to considerable verbal and physical abuse.[55] I recall how North Americans' (including my wife and myself) treatment of servants was resented by neighbors or acquaintances as we gringos were "spoiling a servant for ever serving in a Salvadorian household." The domestic servant's status varies somewhat with the local culture. Significantly, in El Salvador the label for the domestic was *serviente* (servant), in Colombia a *muchacha* (girl), and in Chile an *empleada* (employee) indicating a marginal progression with the level of socioeconomic development. The low pay for domestic service permits the middle and occasionally the lower classes to enjoy this luxury—or "necessity." Socialist governments attack the practice of domestic service—as well as of prostitution—but Cuba with more than forty years of socialism is slightly more successful than was Nicaragua in this respect. Educational programs are offered in Havana and other cities in order to qualify women for white-collar employment. Still, well-situated Havana families demand domestic help. Consequently, an officially sanctioned organization for domestic help, United Family Service, is available with specified wages and benefits.

Beyond Cuba, education as a factor in social class contributes to women's liberation, as, for example, childbearing. Moreover, social class overlaps with ethnicity. For women belonging to an ethnic minority the projection of inferiority is even greater than simply being lower class.

Urbanization and technology possibly have a more negative than positive effect on women. In Brazil, for instance, women more than men leave the countryside for the cities, particularly from the northeast to the southeast. Not surprisingly on reaching the city, women receive even more meager pay than do men, as a study of the textile and garment industry shows.[56] Nonetheless, because of economic necessity, employment of women continues to grow; 20 percent of wives were working in 1980, but 37.6 percent by 1990.[57] Yet, increased employment hardly means better conditions. With a global economy, local capital is seldom able to reach sufficient industrial expansion as noted in inset 6b to offer employment for the flood of immigrants from the countryside. In the expanding informal economy, women even more than men are victims of sweatshop conditions.[58] Mexico's economic crisis of the 1980s deepened the depressed state of most families, forcing many women to seek employment, however meager. As compared to the women entering the employment market during the 1970s oil boom, women entering the workforce in the 1980s had fewer years in school, were older, and more likely to have the care of infants and small children. As another index of the crisis, a Mexican earning the minimum wage in 1970 could cover 96 percent of the diet for a family of five, yet by 1984 it covered a mere 54 percent.[59] Data for the 1990s suggest a negligible improvement in Latin America, but by 2002 Argentine family budgets were at the lowest point for perhaps a century. By 2004 waves of violent protest, especially in the Andean countries from Venezuela to Bolivia, reflect the misery of over half the families in the region.

INSET 6B   AGAINST THE ODDS: EVER THE UNDER FACTOR

The work potential of women is rooted primarily in the dynamics of one, the male–female relationship and two, the class system. This generalization applies not only to a colonial society but also to a rapidly industrializing one, as documented in a far-reaching study by Lourdes Benería and Marta Roldán of Mexico City. Of the 140 women interviewed, only 28 percent had schooling beyond the third year, most were migrants from rural areas. Since subcontracting among multinational firms is the current practice, ever more women work in the *maquila*, a form of subcontracting labor, more precisely a sweatshop, occasionally in the basement of the boss's home or business. Along the U.S. border new factories attest to this growing NAFTA-induced economy employing both men and women—with wage differentials. Employment of women accelerates with the global economy. However, preference for women reflects wage discrimination, ability to work at routine tasks, greater productivity, and lower absenteeism as compared to men (Mondays being most relevant because of weekend drinking). In view of these attributes, notably the low rate of pay and their labor in the home, women are viewed as a subproletariat, whereas their husbands belong to the proletariat. As women are even more involved in the informal economy than are men they have lower status. Better educated women with relatively steady employment can aspire to enter the proletariat or even the petty bourgeoisie.[61]

This interrelationship between gender and class is a complex one with a strong input from the nature of the marital relationship. Men have the advantage of having less drastic shifts during the life cycle—marriage and childrearing. Beyond remunerative work lies an equally large domestic workload.

However questionable the urban situation, the rural scene represents the most agonizing situation for women. Rural labor remains the lowest in the occupational hierarchy, women suffering still more discrimination than men. The drift toward a global free market economy has favored men over women as men find employment in industry and women are left with agricultural roles, as reported for rural Mexico.[60] The grossly unjust system of land tenure and the nature of work status are both central factors. In countries undergoing agrarian reform, except for sandinista Nicaragua, women seldom receive title to the land. As another index of discrimination, a Peruvian survey of 1,169 agricultural workers found only 40 percent of women but 58 percent of men with access to regular employment. Similar data appear for other Andean and Central American countries.

Both for urban and rural women a large segment of employment is in public markets. San Salvador, the capital of El Salvador, has five main retail markets with no less than eight thousand stalls in addition to a wholesale market. Marketing facilities include a rich variety ranging from natural produce and processed foods to pharmaceuticals, textiles, clothing and household products—in addition to ubiquitous bars and cafes. As sellers may outnumber buyers, competition among market women is intense. Even without a national economic crisis, the financial return for these women is painfully minimal. In Nicaragua, as in other countries, political entities regulate and more than occasioznally exploit market personnel. Once more, a socialist government can impose as many roadblocks as can capitalism. Whatever the social system, the struggle for equality is unending.

Is there any final answer to the dilemma of women's role in the economy? One point is clear. Industry increasingly takes women from the home, with little or no reduction in home duties. It is worth noting that in most countries estimates of women's participation in agriculture is at least two times higher than official reports.[62] In reality, the rural workforce declined in the 1970s as the city offered a less forbidding environment, if very marginal work. The decade also saw an expanding commitment to labor organizations. A number of projects emerged in order to improve women's limited horizons in rural areas, particularly in land reform programs.

In spite of marginal improvements, the total outlook is hardly benign. As noted above, the economic disaster of the 1980s drove women even more desperately into the workplace, where they found intense competition with men, who were also adversely affected by the economy. In Mexico women moved from being less than 4.6 percent of the workforce in 1930 to over 24 percent in 1980 (the remaining 11 percent represent miscellaneous types of work), but they remained in a subservient role. Three major areas of women's employment are (1) service, 46 percent (especially domestic, they are also 57 percent of the nation's teachers); (2) commerce and sales, 22 percent; and (3) industry, 21 percent.[63] Interestingly, the most active employment period is between ages 15–24. At the same time, official records point to nearly twice the unemployment rate for women as compared to men.

The *maquiladora* (routine machine processing) industries constitute an assembly-line-export process attached to multinational organizations, especially in border towns such as Ciudad Juárez and industrial sites throughout Mexico. A Mexico City study portrays women's segregated role in industry, even though in at least a third of the families women provide more than half the income—despite depressed wages.[64] A survey of Mexico's *maquiladoras* strongly suggests that globalization has disastrous impact on women in several dimensions. As one instance, in Nogales on the Arizona border, women are relegated to dormitory living.[65]

The status of women is constantly changing, not least the position of women in an urban versus a traditional rural milieu. Anthropologists report that women have more equality in indigenous cultures, notably hunting and gathering societies (where they operate their traditional technology) than they do in the Westernized ones. In the city women have greater entrée to employment, however marginal, as reflected in their higher rate of cityward migration as compared to men. For instance, in Colombia not only do urban as compared to rural women have fewer children but their employability depends on the number they have.[66] A critical dimension in the status and employment of women depends on the economic system. Marxist regimes are scarcely better than capitalist ones as they usually add to the burdens of women, who are expected to follow traditional roles of homemaking in addition to outside employment, for which they are often as underpaid as under capitalism. Sandinista Nicaragua was a possible exception, but the Contra war complicated the advance of women. Cuba presents a favorable, if not ideal, profile. For instance, women represent 66 percent of professional and intermediate level technicians. One goal of the FMC (Federation of Cuban Women) is to promote the education of women, equal rights in the workplace, maternity leaves, and child-care centers. To an appreciable extent, these goals are met, but with questionable success in view of Cuba's economic breakdown following the end of Soviet economic support in 1990 and the seemingly unending U.S. economic blockade. Also, infant and childcare centers are lagging—partly because of lack of construction materials, the responsibility falling on grandmothers and other family members. Despite enormous progress since the pre-Castro regime, gender discrimination still exists. Fidel Castro praised the woman's movement as "a revolution within a revolution," but showed only half-hearted interest in women's welfare.[67] Recent evidence points to an improvement in women's work status. As one statistic indicates, 66 percent of professionals and intermediate level technicians are women.[68]

The research on Latin American women in the workplace generally documents the inferior image that men hold of women and the kind of assignments they receive in industry.[69] The question remains as to why women's labor, particularly in areas of low pay, is highly desired, yet in other areas few career opportunities are available. In contrast to the traditional roles women are expected to fulfill, a sixth to over a third (depending on which country) now work outside the home. Women's adjustment to

Table 6.2   Women as percentage of workforce in given countries

Country	1950	2000
Brazil	20	26
Colombia	19	22
Costa Rica	16	23
Dominican Rep.	10	18
Ecuador	16	20
El Salvador	17	25
Guatemala	12	20
Mexico	15	20

Source: Adapted from Jorge A. Bea, "Population Dynamics
in Latin America," *Population Bulletin*, March 2003, 29.

their various roles—marital, family, and employment—hinges on a number
of cultural, social, and individual factors. In the rural area, progress may
become more visible than in the urban, as it has the longer way to go.

In one summary of the socioeconomic situation facing Latin American
women, three different viewpoints are relevant:[70] (1) The integration thesis
holds that industrialization precedes emancipation and achievement of equality as women take their place in the economic world. (2) Marginalization sees
women as occupying only a subordinate role in industry or services. They
should stay at home and not deprive men of employment. (3) The exploitation thesis points to women as confined to a proletarian role in the capitalist
search for profits. It is possible to rephrase the debate as the articulation versus exclusion arguments—whether women are incorporated into or rejected
from the economic world.[71] So far, integration remains largely a dream, but
women's movements are pledged to bring about equalization of their economic opportunities. As suggested in table 6.2, women have gained wider
employment over the last few decades, even though it lags most of the world.

Women in the Political Scene

From colonial times to the present, women have played a significant role in
sociopolitical change. First of all, the model the conquistadores brought to
the New World was markedly different from indigenous patterns. As an
early challenge to the colonial oligarchy, Baltazara Chuiza led a revolt
against the Spanish in Ecuador. Similar rebellions followed, notably the ill-
fated protest of Tupac Amaru and his wife in Peru. Women also participated
in the War of Independence. In Mexico leadership of women stretched
throughout the nineteenth century.[72] This spirit of resistance glimmered elsewhere in the late nineteenth century, notably the strikes of miners in Chile,
where women maintained *cocinas apagadas* (unlit ovens), that is, withholding
food in order to encourage the men to hold off settlement until their
demands were met.[73] Regardless of their ambiguous social and legal status,
women are slowly and painfully making residual gains. They had little

success in changing their depressed role during the nineteenth century. However, certain social factors sparked a momentum for women to organize.[74] In Argentina at the end of the nineteenth century, several factors— incipient industrialization, waves of European immigrants, diffusion of new political philosophies—became catalysts for women to insist on suffrage and entry into the professions. Argentina finally modified its legal definition in the Woman's Civil Rights Law of 1926 by granting the wife power of attorney. Still, emphasis was on the spouse as wife and mother. Even the support Eva Perón gave through her establishment of the Partido Peronista Feminino was hardly a move toward feminine emancipation. This conservatism persisted into the 1990s.[75] It may be added that Eva was not the only woman to rise to the top of the political orbit (inset 6c).

INSET 6C   WOMEN IN POLITICS: CONTRASTING VIGNETTES

Women occasionally reach the pinnacle on the Latin American political scene. However, vignettes vary markedly. Unquestionably the most flamboyant example is Eva Perón. Born Eva Duarte to a working-class family, she became an actress and a cabaret dancer. Her charm and physical attractiveness brought her at least one powerful admirer.

With ardent support from labor leaders Juan Perón was on the rise, indeed to the point that the military in 1943 banished him to a prison on the island Martín García. Eva along with thousands of *descamisados* (shirtless ones) raucously and successfully demonstrated for his release. Juan and Evita soon married and a year later in a coup he became president establishing a government with the philosophical label of Justicialismo, somewhat patterned after Benito Mussolini's Fascism. The seemingly charismatic Evita shared power with her husband, focusing on women and children concerns, but retaining traditional male dominance. In 1952 Argentines overwhelmingly mourned her untimely death from cancer at age 33. Several leading *porteños* lost little time in not too subtly petitioning the Vatican for her canonization!

However, this was not the end of the feminist Perón dynasty. Although Perón was banished from the country in 1955, uncertainties and rivalries in Argentine economics and politics made for a power vacuum. Peronismo still had its followers. By various adroit moves, Perón was gradually launched on his return to power—an important factor was the negotiations by his new wife Isabel, who was admitted to Argentina before the arrival of her husband. After considerable maneuvers Juan Perón returned to the presidency in 1974. He promptly appointed Isabel as vice president, as he had with Eva. With his illness she became acting president. Anxious to rival the first Señora Perón, Isabel dyed her hair blond and visibly moved among the crowds, but she was not able to capture the charisma of her predecessor. At Perón's death on July 1, 1974, Isabel became president, but not for long. A return of political and economic malaise led to a military dictatorship in 1976.

In a different vein is Violeta Chamorro, president of Nicaragua, 1990–95. Born Violetaa Barrios in 1929, she was the daughter of wealthy landowners, and was educated in Nicaragua and the United States. In 1950 she married Pedro Chamorro, descendent of a distinguished political family. Pedro inherited his father's role as publisher of *La Prensa*, an independent newspaper. He soon joined the opposition to the brutal Somoza regime (which was sedulously supported by the United States).

Imprisoned, later released, again threatened, Pedro fled underground to Costa Rica, Violeta and her four children joining him in San José. After an amnesty was proclaimed Pedro returned to editorial and political activities in Managua. But as Anasasio Somoza became even more tyrannical, assassins were ordered in 1978 to kill Pedro.

As with most Nicaraguans, Violeta was caught up in the Sandinista revolt. She consented to join a provisional junta, the dominant figure being Daniel Ortega. As she gradually felt that militarism, press censorship, and nondemocratic themes were unacceptable, she resigned after nine months. Her children, now adults, moved in both pro- and anti-Sandinista directions as did other Chamorro kin. Through the 1980s conflict was rampant between the Sandinistas and the National Opposition Union (UNO) or the Contras. Violeta opposed the violence on both sides. As compromise and peace came in 1989, an election was scheduled for early 1990. Violeta chose to run for president under the banner of the UNO, now rid of the more unsavory Contra element.

Violeta Chamorro's narrow victory over Daniel Ortega set the stage for a continuing accommodation to Sandinistas. For example, she appointed Daniel's brother Humberto Ortega, as defense minister. A most difficult task was negotiating with the Contras for their disarmament. She also reduced the numerical strength of the Sandinista army and finally ended conscription.

Economic problems continued to plague the nation. Accordingly, Violeta became the first Nicaraguan president to visit Washington. Her 1990 visit with President Bush resulted in promises of help, but nothing specific. As her term of office wore on, health problems increased. More and more, she turned to her son-in-law for decision making. Still, her image remained a model to probably most of her fellow citizens.[76]

In a different social climate, the Mexican Revolution was the prelude to the 1916 Congreso Feminina in Yucatán. Mexico, along with Chile and Uruguay, are among those nations with de jure, if not de facto, gender equality. By the 1960s women's organizations and publications such as Fem prompted a nucleus of women to fight for *reivindicación* or reclamation of their rights through political action (table 6.3).[77] The vote was also extended to women in the twentieth century, Ecuador, scarcely rated as an enlightened country in Hispanic America, was the first to give women voting rights in 1929. Brazil, Uruguay, and Cuba came along a few years later. Despite high literacy rates Argentina and Chile lagged until the late 1940s, decades before Switzerland! Opposition to empowering women with the vote was based on the same stereotypes as European and North American males uttered a generation earlier—women are less intelligent, therefore not suited to the task of choosing officials. Also, many feared that they would be too much influenced by the church.

On this frontier of women's rights the twenty republics display a diversity of aspirations and chronology, even though the fundamental desire is for recognition of their status and the end of gender discrimination, if not actual persecution. Improvement is occasionally visible as in the declaration of gender equality in Ecuador's 1945 constitution, even though its implementation was long delayed. In part because of the power of the Roman Catholic Church in Colombia, women's marital rights were forestalled, but a 1974 law acknowledged men's adultery to be almost as reprehensible as women's.

Table 6.3   Women's electoral rights

Country	Right to vote	Stand for election
Argentina	1947	1947
Bolivia	1938–52*	1948
Brazil	1934	1934
Chile	1939–48	1931–49**
Colombia	1954	1954
Costa Rica	1949	1949
Cuba	1934	1934
Dominican Rep.	1942	1942
Ecuador	1929–67*	1929–67*
El Salvador	1939	1961
Guatemala	1946	1946
Haiti	1939	1950
Honduras	1955	1955
Mexico	1947	1953
Nicaragua	1955	1956
Panama	1941	1962
Paraguay	1951	1961
Peru	1955	1955
Uruguay	1932	1932
Venezuela	1946	1946

* In certain countries the franchise or similar rights came in two or more stages, e.g., the vote was first granted to all literate women, later to all women.
** Adapted from Latin American Center, *Statistical Abstract of Latin America* (Los Angeles: University of California, 2002), p. 181.

Even in an advanced country such as Uruguay, demands for the end of wife beating are producing shelters—the novel idea of making violence against women a crime.[78] Brazil in 1995 officially set up the National Council for Women's Rights (CNDM), which monitors cases of physical abuse, economic exploitation, and provides for childcare centers. No less than ten nations have adopted similar provisions. Women's rights, in addition to freedom of sexual behavior and gender choice, are recognized in Brazil's 1988 constitution.[79]

The Impact of Women's Movements

Women's movements have been an intriguing aspect of Latin America since the 1970s. Jane Jaquette looks at women's movements as falling into three types: (1) human rights movements with a focus on violation of civil liberties in the climate of authoritarian regimes; (2) feminist groups arising from disaffection with the failure of political parties to recognize women's needs—a movement beginning with consciousness-raising, developing into active organizations; and (3) mobilizing impoverished urban women, which began as a grassroots movements spurred by the church, political organizations, and international foundations. In a comparison of five countries (Argentina, Brazil, Chile, Peru, and Uruguay), intersection of these three types of movements varied widely from parallel to integrated activities and in the degree of support offered by political, religious, and economic structures.[80]

Movements take place in different periods, but the 1970s and 1980s were crucial, a score of organizations appearing in Venezuela in those decades.[81] On the other hand, in Paraguay, after thirty-five years (1954–89) under the Alfredo Stroessner dictatorship, a woman's movement finally sprung up in the 1990s. Significantly, eight of the fifteen members of the national executive committee of the MCP (Paraguayan Peasants Movement) are women. Among its objectives are legal recognition of women and meeting various practical needs—limiting family size, specifically the use of contraceptives, still discouraged by the church. Throughout Latin America women's voices are heard on a broad spectrum of issues—wage differentials, protection of the rights of indigenous women, decriminalization of abortion, and most critical, ending verbal and physical abuse by men, particularly spouses. Prostitutes are also demanding legal recognition. Movements may grow out of social organizations, as they did in Santa María Tzejá, a Guatemalan community caught in the army's campaign of destruction of Mayan peasants. A number of women joined the guerrillas, others aided the refugees, still others organized the rebuilding of the community after the armed conflict ended. These groups and organizations, often in liaison with NGOs, led into national and international movements.[82]

Women's movements abound in every country with the possible exception of Haiti. They emerge from middle or lower class, urban or rural, national or local orientation. Movements often develop out of a favorable political moment as when a progressive regime achieves power or as an underground phenomenon in a regressive military takeover. For instance, with demilitarization of the mid-1990s, women in El Salvador are finding for the first time in their history a significant opening, though modest, into politics—29 of the 262 mayors are women.[83] After Scandinavia and the Netherlands, Latin America ranks next in the ratio of women in the legislature, Argentina and Cuba leading the list, both with 27.6 percent.[84] In Mexico a coalition of women's organizations include in their agenda: guaranteeing worker's rights in the *maquiladoras*, declaring rape a public crime, requiring men to prove their non-paternity if they are not to pay for child support, and protecting women in the penal code. Chile represents a special case in state support of women's rights. In 1991 the government established the National Service of Women (SERNAM), which is designed to assure the same rights to women as apply to men. While the law has worked less than adequately it gives a certain degree of stability to gender rights—a security absent in certain countries, especially those subject to political and economic vicissitudes. At the same time, the effect of the law reduces the militancy that women may feel in other nations.[85]

In other words, movements develop in varying stages and express differing meanings. Among middle-class intellectuals feminism is more ideological as it stems from debates among left-wing students and professors, inspired by ideas from Europe and North America. In contrast are working-class women's movements rooted in the *barriadas* (squatter settlements) and trade unions, usually dedicated to soup kitchens and similar tasks. Still, its mobilization is oriented to a political program pledged to reform reproductive rights and reduce violence against women. True feminists feel that such projects as soup kitchens only underline the passive, exploitive destiny of women. In the ever-changing scene of political, economic, and ideological pressures a sinuous line of both bifurcation and merging are apparent in feminist movements.[86] Fundamentally Latin American women are determined to end the mythology surrounding gender and widen the democratic channels for a fuller participation.

The Intricacies of Political Participation

As implied earlier, the twentieth century was marked by only a fragmentary advance in women's rights. As one example, despite the major role women played in the Mexican Revolution, the vote did not come to women until 1953. For Latin America in general, social and political reforms appeared in the brief democratic wave of the 1960s, only to be reversed in the dictatorships of the 1970s and 1980s. The impetus to address women's problems yielded significant gains only in the last fifteen years of the century as political channels became more open. Whatever the image of women as politically conservative, their inferior social status often leads them to more liberal attitudes as compared to men.

The supposed ambivalence of women about participating in the political process is contradicted by voting statistics since the late 1980s. Further, the changes in gender ideology lead to various effects. In Buena Vista, a community in Morelos, Mexico, women organized in order to initiate sociopolitical change in municipal services—land, health, education, and water supply. In part the movement was inspired by the theme of "mother as redeemer."[87] Similarly, women's groups in Mexico City perform a number of services from caring for the ill to getting out the vote and pressing for a better community, including housing.[88] Studies of social movements point to the dedication and relative success of women over men in various organizations, particularly ecological ones—all the more impressive considering the commitment they make in spite of the work load both inside and outside the home.

In the political arena women meet resistance from both the right because of their interest in social action and from the left because of their supposed conservatism. As noted earlier, women's involvement has a long history. They were active in the formation of the Alianza Popular Revolucionaria Americana (APRA) in Peru during the 1920s. Similarly, they held key positions in opposing dictatorship and supporting socialist movements in Argentina, Brazil, Cuba, Chile, and Nicaragua, in addition to giving up their

sons and husbands to the armed struggle.[89] Women were very supportive of
the Cuban Revolution, followed by emergence of the Federation of Cuban
Women (FMC) and extension of work opportunities to women, even
though they seldom reach high government positions. By 1974 women
composed only 13 percent of the membership of the Communist Party.

Chile offers another reminder of the inability of socialism to root out
machismo. Women were divided, usually by social class, in their loyalty
to Popular Unity under Salvador Allende (1970–73), who appointed
only one woman to his cabinet and then only for a brief period. During
the Pinochet (1973–89) dictatorship arose the organization Women For
Life (many of its members had lost sons, brothers, and husbands in the reign
of terror). Their demonstrations were greeted by water cannons and other
violent action, but the organization ushered in the twilight of the Pinochet
rule, which ended in 1989. In the overthrow of Somoza in Nicaragua
30 percent of guerrilla forces are reported to be women. Besides playing a
leading role in health care, women demand greater access to birth control
methods and to job training courses, rising to positions of power in labor
organizations.

A comparison of women's mobilization in five countries sheds some light
on the rising influence of women in national struggles. In Colombian and
Cuban guerrilla activity of the 1950s and 1960s women had little direct
involvement, whereas in the movements against authoritarian rule in the
1970s—El Salvador, Nicaragua, and Uruguay—feminism had its impact.
Both men and women were trying to change their universe toward more
equalitarianism and democracy. Salvadorans, Sandinistas, and Tupamaras
articulated more explicit protest to patriarchalism, pushing for feminist
long-term goals such as ending discrimination in education and the labor
force. Second, short-term goals (child care, health clinics, and literacy) were
to facilitate women's role in family functioning.[90]

The Heroic: Violence, Martyrs, and Change Agents

The suppression of women is sanctioned by a traditional society, as in the
patria potestas, institutionalized by the Romans and accepted in Western
cultures well into the nineteenth century. It lingers on in several quarters of
Latin America. Also, violence is meted out to both genders by authoritarian
regimes. This terrorism falls more on men than women but the violence
assumes more traumatic methods and effects on women—and on children.
This form of violence characterized Argentina (inset 6d), Chile, and
Uruguay during their military governments of the 1970s and early 1980s. In
addition is a persistent siege of violence in Central America, notably El
Salvador, Guatemala, and Nicaragua, which only in the 1990s showed some
abatement.

INSET 6D   WOMEN CONFRONT VIOLENCE

Even though the cone countries represent a relatively advanced area, they are cited for their terrorism. Argentina is the most publicized case. The "dirty war" from 1976 to 1983 was probably the most tragic episode in Argentine history. The ruling generals were determined to suppress even the suspicion of insurrection. Some fifteen thousand citizens of various ages, particularly the young, were sequestered, kidnapped, or simply disappeared. Among the resistance groups none was more enduring and dedicated than Las Madres de la Plaza de Mayo (the mothers of the Plaza de Mayo). On April 13, 1977 on the plaza appeared fourteen mothers. These women had come from the Junta headquarters inquiring after missing sons and daughters, only to be told "your son probably ran off with someone," or "your daughter was killed by a terrorist colleague."[91] A week later, twenty women, and by September some three hundred were circling the Pyramid two by two, wearing white headscarves. Tear gas, arrests, and kidnappings soon followed. As nuns were among the protesters, the government had little pretext to claim that the demonstrations were a form of subversion.

Although over half the victims were completely innocent of what might be interpreted as political dissent, terrorism became ever more far-reaching. Simply being in the address book of a suspect was sufficient grounds for arrest. Another savage form of duress was the disappearance of infants and children who were sent to neighboring countries or adopted by childless government officials. At the World Cup in June 1978, women paraded with signs "We have no news of our children, our grandchildren, or our parents." International attention seemed only to aggravate the terror. Attorneys defending the Commission of Relatives joined the lists of the disappeared. Laws were passed in order to facilitate the certification that the victims had died, all of which was to discourage inquiry into their fate. This Argentine episode has many facets; one telling aspect of these demonstrations is the portrait of citizens moving beyond grief to make a political statement. Apparently the action of these women failed to move the military and political elites, but their relatives, friends, neighbors, and eventually the public at large understood their position.[92] Although politically marginalized they were indirectly promoting a democratic nation. Further, they were transforming public space as they occupied the Plaza de Mayo—long the symbol of masculinity dating from the Inquisition to the twentieth century. The event also signified "an alternate vision in which 'family and justice' are not sacrificed for 'security.' "[93]

Despite a relatively democratic tradition in Chile, terrorism under Pinochet was no less a tragedy. The 1973 right-wing coup against the Unidad Popular of Allende ended three years of socialism. Although estimates vary, presumably over three thousand were killed in the first several months of the Pinochet dictatorship.[94] This violence fell upon both women and men, especially in the working class and rural population. Although the principal targets were men, women also became victims of imprisonment, kidnapping, and rape. As in other militarist regimes (Argentina, Uruguay, El Salvador, and Guatemala, among others) women were subject to the most brutalized method of torture as "required" for state "security." The techniques, seemingly used by the military and local police throughout Latin America, reach various levels of depravity, often in the form of "family torture," which is an attempt to extract information from an uncooperative male detainee. Rape and torture are often used in front of the husband to force his confession. Often the same basic methods are applied to both sexes, but for males torture is to destroy their "sexual confidence," for females torture is "systematically directed to their sexual identity and anatomy."[95]

Similar histories characterize much of Latin America. Physical and mental punishment meted out to women as well as men in Central America is documented in a number of volumes. Nor are democracies immune to seizure, disappearances, and torture. A report from Mexico tells of how mothers, led by the leftist political leader Rosario Ibarra, have a campaign to find out the fate of their sons and other relatives. Inspired by the Madres de la Plaza de Mayo, this movement was successful in liberating 148 of 522 disappeared persons, many of whom were hunted down by government agents and the police.[96] To place this analysis in a political context, women appear to follow at least two paths toward sociopolitical action. One is overtly ideological, particularly among more educated women, as they work through political parties and other organizations, more often on a nonviolent than violent basis. The other occurs more during dictatorships and may be an expression of *marianismo* as they view themselves as the suffering mother of Christ. This was most visible in the Mothers of the Plaza de Mayo during the generals' reign in Argentina and the Families of the Disappeared in Pinochet's Chile. Both these movements became models for Latin American women to organize for an equalitarian order—usually socialist.[97] Yet, any generalization is fraught with complexities. Many Peruvian women in Ayacucho wrestled with pressures from family, church, government in their decision as to whether to join or not join the Sendero Luminoso.[98] In the Zapatista National Liberation Army (EZLN) women are playing an important role in mobilizing both peasants and townspeople.[99]

Conclusions

Socialization and gender relations are structural in Latin America, yet are undergoing change, notably in those sectors of society most affected by trends set in advanced Western cultures. In particular, machismo is threatened, but will persist as males are seldom enthusiastic about giving up their domination.

What is the future of feminism in Latin America? The profile of women's needs remains problematic. Among the issues is the feminization of poverty—also a problem in the United States. The situation is more severe in developing nations. One hopeful note in Latin America is the increase in educational opportunities, employment of women, and a growing consciousness and communication of their disadvantage. Moreover, women in the feminist movement stress that their style is different from that of their North American cousins, at least those in organizations such as NOW

(National Organization of Women), but the issues and goals are similar—adequate pay, child care, legal equality. Increasingly, modernity became linked to gender concerns. Research points to societal factors as predictive of progressive attitudes of women. The twenty-first century will undoubtedly bring major shifts in the global economy and in the international power structure. It is likely that Latin America will be an increasingly active player in this scheme. As the region is generally closer to the Western cultural heritage than are Africa and Asia, we may see improved societal relations, including women's participation in whatever benefits accrue. Moreover, the political atmosphere with de jure if not de facto democracies allows Latin American women to seek new avenues of autonomy.[100] As with their Western sisters, Latin American women now understand value structures and power relationships that stand in their way. Women are gradually learning the power game and how to avoid roadblocks of the past.

Chapter Seven

Marriage and the Family

In all societies, even the most primitive, two institutions are indispensable—the economy for insuring physical survival and the family for providing inter-generational continuity. For all societies the family remains the integrative institution providing a number of functions—assignment of statuses, control of the sex urge, reproduction of a new generation, and socialization of the young. Family and kinship constitute the most important socializing agency. Through nearly five centuries of its history—preceded by well over a millennium of an indigenous heritage—the Latin American family transmitted the culture, and even in the urban society of today it remains the individual's anchorage in a world of uncertainty. For most Latin Americans, the family is the one institution upon which adults and children rely for their identity and survival. Other institutions have robbed the family of its functions in the maintenance and socialization of its members, but to a lesser degree than in most Western nations. Even the word for marrying, *casarse*, means the setting up of a home. As a kind of infrastructure, the family gives continuity to society and influences the other institutions.[1]

Probably only Latin Americans can understand the depth of their family culture; an outside observer can hardly appreciate its many meanings. Family forms are highly diverse as based on historical and regional traditions, or on the heights and depths of class and caste. One has only to compare the Inca heritage on the *altiplano* with the culture of the Afro-Caribbean, the *casa grande* and shacks of the fazenda with the urban dwellers of São Paulo. Even within the mestizo and European family the marriage system differs by class and the urban–rural cleavage. Yet, as the landed elite relics on family and kinship to hold together their wealth, shanty dwellers or agricultural migrants find the family their major psychological and economic resource.

In addition to the effect of subcultures and historical traditions, contemporary events are changing family norms. Technological culture and urbanization have an inevitable impact on the family as they have in all societies affected by Western industrialization. Also, economic and political events set in motion changes in the family structure and function.

This chapter begins with a brief note on several theoretical bases of the family. We then examine the forms of marriage and the roles played by husband and wife in the marital system. An analysis of family and kinship structure in the life cycle along with the problems of marital and family instability will follow.

Theories of the Family

At least four theoretical frameworks are brought to bear on marriage and the family. One is the structural–functional, which serves to explain the checks and balances of family functioning—an equilibrium of individual needs with family stability. *Functionalism*, as the theory is often identified, fits into the concern for modernization with its stress on the nuclear family but recognizes the functions of the extended family, which is salient in Latin America. However, the theory proves to be inadequate in explaining family breakdown and the crises of families in a society undergoing socioeconomic privation. Second, in the sociopsychological tradition is *symbolic interactionism*, which stresses role relationships, role conflict and strain, communication networks, and interpersonal competence.[2] Third, also focusing on interpersonal processes is *exchange* theory and its theme of costs/rewards along with bargaining and complementary needs. Both these theories are more relevant to family processes than is functionalism, but seem somewhat more suited to the family in a less economically impoverished culture than Latin America. These models should be supplemented with an explanation of the societal distortions.

Fourth, *conflict* theory assumes different forms, but fundamental is the idea that society is marked by grossly unequal resources. This viewpoint emphasizes competition, aggression, negotiation, stress and stressors. Further, it shares several concepts with exchange theory, that is, the trade-off of benefits. In reference to the family, two major interpretations of conflict have emerged: (1) classical Marxian—society basically represents the conflict between property owners and workers. Patriarchalism derives from male dominance in a capitalist society. The male's participation in industry permits exploitation of women in the home. In Latin America approximately 30 percent, with regional variations among specific countries, of women are employed outside the home—the lowest ratio in the world, even less than the Middle East.[3] (2) Feminist theories, as explained in chapter 6, find the basic conflict to be the traditional male dominance, which goes beyond economic systems.

In another respect, conflict seems endemic in the Latin American family and the culture in general. Families or family members struggle with each other for limited resources, whether space, attention, affection, sex, power, money, or food—often ending as trade-offs. Individuals and families display differing techniques of suppressing or regulating conflict. Conflict can even be functional within the family.[4] For instance, in many Latin American families, contested members may somehow find the means for establishing their status or identity. In still other families, harmony may prevail. The cultural pattern may set the stage for marital and familial conflict. For example, marriage in Latin America is permeated with a dualism emerging from both a moralistic Catholic outlook and a materialistic, opportunistic streak. Presumably the transition from paternalistic feudalism to capitalism has not changed but possibly magnified this asymmetrical relationship. The hierarchy based on gender provides for an intense opposition of male and female.

All the theories mentioned earlier have some validity in analyzing family processes. Whatever the model, families vary in patterns of cohesion and adaptability, levels of separation and connection, engagement and interaction.[5] As a social institution the family involves a set of norms according to which its members supposedly behave. However, decision making, negotiation, and resources favor the male in Western society as it developed from its classical Mediterranean sources.[6] The Latin American family is possibly more deeply rooted in the past than even most family systems of the West. Have technological advances and democratic processes provided more—or less—of a climate of change to Latin America than to Europe or Anglo-America? It would be difficult to assess this question. We do know that the rapid rate of urbanization along with the democratization of the mid-1980s brought a spirit of innovation into Latin America, but economic realities set boundaries to the degree of change. Regional and class variations are exemplified in inset 7a.

INSET 7A   CONTRASTING VIGNETTES OF FAMILY DYNAMICS

Sixty miles north of Asunción, capital of Paraguay, a six-hour trip on a steam train in addition to over an hour by foot brings one to the Burges' home in the village of Caballero, where Simeón and Catalina live with their four children ranging in age from eight years to seven months. The family lives on ten acres of land owned by Simeón's father. Beyond the woods one finds a garden and over a score of animals (chickens, a cow, a pig). The only source of money is chopping wood, and occasionally Simeón is forced to migrate to the city in search of construction work.

The acreage also accommodates other family members—aunt, uncle, and cousin, most of them occupying a dual structure, a family room with a detached kitchen. Catalina endures a heavy schedule with care of the vegetable garden and animals, the children helping with various chores. The immediate or nuclear family maintains a close relationship with the extended family. There is no question about the father's dominance. When Simeón returns home in the evening a quiet respect prevails, neither crying nor dissension is permitted. Moreover, Catalina is grateful for his fidelity. It may be added that the enormous loss of men in the 1870s War of the Triple Alliance and the Chaco War with Bolivia (1932–35) left Paraguay with a strong tradition of the male's importance—and an entrenched machismo.

In contrast is the Arantanha family—Pedro and Efigenia (both aged thirty), their two children Pedrinho (three years) and Carola (eighteen months). Pedro, like his father, a naval physician, dashes off in the early morning into Rio de Janeiro's traffic on his motorcycle. Efigenia, a pediatrician in a clinic, drives to work in her car. A maid cares for the apartment and children, who periodically go to a day center. Conveniently, their high-rise apartment complex provides, among other amenities, a pool to which they retreat most late afternoons.

Although Efigenia dreamed of a large family, after the birth of Carola she realized the demand of children on time and energy and decided with Pedro's approval on a tubal ligation, assuring continuance of a middle-class lifestyle.[7]

The Dynamics of Marriage

Marriage must be viewed in the context of the family, both nuclear and extended. As one Mexican psychologist reports—if somewhat simplistically—from his survey, North Americans think first of the individual, Latin Americans think of the family.[8] More than in the West, variability is inherent in the marriage pattern. First of all, four types of marriage are found:

1. marriage by a civil procedure followed by a religious ceremony, as preferred by the middle class;
2. civil ceremony only, universally recognized in Latin America but customary in parts of Brazil and in contemporary Cuba;
3. religious ceremony as found in rural society, or in the few areas where no civil marriage is required; and
4. consensual union (*unión libre, unión de hecho*), a widespread practice among the urban lower class and in rural areas, where access to civil or religious authorities is limited either by low demographic density or finances.

Both the consensual union and religious ceremony have lost ground in countries requiring married applicants for social security to prove they have a civil marriage.

The influence of region and social class is evident. For instance, in my Central American samples, twice as many of the middle class as of the lower class in San Salvador had the legal–religious union, whereas in more orthodox San José both classes elected this route. A formalized religious ceremony is an index of upward mobility. Middle-class respondents reported a smaller number of consensual unions for themselves than for their parents, whereas little generational difference was found for the lower class. The particular national climate is also relevant. Europeanized countries such as Argentina, Uruguay, and Chile are oriented to more conventional Western norms. By the 1980s the urban middle class in nearly all countries began to accept the idea of cohabitation, but not to be accompanied by reproduction as in the *clase popular* (working class).[9] This liberalization probably accounts for the gradual rise of the *unión de hecho* in Costa Rica from 8.5 percent in 1970 to 10.3 percent in 1980.[10] Although it is difficult to determine the number of common-law marriages (consensual unions), they are reported to vary from 2.4 percent in Chile and 7 percent in Argentina to 43 percent in El Salvador, suggesting a dichotomy between European and indigenous–mestizo traditions.[11] Estimates of consensual unions in rural areas are much higher, for instance, 66 percent in Brazil.[12] Throughout Latin America, turnover is three–seven times higher (depending on the country) than legal unions, at least half eventually dissolving. The number of individuals who have a *unión libre* at some time in their life is much higher than those legally married.[13]

In this context, an intriguing question is the difference between the two genders as to what constitutes the married state. In the 1990 Mexican census, 16.1 million women listed themselves as married but only 14.7 million men so responded. The tradition of the *casa chica* or dual households is one explanation of why more women than men consider themselves as married. Another aspect of this marital identity is the large number of male emigrants who leave their wives at home.[14]

For all Latin America the ratio of individuals married, whether legally or not, is less than in the United States or Western Europe. As compared to the annual rate of 9.1 marriages per 1,000 population in the United States in 1992, Mexico had a rate of 7.4, Argentina 6.0 and several countries remained below 5.0. Marriage statistics are probably no more reliable than fertility rates or other demographic data; still, the prevalence of *unión libre* remains in most mestizo and indigenous areas along with the liberated urban middle class. Also, in view of differential migration—men to agriculture and extractive industries, women to the city as servants and menial white-collar occupations—a number of individuals do not find mates.

The consensual union is partially an outgrowth of the less formalized marriage systems in certain indigenous societies. For example, in the Andean village of Huaylas a "wait-and-see" attitude persists. Cohabitation may begin at age twenty for the boy and eighteen for the girl, with formal marriage deferred for perhaps a decade.[15] Since the woman is often forced to support the husband, she is likely to hesitate about becoming formally married. Nor is the male anxious to assume marital responsibility, even though he has less to lose. Finally, the free union has little stigma, at least for the bulk of the population. In an urban Colombian sample, women gave two reasons for preferring the free union: the need for individuality and freedom and the rejection of formalism.[16] Most rural unions are highly stable. Many result in marriage when an itinerant priest appears in the locality, or during Holy Week, when the fee for a church marriage is dropped.

Courtship and Mate Selection

Mate selection varies greatly by ethnic strain and social class. Latin Americans traditionally guarded a girl's activities with the opposite sex. In rural communities boys and girls carry on flirtations at fiestas or *serenatas* or possibly at school. In a small town a furtive glance during a *serenata* may become the opportunity to establish a later meeting. Dating as known in North America is largely confined to liberated segments of the urban middle and upper classes. Traditionally, the upper class demanded more circumscribed behavior than the lower classes, in which permissiveness is generally accepted.[17] However, as in Europe and North America, class boundaries are less firm today.

As virginity at marriage was theoretically binding in the middle class, chaperonage was the institution calculated to discourage intimacy with the opposite sex. An older sister, cousin, or aunt accompanied teenage girls and young women on "dates" or social events, whether at the movies or the country club, but the system virtually disappeared after the 1960s.

Serious courtship is launched when the parents become aware of the young man's intentions. In the middle and upper classes both families are understandably concerned about the desirability of the marriage. Marriage can be a means of upward mobility. For example, migrants to Buenos Aires in the nineteenth century would try to seek unions for their sons and daughters with the older aristocracy, usually descendents of the colonial merchants.[18] Chaperonage was relaxed in later stages of courtship or at engagement. In recent decades courtship has assumed far less formality—chaperonage, serenading, and approval by the extended family—in urban culture, whether cinemas, automobiles, and secondary group life. Yet approval by the immediate family is generally requisite.

In most village and peasant cultures these norms are arbitrary. In many communities mate selection proceeds by informal contacts early in life. For instance, in a mestizo community on Margarita, an island off the Venezuelan coast, sexual contacts occur in the early teens. More often the girl is invited to live with the family of a boy of her choice, and a nuclear household is established later. A civil marriage, or more rarely a church wedding, can occur. Polygynous relationships may or may not follow.[19] Similarly in Aritama, a Colombian lowlands mestizo village, mate selection falls short of traditional norms. An undercover rendezvous, elopement, and some trauma to both the girl and her family characterize the choice of a mate.[20] As always, folkways and mores are conditioned by the particular community and occasional influence from urban areas. Less diversity is found in urban middle classes; even so, the style differs between Mexico City, San Juan, and Santiago.

The Marriage Event

The marriage ceremony is, of course, shaped by region, ethnicity, and class. Marriage may have to wait until the couple's parents and godparents accumulate sufficient funds to provide a few essential furnishings and the makings of a fiesta. As a religious ceremony in the city can be an expensive affair, couples sometimes elect a civil ceremony, which carries less social obligations. Elopement is also a way out, but marriages without parental consent could mean disinheritance as well as weakening family solidarity—too important a commodity for a young person to ignore. Nonetheless, elopement can occur.

At one time in rural areas, physical proof of the bride's virginity was requisite. However, ideal and reality are not the same thing. Because of the mobility accompanying urbanization, traditional values break down. In theory, premarital chastity is still an ideal, but employment outside the home, mass media, and new norms of sexual freedom, among other changes, are bringing the mores into question in the city.

Marital and Family Structure

The average age of first marriage is somewhat later than in the United States but slightly earlier than in northwestern Europe. The age varies from nations such as Argentina, where men marry at twenty-seven and women at twenty-four to Central American countries, where it is close to twenty-four and twenty, respectively. The *ideal* age at marriage appears to be older than the actual age. Responses from women in six major Latin American cities show a preference for age twenty-one to twenty-four years (Caracas representing the lower age and Buenos Aires the upper age). Urban residence and educational attainment are also positively correlated with a later age of marriage.[21]

In respect to the relatively small number of persons married, Latin America stands below a postindustrial nation such as the United States, where over 85 percent of the population is married before the age of fifty. This lag runs counter to the socioeconomic enhancement of marriage in Western culture and the strong norms supporting marriage in other developing areas of the world. Still, of the large percentage of the population not married, a number have had a consensual union.

As follows from higher fertility, average size of the Latin American nuclear family is larger, although declining, than in most Western nations. Also, even in low fertility nations, rural family size is above the urban. The high ratio of single individuals, new households, widowed and separated is balanced by large families, with wide regional differences. It is relevant that in 1998 the birthrate varied from 3.7 in Guatemala to 1.8 in Uruguay.[22] Whatever its size the dominant form of the family continues to be nuclear, that is, only a minority includes a grandparent or other relative. Actually, nuclear families are occasionally larger than extended households. In Peru it is estimated that 57 percent of rural families and 51 percent of urban are nuclear. Roughly a fourth are extended, a tenth are single person households, and the remainder are too complex to classify.[23] In Santo Domingo, a *clase popular* district (*colonia*) in Mexico City, the nuclear family prevails with both parents highly child-centered, in contrast to the rural pattern of patriarchal, extended family culture they left behind in rural states such as Hidalgo or Oaxaca.[24]

The practice of *adoption* is widespread, not only because of the high rate of illegitimacy, but because of the ravages of malaria and other diseases. Also, poverty and migration produce broken homes. For instance, in Gurupá, an Amazon community, families of higher economic status adopt children of their more unfortunate neighbors. The adopted child is often the one expected to do a large portion of the chores, but generally he or she is treated as a member of the family, and is loved as much as the parents' own children.[25]

Instability and Divorce

A major aspect of the marriage system is its instability, not only for Latin America. In the urban lower class over a third of families are incomplete. A major cause of instability is economic insecurity or the inability of the male to support his family. Another factor is the male-oriented nature of marriage itself. Many men have a *casa chica*, which borders on concubinage. In a Cuernavaca, Mexico, sample, three types of extramarital relations are found: (1) casual episodes with a fellow employee, who he might take rather than his wife to the company Christmas dance, (2) overnight stays with another women, often older and more experienced than his wife, and (3) setting up a household with the other woman, although time and finances are likely divided between the two.[26]

An urban lower-class woman may have children from three or four unions. As women remain in a submerged position after marriage, rural marriages are not altogether stable. A portrait of the Brazilian coffee subculture reveals how women are abused and abandoned by alcoholic husbands as well as forced to rear children with severe food shortages. The transition from *colono* to wage earner is more destabilizing to the family and adds to the wife's burden as she is forced into work in the fields. Also, among intact unions she must defer to her mate and his family's and parents' needs.[27] No less than urban marriages, the rural suffer from instability because of ethnic, economic, and relocation stresses, as in the case of Carolina Maria de Jesus (inset 7b).

INSET 7B   FAMILY LIFE IN THE BACKLANDS

Carolina Maria de Jesus wrote one of the most stirring chronicles of life in a São Paulo *favela* (slum). It was to become a best seller.

Carolina was born in 1914 in Sacramento, a remote town in Minas Gerais. Daughter of a wandering musician, who later abandoned her mother, her life was one of almost constant anxiety, at least in her younger years. Among the townspeople Carolina and most of her family were regarded as pariahs as they were of very dark complexion—"so black that the skin seemed blue"—in a nation where skin color still determines status. Even lighter-skinned cousins (mulattos) spurned her. Besides her color, she was ostracized for being illegitimate and poor—poor even by local standards. In fact, the family could hardly buy any commodity; they even made their own soap.

Due to the charity of a landowner's wife (who seemingly was trying to atone for her ancestors having owned slaves), Carolina was able to attend school. Mocked by other children because of her low status and being somewhat rebellious, she first hated school, her mother beating her almost daily in order to force her to attend. Later she loved school, devouring every book her teacher could find for her.

After only two years at school the family moved to a *fazenda* (ranch), where her "stepfather" found a job, and soon Carolina, not yet ten, was also forced to work in the *fazendeiro*'s house, temporarily ending her schooling. Relocations of the family followed according to the uncertain world of rural labor. Eventually the family returned to Sacramento, where further duress was to prevail. For one, Carolina's mother once tried to stop a neighbor from beating his wife. He then attacked even more violently. When the police arrived, both attacker and interceder were arrested. This and other injustices made an indelible impression on Carolina, who not only had a keen awareness, but learned to live in the universe of the mind. She was especially influenced by her grandfather, who, though denied a formal education, regularly read the newspaper *O Estado de São Paulo*, relating the contents to his granddaughter. As school again became a reality, her horizons broadened as she managed to find books that became an escape from the vicissitudes plaguing her, including skin diseases that appeared intermittently.

As she neared adulthood her employment improved as did her social life. She had a series of liaisons with men of lighter skin than herself—Brazilian society is more tolerant of encounters between a dark female and a lighter-skinned male than the reverse. In 1947 she was married and moved into a São Paulo *favela*. Years later with the publication of her diary Carolina became famous, Yet, she was never quite accepted by the São Paulo establishment.[28]

As in other Roman Catholic countries, divorce has only recently been institutionalized. Formerly it was only available in Middle America, but is now obtainable in every Latin American country, even though the procedure is of varying difficulty. In Mexico, still influenced by its anticlerical past, divorce is granted on grounds similar to those in the United States. Among those are adultery on the part of either spouse (in theory at least), desertion for more than six months without just cause, cruelty or injury, conviction of a crime that carried two years imprisonment, character and mental disorders, and mutual consent. Mexicans took less advantage of these provisions than did itinerant North Americans (who could obtain a divorce in a border town during a twenty-four-hour visit, until 1971 when the government lengthened the waiting period). Except for Cuba, Venezuela has the highest rate of divorce for Latin America, if less than a third of the U.S. rate.[29] Grounds are comparable in the Caribbean and in Central American countries, where Catholicism is relatively weak. Even before the legal acceptance of divorce, dissolution of a marriage acquired elsewhere is recognized by the local government. The vulnerability of a marriage to dissolution is related to the type of marital ceremony or initiation. In Mexico the probability of ending a first consensual union is 46 percent, still higher for relatively late unions, but only 19 percent for marriages beginning with a civil ceremony, and 6.4 percent for those with both a civil and religious ceremony. Comparable figures are found for most Western countries.[30]

The rise of divorce since the 1980s has led to a good deal of breast beating. Various reasons are given—from the disequilibrium in gender relations to economic problems, from early marriage to the effect of the mass media, from supermaterialism to the breakdown of religious values.[31] Other factors may apply, for instance, the anticipated costs of divorce combined with the threat of AIDS may explain the increase of 10 percent in the number of married couples in Brazil between 1983 and 1985s; divorce occurs primarily in the early years of marriage.[32] Other voices stress the need for more family counseling, crisis centers, and providing family education in the church and schools.[33]

Role Behavior in Marriage

The Latin American family is patriarchal and the rights of the husband and father are clearly set in the legal system. Women's primary function is motherhood, as revealed in the literature on feminine roles. The status of women in the economic and political spheres is only with difficulty detached from maternity.[34] At the same time, the number of births are declining as revealed by the population growth (table 7.1). Still, motherhood is probably more salient in Latin America than in Europe or North America.

The activities of the Latin American wife are largely circumscribed. In many villages she is not free to wander about the streets lest this be interpreted as a lack of respect for her husband or not meeting the norm of a faithful wife. The wife could easily develop a martyr complex, but her mother has undoubtedly prepared her psychologically for the marital role. Interview data reveal that most wives regard their role as subservient. According to my research, marital frustration was present to a significantly higher degree with the lower than with the middle class in both San Salvador,

Table 7.1 Average growth rates in percentages for selected countries, 1961–99

Country	1961–70	1991–99
Argentina	1.5	1.2
Bolivia	2.4	2.1
Brazil	2.8	1.2
Chile	2.2	1.3
Colombia	3.0	1.7
Costa Rica	3.4	2.5
Ecuador	3.2	1.9
Haiti	2.1	1.5
Honduras	3.1	2.5
Mexico	3.3	1.5
Peru	2.9	1.6
Uruguay	1.0	0.6
Venezuela	3.5	1.9
Latin America	2.8	1.5

Source: Latin American Center, Statistical Abstract of Latin America (Los Angeles: University of California, 2002), p. 525.

El Salvador, and San José, Costa Rica. In countries oriented to the European lifestyle, Chile for example, the male–female axis approaches but does not reach equalitarianism—or does it anywhere? However, as noted in chapter 6, this freedom is changing because of women's movements and increasing the wife's employment, as a study of Colonia Santo Domingo in Mexico City documents. A reverberation of the 1982 economic crisis is that working-class fathers have become more involved in their children than are their cohorts in the middle class, for whom servants are affordable.[35]

The Husband–Wife Relationship

Often a central issue in the marital situation is the question of male infidelity, as described in chapter 6. Especially in mestizo cultures, the husband is free to carry on not only his extramarital ventures but to "hell-around" with male peers. This "amigo system" alienates many wives. A "good" wife should never question the hours her husband keeps, or the women he may see, but she complains if his attentions to another woman become too public. Nonetheless, this macho, patriarchal tradition has been losing force in Brazil since colonial times.[36] The 1988 Constitution gives women and men the same rights, that is, the idea of male as head of the family is gradually dying. At the same time, equalitarianism is more apparent among those under than over age thirty-five, and is more visible in São Paulo than in a village.[37] Still, machismo, of course, appears in many forms. In the Peruvian highlands, mestizos—not Indians—like to brag about the size of their progeny, seemingly as one more symbol of machismo.[38]

The articulation of expressive and instrumental roles by gender is seemingly universal, but overlapping between the sexes occurs. In the *clase popular* (working class), when at home the husband may sit alone at the table and is usually served first. He may or may not control the purse strings and in a few instances does the basic shopping for the family. The community's social climate in part determines whether or not the wife can be trusted with money. By tradition men cannot be bothered with household chores. Yet, in spite of her reduced status the wife is aware of her strength in the home as she provides nurturance to the family. Further, in the rural sector economic change modifies gender roles. Corporate agriculture or cooperative farms involve women as well as men, as a study of a central Mexican community shows. Also, the husband's migration to the city or to the United States can insure financial survival and permits a somewhat more even relationship between the spouses.[39]

Educational level determines the wife's degree of power and affects her marital role, including the decision on maternity. Also, occupational roles relate to marital roles; they overlap on the farm. In most peasant communities, women have a greater skill repertory than men. Both the *india* (indigenous) and *chola* (mestizo) in Bolivia, for instance, would rival their counterparts in any society in norms of female productivity. The economic role of women is well known, as the village marketplace constitutes a unique entrepreneurial world. In fact, women who are permitted by their mate to engage in market operations are economic assets, besides being strategically located in the village communication network. In this connection, the husband's dominant position in the conjugal relationship is shaped by economic status as well as regional, ethnic, and family tradition. Life history and individual needs also determine male dominance. In Oscar Lewis' *Five Families* (1959), two husbands did not adhere to the strong male syndrome because of impotence and latent homosexuality.[40]

To what degree does the equalitarian movement infiltrate Latin American marriage? As noted in chapter 6, evidence suggests that change is taking place. Especially among the younger upper-educated in the city, more mutuality appears in the marital relationship as well as between parent and child. Recent research literature suggests that Latin American women are questioning the traditional values and norms, including motherhood, which are being redefined as they look at their role in society in general and in the marriage in particular.[41] According to at least one Mexican social critic, the invasion of North American cultural norms encourages liberation of women and a quasi-egalitarian courtship and marriage.[42] On that point, movie *aficionados*, for instance, can hardly escape the different standards in the gender-role relationship as seen on the screen from what is traditionally prescribed. Still, Latin Americans are exposed to new gender models and freer sex expression in their films produced since the 1970s.

Marital Tensions and Violence

If one judges by the situation in Aritama, conflict marks the husband–wife relationship. Continuous anxiety exists as to whether the husband will be employed and mete out enough of his earnings to insure basic family needs.[43] Communication may not move beyond essentials. The family is often in a more tranquil state when the "husband" is absent. It is not surprising that less than half of the lower class but two-thirds of the middle-class spouses (more often women than men) in my San Salvador sample said they "would marry the same person if they had their life to live over again."

Family violence came under increasing scrutiny after the 1970s. The general climate of violence is set by societal distortions, especially poverty. The husband–father's abandonment itself is a form of violence and in the case of Peru affects nearly 30 percent of families. Beatings of wives and children are legion.[44] Ethnic relations are a factor. Indian women are especially subject to rape by mestizos, who are traditionally not held accountable. Brazilian newspapers continually report domestic violence, including murders, with the exoneration of husbands as they are acting in "legitimate defense of their honor." The outrage of feminists led to the creation of shelters for battered wives.[45]

Coital relationships themselves can be a source of tension, at least in the lower class. Many husbands believe that any arousal of sexual desire or making the sex act more enjoyable drives the wife to seek relationships outside the marriage. Caressing and kissing are often absent in coitus and clothing is not always removed. Sexual experience is hardly ideal when the entire family sleeps in the same room. Even when the couple goes outside the hut, coitus tends to be hurried, and is usually a release only to the male.

In the context of exchange theory, the marriage relationship represents an equilibrium between costs and resources. Also, rewards should be proportional to the investment.[46] The Latin American woman has to judge the benefits/costs she finds in the marriage, whether legal or consensual. Among other inputs, she has to consider the alternatives, degree of satisfaction or dissatisfaction, barriers to the dissolution of the relationship, all within her economic and cultural realities. The cost of separation—or divorce to the point that it is possible—may be too high. These decisions become all the more difficult, even terrifying, in the context of a severely stratified society wrought with unemployment, hunger, and violence.[47] Marital and family tensions are also compounded by economic demands, notably the urge for cityward migration and displacement of family members.[48] Most disturbing were the effects of the economic crisis of the 1980s, which precipitated a rise in marital breakup and divorce.[49] Most devastating to marriages and the family are the waves of political violence—dictatorships in Argentina and Chile as well as civil warfare affecting much of Central America. As one instance, in the *contra* warfare against the Sandinista regime families were torn apart with the death of many males.[50]

A Mexican study suggests that both modernization and marginalization are relevant to marital power and violence. On one side are the effects of education, communication, and formal employment in the urban setting on mutual decision making and reduction of marital violence. Even so, 19 percent of the women reported being struck violently by their mates, especially in consensual unions. Unemployment and excessive fertility are contributing factors.[51] Evidence of the changing scene in Mexico is seen in Colonia Santo Domingo. As elsewhere, marital violence is associated with parental models— the burden of children and the effect of alcohol. The number of women beaten by their mates prompted the authorities to establish in 1992 the Centro de Atención a la Violencia Intrafamiliar (CAVI) in which groups of men are to meet with a psychologist once a week for three months in order to analyze the meaning of domestic violence. According to the files, physical violence accounts for 81 percent of the reports, sexual violence for 18 percent, and psychological violence for 1 percent. In 1992 over six thousand complaints were filed, a 10 percent increase over previous years. Men were aggressors in 88 percent, women in 12. Clearly these reported cases are only a fraction of marital violence, but the monitoring is an indication of a changing scene.[52]

Various surveys indicate that between a fourth and half of Latin American women are victims of domestic violence. Even in an advanced country such as Chile over half of women who live with their partners suffer from either physical or psychological abuse and more than 10 percent are subjected to severe physical violence. In Argentina 37 percent of the women who are physically attacked by their husbands have endured this kind of abuse for more than twenty years. Domestic violence also has effects on the children—a noticeably higher dropout rate from school for those who experience or witness abuse. A study of Guatemala City lower-class *barrios* reveals the extent of abuse, mostly beating of children. Another form of violence is sexual abuse of daughters, principally stepfathers against stepdaughters. Often the daughters are afraid to tell their mothers and often the mothers refuse to believe their daughters or are too frightened of the husband to take any action.[53] Studies also stress the importance of outside employment as a preventative of violence against the wife.[54]

Single Parenthood

Throughout Latin America, as in other parts of the world, mothers—and occasionally fathers—assume the role of single head of the family. In Costa Rica the number of single mothers rose 150 percent between 1973 and 1992.[55] Similar increases appear in other countries. Accompanying this change is the rise of teenage pregnancy; in Chile births by adolescents increased 38 percent between 1990 and 1993.[56] These statistics reflect a number of social changes—secularism, mass media, peer pressure, and new sexual mores. Moreover, economic, physical, and mental pressures on a woman coping with children in both urban and rural settings are well documented. Also, as single mothers are usually condemned to poverty they are further marginalized.[57] Even in relatively equalitarian societies, depression is more common among single than married mothers.[58]

With a rising rate of liberated women and divorce, men are also inheriting single parenthood. As Gutmann notes in Santo Domingo, "*estoy de Kramer*" ("I'm another Kramer," based on the movie *Kramer versus Kramer*, a popular film on the divorced husband caring for the child) was an occasional response of males. But generally it is the wife who inherits the children after a breakup. Painful though her situation may be, she is generally more socially accepted than she was in the past, and is free to call on the help of relatives, likely choosing to live with her parents as well as her children.

Family and Ceremonial Kinship

A closer affinity is found between relatives and within the family than in the urban culture of Northwestern Europe and North America. Whether by blood or adoption, kinship is highly ritualistic. Moreover, since the supply of potential mates is limited in rural cultures, a higher amount of intermarriage occurs. An individual may be related, if distantly, to many members of the community. These ties are one more reminder of the sacredness of the family. In Brazil, as elsewhere, it is the kinship system that prepares one for the "*luta pela vida*" (struggle for life).[59]

Compadrazgo (literally, godparenthood) is institutionalized in all Latin cultures and has its roots in indigenous societies. One variation is among Brazilian Indian groups who have a relatively vast network of relationships, known as *compadrismo* or "in-lawism."[60] In addition to her or his own parents, the child has a *padrino* (godfather) and a *madrina* (godmother), who will function at baptism, confirmation, and marriage. In several societies different godparents are appointed for these specific events. Godparents are known as *compadre* and *comadre* to the parents, who may feel closer to them than to their own siblings. The *compadre* relationship is usually more binding when socioeconomic status is symmetrical between the parties.[61]

In the urban middle class, godparents are selected most frequently among relatives, in the lower class perhaps a neighbor or a close friend. However, the middle class in Popoyán, Colombia, choose *padrinos* among their circles of friends and only rarely would the candidate refuse.[62] In Guatemala, as in several Andean countries, Indian parents often select *ladinos* to be the *padrinos* of their children but the *ladinos* seldom reciprocate. The *ladino* may expect his godchild (*ahijado*) to perform various chores, and in case of the parents' death might adopt the child as a *hijo de casa* (member of the family).

In societies where indigenous folkways are blended with those from Latin cultures, *compadrazgo* assumes special meanings and rituals For instance, in a number of Colombian villages, Chibcha (a pre-Columbian culture) influence is reflected in the *compadres de suta*, or the cutting of the hair and fingernails. This multiplication of *compadres* for specific occasions also appears in several Peruvian subcultures.[63] But it is the *padrinos* of birth or baptism who really count. They are expected to confer presents on the child periodically. Most important, they are responsible for the child's spiritual and physical welfare. The *madrina de bautismo* (original godmother) has special significance for the boy in Margarita; he honors her as he does his own mother, and is obligated to work for her on various occasions without pay.[64] In remote cultures *compadrazgo* is more indispensable, if for different reasons, than it is in the upper class.

The *compadre* system may be extended to close friendships that have nothing to do with parenthood—*fictitious* as opposed to *kinship compadrazgo*. In Brazilian small towns are the *compadres de fogueria* (co-fathers of the fire). Bonfires are lit at certain festivals, especially St. John's Day. As they burn down, jumping over the fire along with other ceremonials is a means of establishing a *compadre* relationship to a close friend. While this tie is not binding, it may be useful to the individual later.[65] In other words, *compadrazgo* becomes a means of enhancing upward mobility. The relationship is more expandable than is kinship. In Guatemala the goal is to choose a *hacendado* or at least a *ladino* for a *compadre*. The latent function of the system permits an escape from kinship based on ascribed status. It is not certain as to what degree the institution is weakened or strengthened by urbanization. According to one study, economic pressures seem to destroy something of ritual kinship, for others it is an instrument of social belongingness in a growing, complex economy.[66]

The Extended Family and Beyond

Besides involvement with godparenthood, various levels of kin, grandparents, aunts and uncles, and cousins have a significance well beyond the nuclear family subculture of Anglo-America or Northwest Europe. In-laws, too, occupy a special meaning since individuals still belong to their family of orientation (family in which they grow up).

The extended family can be richly nuanced, Brazil being an interesting example (inset 7c). In that context, special relationships involving both kin and friends are reflected in the nomenclature from Margarita, Venezuela, as seen in table 7.1. The particular terms vary between communities, but protocol is of special concern. Social relationships are divided between those to whom one may use the familiar or the formal second person, "*tu*" and "*Usted*," respectively. Moreover, the boundary between nuclear and extended is not sharply defined and differs with the regional culture. In one survey, children were asked as to which family they belonged. Of the total Latin American sample 64.5 percent answered the nuclear family and 28 percent thought of the extended family (the remainder were undecided); the respective percentages for Chile were 60.5 and 19, for Guatemala 44 and 48—a reminder of the importance in the indigenous heritage of a large and often complex family grouping.[67] As another instance, in a Mexican shantytown, couples after their marriage tend to live with the husband's family rather than the mother's, but neolocalism is preferred once the financial base is assured. Whether of lower or upper class, conflict between the two extended families is evident, only the establishment of a new "grand-family" (combining both families) resolves the issue.[68]

It is, above all, the extended family to which many Brazilians belong. It has an almost mystical bond. In the classical *casa grande* subculture the names of distinguished ancestors are carried on by the newer members of the family. Also, in order to consolidate wealth, inbreeding through marriages to cousins prevailed in the nineteenth century, in addition to the sexual liaisons of brothers with half-sisters who were the offspring of the father and an Afro-Brazilian or mulatto concubine. Still, the sanction against coitus was as strong between *compadres* as between siblings.[69]

INSET 7C   THE ULTIMATE IN FAMILY BELONGINGNESS

Latin American cultural history is replete with examples of families, whether upper, middle, or lower class, who provide continuity in the struggle for survival. One case is the Gómez family in Mexico. With various social networks over successive generations, members draw on both private and public sectors in order to succeed in business operations. Those with lesser income lean on more affluent relatives. Although consanguineous counts more than fictive kinship, a flexible interlocking usually permits access to resources. Genealogical, social, and economic distances might weaken ties, but the family remains as a functioning unit.[70] Similarly, in Monterrey the Garza-Sada and the Reyes families operated as a powerful network with marriage confined to extended kinship.[71] At the same time, political feuds can occur in families. Among the Bonfims, a high-status family in Minas Velhas, a Brazilian mining community, support of opposing political allegiances reached the point that some sons no longer spoke to each other. That is, an important function of the *compadrazgo* system is the integration it gives to family and kinship, not least in the occupational sphere. Throughout Latin America relatives are brought into one's business, whether an industrial enterprise or a local *tienda* (shop). Nepotism does not raise the ethical questions it does in Europe or the United States.

Brazilians refer to their system of extended family as the *parentela*. Various interpretations can be applied to this institution. Often this relationship goes beyond first and second cousins, and the terms of uncle and aunt are liberally applied. For upper-class families the *parentela* can be an extensive affair and calls on a good deal of genealogical sophistication. The *parentela* may also involve peripheral members of the family such as relations by concubinage, including varied ethnic strains. The *compadrazgo* (*compadresco* in Portuguese) extends the *parentela* still further. As elsewhere in Latin America, the system brings status to the family. The *parentela* can include close friends, not least as a means of the person's upward mobility. Whereas it was once a large network among the upper class, it is now smaller and more loosely structured. Seemingly, urbanization is breaking down the effect of the *parentela* and in some cities it only marginally exists.[72] As a country of extreme diversity, institutions in Brazil are under constant revision, at least in the city.[73] In greater Manaus (major city on the Amazon), the *parentela* (which characterizes 32 percent of urban as compared to 24 percent of rural households) may be more of a survivability strategy or a "temporary artifact" of the family life cycle; a quasi-*parentela* appears when the son or daughter marries or a relative migrates.[74]

Historically, family sociologists point to the trend toward nuclearization of the family as an adjustment to modernization. The extended family traditionally is itself a production unit and therefore is not truly functional in an industrial economy. According to a study of six Latin American nations (Colombia, Costa Rica, Dominican Republic, Mexico, Panama, and Peru), migration to the city does not necessarily dissolve extended ties.[75] Rural migrants to the city depend on kin, but it is equally predictable that over time they partially break these ties. Still, it is improbable that kinship, whether blood or fictive, continues in its traditional form, notably in the city. The lower class cannot afford the expense of maintaining kin relationships and the middle class is too preoccupied with problems such as making a living, coping with inflation, securing education for their children, and other aspects of upward mobility.

The Life Cycle

As in all societies, stages of life are marked by certain role expectancies. As the level of aspiration is lower, expected roles proceed in Latin America with relative smoothness from one phase to the next; in other ways the transition may be more abrupt. One adjusts to the demands of the aging process as well as to societal change, but economic crises can be devastating. For instance, fluctuations in the international coffee market in Brazil and Central America can uproot farm families, who are forced to react to low prices by resorting to wage labor, migrating to the city, finding new economic roles, among other adjustments. More roles change in response to individual and family needs, parenthood for one.

Birth and Infancy

Pregnancy and birth are surrounded by varied folkways and magical practices. Attempts at abortion are made with laxatives, local herbs, and special medicines. Pregnancy is attended by certain taboos in rural areas, all kinds of omens favorable and unfavorable are present. When labor approaches the woman goes into seclusion. In Puerto Rico the expectant mother is given a forty-day period (*la cuarentena*) in which she is relieved of household duties. The midwife is skilled in delivery practices, if not always in antiseptic measures. Today, urban women, including the lower class, are going to a public or private clinic for childbirth.

After birth, taboos and ritualistic practices continue, notably in indigenous areas. For one, the umbilical cord must be ceremoniously disposed of, either to rest in the family album or planted in the family garden. In Minas Velhas the child must be kept in a darkened room for a week to avoid the "seven-day sickness"—a kind of tetanus infection. Breast-feeding is the norm. In Aritama, feeding is mostly on demand, perhaps nine–twelve times a day, and weaning occurs when the mother's work or health make it mandatory. Patterns of discipline and toilet training vary with the locality. As in most cultures, affection and rejection alternate. Sooner or later, care of the child falls to an older daughter or another relative as the next child arrives.

Childhood and Adolescence

Two basic principles underline socialization of the young. As explained in chapter 6, most important is gender differentiation. Dress varies for the sexes; the girl must be more fully clothed. The kind of games children play are controlled. This difference is accentuated as adolescence approaches. The other major object of socialization is the idea of *respeto* for the father. Whereas the mother usually wavers between indulgence and discipline, the father more likely imposes strict controls to the degree he is in the home. By the late teens the son may enjoy sufficient status to expect family members to submit to his authority. It is anticipated he will assume the father role. In a semantic differential scale I discovered Chilean adolescents to have a greater idealization of both parents than was found among United States and German cohorts.[76] In a study of Venezuelan elites, mothers were idealized and fathers respected.[77] Mothers try to inculcate a loyalty in the children as they look forward to their material and emotional support in the upper years. For one Mexican community, extent of women's political action depends on the possible effects for their children, whereas men engage in similar activity with little concern for the effects on the children or wives.[78]

As with adults, children are affected by stimuli beyond the home and may experience conflict in bridging the gap between these worlds, particularly as economic injustices sharply curtail the potential for infants and children. Consequently, social reformers place a high priority on aid programs for the single mother and her children.[79] A serious issue is the ability of municipal authorities and international programs to meet the standards set by the United Nations Convention on the Rights of the Child. One study focused on the plight of alienated children and adolescents in the poorer *barrios* of Managua, Nicaragua. The Sandinistas were sensitive to this issue, but in the wake of the 1990 election NGOs became increasingly involved in trying to resolve the problem.[80]

Again, social change is evident in the parent–child relationship, particularly among the rising middle class. Parents from the Rio Grande to Patagonia are rethinking the folklore of yesteryear on socialization practices.[81] Another ubiquitous factor shaping imagery and values is television, which disproportionately exhibits North American programs. However, since the 1990s in larger countries such as Brazil and Mexico over half the programs are domestic. Moreover, generational differences hold, claims a report from Venezuela, where 90 percent of the homes have at least one set, and children no less than adults are transfixed to the screen. In the working class, viewing is usually with the whole family in the one all-purpose room; whereas in the more private world of the middle and upper classes the servant may remind children when their program is on.[82]

Not only urban areas but also rural communities are affected by changes in education, the mass media, migration, and the national consciousness. For example, in Pocoata, a rural community in Bolivia, adolescent girls find a compromise in negotiating their identity between a Quechua past and a national present in their choice of clothing, language, and social interaction. This transition is a means of preparation for adulthood.[83]

Old Age

Aging was not an acute problem in Latin America simply because until the late twentieth century, relatively few reached the upper years. In view of new sanitation and health standards in the last three decades of the twentieth century longevity has risen remarkably, as shown in table 7.2. A struggle continues to deliver sustenance to the aged. Fiscal problems inhibit the difficulty of reaching the goal of social security for retirees.

Table 7.2 Life expectancy for given periods and countries

Country	1950–55	1995–2000
Argentina	62.7	73.1
Bolivia	40.4	61.4
Brazil	51.4	67.9
Chile	54.8	75.2
Colombia	50.6	70.2
Cuba	59.5	76.0
Haiti	37.6	57.2
Mexico	50.7	72.4
Peru	43.9	68.3
Uruguay	66.3	69.7
Latin America	51.8	70.0

Source: Adapted from *Statistical Abstract of Latin America* (Los Angeles: University of California, 2002), table 700.

As a result of the status accorded to age, some persons in Tobatí, a Paraguayan village, exaggerated their age in the hope of gaining attention. The degree of support children give their aged parents varies individually and subculturally. Still, traditional adherence to the large family is based on the probability of assuring help in later years. In Cuernavaca the "successful elderly" could live off their income, but the poorer ones felt they had to make themselves useful to their adult children in order to feel secure about receiving help.[84] Urbanization and decline of rural areas aggravate the problem of the aged. Psychologists and other professionals advocate a higher governmental priority for the elderly.[85]

Death of the aged is not accompanied by the same grief as it is for the young, but custom demands the usual *velorio* or wake, an all-night affair consisting of a mixture of sorrow and festivity. Traditionally, the *novena* follows, an open house following each rosary for the nine evenings. The *velorio* and the *novena* are institutionalized, usually with a background of white draperies, pictures and figurines of saints, and lighted candles. Ideally but far from universally, the ceremony is repeated a month later and on the first anniversary of the death and annually for a given number of years. Generally, rituals exist to perpetuate and yet neutralize death.[86]

Family and Society

In summary, the family can be examined for the structure and function it confers to the society. Three dimensions are salient. For one, the nuclear and the extended family: throughout Latin America the nuclear is the preferred orientation. However, in view of the importance of consanguineous and fictive kinship, the extended family is more viable than in most Western societies. As in most cultures, the extended family is more meaningful in the middle and upper classes than in the lower class. In regions such as rural Cunha in the state of São Paulo, kinship is highly valued; a given canyon or other locality may be occupied by one extended family, but residence is nuclear.[87] The extended family functions as an auxiliary system when the nuclear family is not operative, that is, for Cunha migrants who trek to the city.

Two, an instrumental–expressive axis operates to varying degrees in all families. On the surface, these functions are clearly differentiated between the two sexes. The husband and father enacts the instrumental role of economic and power relationships, the mother fulfills the emotional and socializing tasks. For over a fourth of Latin American families, the mother must serve both roles.

Three, decision making is divided between patriarchal and matriarchal patterns. The father's rule is final in the family, although it is the wife or mother who makes decisions on a day-to-day basis, especially in a consensual union. As another variant, equalitarianism is the norm in the new professional middle class. It is noteworthy that the individual in Iberian societies carries the family name of both the father and the mother; however, the father's name has priority as in the name of José García (father) Gómez (mother). Idealization of the mother and respect for the father underscore this importance of matriarchal–expressive and patriarchal–instrumental functions.

As in other societal phenomena we see the complexity of many variables. The functionalist approach reveals the intermeshing of the family with other institutions—economic, political, religious, health, and recreational. Families vary from being functional to dysfunctional; many exhibit a capacity to adjust their behavior to ongoing crises. Symbolic interaction is relevant as individuals implicitly define their situation, in which traditions, especially an entrenched patriarchalism, limit the options to participants, both female and male. When one considers the economic and social deprivation in which most families are forced to compete, conflict theory is most relevant.

Because of the intricacy of family functioning, the search for a definitive theory seems premature. For one problem, not all actors in the family system can verbalize their situation to a sufficient degree in order to define specifically their roles and interactions, even less to detach themselves from their limited reality. Likely, the theories mentioned earlier combined with a systems approach offer a means of analyzing the changing Latin American family. From the family it is appropriate to turn to religion and education as other socializing agencies in the next two chapters.

Chapter Eight

Religion at the Crossroads

For the mass of Latin Americans, two social institutions are central in emotional commitment: the family and the church. For 85 percent of Latin Americans religion means Roman Catholicism, although the significance of this religion for both individual and society is sometimes baffling. Several questions are relevant: How do historic and regional differences account for varying religious developments among Latin American nations? Has or will religion become a regressive—or progressive—influence? How does religion meet the challenge of other ideologies, such as nationalism, modernization, capitalism, and socialism? What is the effect of the growing Protestant community? To what degree can the church bring about a revitalized social ethic?

Regarding the church's need to embody a humanitarian philosophy and the willingness to implement this determination, the historical evidence was until the 1960s largely negative. On the other hand, what social consciousness appeared intermittently during the first centuries of Latin America came predominantly from Roman Catholicism. As the region accounts for nearly half of the Roman Catholics in the world, the status of Catholicism as a progressive force can rise or fall according to what happens here. The definition of what constitutes the mission of Catholicism is subject to an ever-sharpening debate between what is the "New Church" and the "Old Church."

Functionality or Conflict?

Religion can be defined as a system of beliefs and symbols that give meaning and purpose to life. In other words, religion is directed to the unknown, relieves frustration, and provides identity and continuity. That is, religion has a "reality-maintaining" task; it legitimates our world.[1] As Emile Durkheim, the first sociologist to approach the subject scientifically, points out, religion focuses on the sacred as opposed to the profane, and through ritual relates the individual to the sacred. According to Durkheim, society constructs a religious institution in order to secure its legitimacy and unite its members. From the sixteenth century to the present, the Latin American church has had an uneven profile in fulfilling this function. Even social protest reflects the individual's need for ritual. For instance, the fiesta celebrating the Virgen del Carmen in a Lima *barrio* lends identity to the migrant *cholos* (recent Indian migrants to the city) as the church incorporates indigenous, Moorish, and Christian images.[2] Nor could the Mexican Revolution deter the ritual of pilgrimages to the shrine of Guadalupe.

Sociologist Max Weber approached religion from a functionalist viewpoint as a close relationship among ideas, social structures, and organizational development, as shown in his "ideal type" of the Protestant ethic and its positive effect on the rise of capitalism in Western Europe. In many societies, including Latin America, the church functions as a kind of monopolistic, bureaucratic, hierarchical structure, interacting with the state and other institutions, and committed to enlarging its power by increasing its membership. Not only does religion rationalize the given society, it also provides values and norms as well as a sense of immortality and salvation to its adherents. In addition to its manifest functions the church may also exhibit some latent functions. For instance, Protestants in Latin America speak of the consciousness of self-reliance, thrift, and hope for upper mobility they find in their new religion—giving limited credibility to Weber's thesis.

Few observers would question the functionality that religion gives the social order and the individual, but it is interlaced with contradictions and conflict. A cursory glance at the daily press reveals religious loyalties as the basis of disorders in places as diverse as Northern Ireland, the Balkans, the Middle East, and India, not to mention the historic conflicts of Christians, Jews, and Moslems. Various beliefs, whether Christianity, Marxism, or other symbolic systems, compete against each other for scarce resources—allegiance and power. The history of Catholicism in Latin America can be portrayed as the unfolding of conflict interspersed with episodes of integration, as witness the inquisition, state–church friction, alienation of marginal populations, at least until the 1960s, when under the inspiration of Pope John XXIII (1958–63) the church took on a deeper social conscience.

Socialization both in religious and secular life can be considered as a variety of coercion—still another aspect of conflict. That is, religion is a means of establishing order within a society by determining goals and enforcing means with a system of rewards and punishments. In this context, exchange theory is as relevant to conflict as to functionalism. Human beings are motivated by a system of rewards as balanced with costs. Religion functions in many cultures by offering rewards, among others, immortality. Since these rewards are not readily obtainable, at least in the here-and-now, the individual accepts a kind of promissory note of future benefit. Perhaps because of the ritual surrounding death in preliterate civilizations, the mystique of death in Catholicism has special significance to Latin Americans.

Conflict assumes a number of social and psychological nuances. Marxism itself may represent a quasi-Christian rejection of the quest for economic self-interest. Conflict is also a source of change; tensions within the church and between the church and civil society in Latin America can promote significant change, as we will find in "Liberation theology" later in the chapter. Additional

Table 8.1 Indices of Christianity in Latin America*

Country	Practicing Christians (%) in 1900	Nonpracticing Christians (%)		Priests per 1,000 inhabitants 1990
		1900	1995	
Argentina	90	10	55	5.1
Brazil	80	20	57	9.2
Chile	80	20	56	4.9
Mexico	80	20	33	7.1
Paraguay	70	30	41	6.7
Peru	95	54	20	8.3
Uruguay	80	20	34	4.4

* As measured by survey data.
Source: Adapted from Latin American Center, *Statistical Abstract of Latin America* (Los Angeles: University of California, 2002), p. 381.

conflict arose as Roman Catholicism returned to a new conservatism during the papacy of John Paul II (1978–2005), who held on to a restricted view of sociopolitical action by the church.

Another dimension of conflict is the question of morality. A discrepancy is almost universally found between the norms prescribed by religious authority, whether Christianity, Judaism, or Islam, and various expressions of these norms in day-to-day behavior. Secularization in most societies seldom reduces conflict between ideal and real norms. Possibly more than Protestantism, Roman Catholicism has difficulty determining the boundary between church ritual and everyday morality. A moral secularism with Marxist overtones pervades Cuba and (and to a lesser extent Mexico and Brazil, Chile, and Venezuela) in an anticlerical heritage.

Once again we see a social institution representing a mix of functionalism and conflict. Through most of its history the church administered to economic elites. Debate continues on the accomplishments and validity of the "popular theology" of recent decades. At the same time, throughout the West, religious authority is on the decline, even though this drift toward secularization varies in time and place. Apparently, upper educated people are moving away from traditional religious values, whereas the less educated primarily seek emotional and cognitive closure in their religion, but this distinction is subject to investigation. As noted in table 8.1, allegiance to religion, specifically Christianity, has been losing in recent decades.

The Historical Fabric

Closely on the heels of the conquistadors came the clergy. Catholicism enjoyed great power and prestige at home. By the sixteenth century the church had driven Moslems and Jews from Spain, and Charles V and Philip II waged holy war against the enemies of Catholicism from as far away as England and Turkey. The clergy was filled with a zeal for conquest and conversion. Moreover, the Vatican and the monarchy lost little time in conferring a number of privileges on churchmen who went to the New World. A papal bull in 1494 divided the Americas between Spain and Portugal with rights to conversion of all the natives of the respective areas as well as setting a line of demarcation between Luso and Hispanic America.

Concern with purity of faith also preoccupied the colonial church. Threat of the inquisition kept the clergy and general population under strict control and acted as a kind of "police court for tracking down bigamists, robbers, seducers of youth, and other undesirable people."[3] Censorship of the press, theater, and what passed for education was assured. The inquisition in Iberia and its colonies was little worse than the religious intolerance in most of European civilization, including the Salem witch burnings in the late seventeenth century.

The profile of the church was by no means all negative. The church had the difficulty of Christianizing a vast continent and a half, but first it had to establish a coherent structure by meeting a number of challenges—finding income, developing agriculture, fighting disease, and organizing clergy and parishes, as the history of Mexico in the sixteenth century shows.[4] The zeal for conversion brought Franciscans, Dominicans, Augustinians, and Jesuits (see inset 8a) to the New World. Although motivation of these missionaries may be questioned, many were devoted to the needs of the Indians. For instance, in Hispaniola as early at 1515 the church established missions in order to protect the rapidly vanishing natives. Vasco de Quiroga founded the "community of innocents" in Michoacán, Mexico. But most dedicated to the indigenous was Bartolomé de las Casas, who endlessly pleaded for better treatment.

INSET 8A   A JESUIT COMMUNE IN THE WILDERNESS

Possibly the most extraordinary venture into a socialized community was the Jesuit organization of twenty *reducciones* (colonies or compounds) encompassing perhaps a hundred thousand Guaranís and other indigenous groups in Paraguay, Uruguay, and northern Argentina (1637–1767). Jesuits first came to Latin America in 1573 and by the time of their expulsion two centuries later two thousand had been dispatched to various corners of the hemisphere.

By the mid-sixteenth century, stirrings for better treatment of the indigenous became a priority of the church as symbolized by Pope Paul II's edicts in 1537. Particularly vulnerable was the brutal *encomienda* system of forced labor.

For over a century, highly committed Jesuits brought education, crafts, and agriculture to this relatively forbidding terrain. Moreover, the Jesuit missionaries insisted on a European moral code, attacking, among other customs, the practice of polygyny. A simple lifestyle (including dress and habitat), a strict code of ethics, and religious dedication were the themes. The missions displayed a mixture of benevolence, productivity, and despotism. The Jesuits demanded as much of the Indians as they did of themselves.

Unfortunately the colonies were within reach of the *mamelucos* (gangs of mixed blood Dutch and Portuguese) who were engaged in enslaving the indigenous and selling them to São Paulo planters. Raids in the late seventeenth century decimated several *reducciones*, but as the Jesuits became more militant, arming Guaranís as well as themselves, order was restored.

By 1700 no less than thirty colonies were enjoying a secure existence. In addition to food crops, they were exporting a variety of products—cotton and linen fabrics, tobacco, and *yerba mate*, a tea-like beverage popular today in the Plata region. Also, the Guaranís were useful to the Spanish when enlisted into the military for combating foreign (English, French, Portuguese) invasions.

Unfortunately, during the eighteenth century, settlers and traders, jealous of the *reducciones'* success and wealth, increasingly encroached on Jesuit territory. Responding to these pressures the Crown became disenchanted with the ever-growing strength of the Jesuits and Carlos III expelled the Jesuits in 1767. Sadly the colonies gradually fell into decay, and little remained of the Jesuit heritage. Notwithstanding this dominance and the near obliteration of an indigenous culture, one might argue that these some half-million indigenous experienced a higher standard of living, including a primitive form of hospital service, than they were to know for generations. Also, the contrast between quasi-utopian societies stands in contrast to the exploitation of the planters who surrounded them.[5]

Although aboriginal cults were woven into the façade of Catholic rituals, Africans as well as Indians were eventually "Christianized"—and exploited. No one did more for the rehabilitation of the Afro-American than Father Pedro Claver of seventeenth-century Cartagena, Colombia. Franciscans as well as Jesuits became concerned about both the indigenous and Afro-Americans, but were only partially successful in finding a meaningful mode of life for these people in the harsh world of European settlers. Moreover, the flight of many indigenous from the missions underscored the difficulty of incorporating natives into this semblance of a European culture.

The Church in the Independence Movement: Aspirations and Reality

Through most of the colonial period the church rivaled the secular government in power. In conducting schools and hospitals, the church was the refuge of the aged, sick, and abandoned. By the time of independence the church had become an enormous superstructure as it owned nearly half the wealth in the colonies, with inequalities of a feudal system; an archbishop would have an income a thousand times larger than would a parish priest. Also, the church was almost entirely a Spanish operation; only toward the end of colonial times were creoles accepted as clerics. To most Latin Americans this preference for *peninsulares* (Spanish born) could only be a direct affront.

Church officialdom sharply opposed independence. Still, lower rungs of the clergy participated in the movement. Among others, two mestizo priests,

Miguel Hidalgo and José María Morelos in Mexico, played key roles in this struggle and paid for it with their lives. Morelos was far more than a revolutionary warrior. In his writings he opposed compulsory collection of tithes and pleaded for a redistribution of large haciendas to the Mexican people. Unable to convince his contemporaries of the value of his social philosophy he was considered a dangerous radical for over a generation after his death. What he and other Mexican revolutionaries stood for was lost when General Iturbide, through some devious maneuvering, arranged independence on his own terms. In Venezuela, Argentina, and Chile the clergy disobeyed their bishops and worked for independence. Uruguayan priests followed José Gervasio Artigas into battle against the Spanish. Priests of mestizo background were well aware of the inequities in the Spanish regime.

After independence the church languished as the hierarchy returned to Spain or was alienated from the new power structure in the former colonies. The church could no longer continue to be the central arbiter in the secular world. Since the Vatican refused to recognize the new governments, continuity of clerical offices became a serious problem.

The Struggle over Anticlericalism

The wars of independence left the new republics in no mood to return to the old order of religious domination over civil life. On almost every issue the church continued to fight against liberalization of land policy, raising the status of mestizos and nonwhites, and extension of civil rights. Also, the clergy had a reputation for something less than savory conduct. Even though the Vatican insisted on clerical chastity, concubinage among the Brazilian clergy in the mid-nineteenth century reached the point that the government veered toward the idea of establishing a Brazilian Catholic Church. In addition, linkage of many clergymen with the Masonic orders weakened ties with Rome.

In almost every country, civil authorities removed public education from the church and demanded civil registration of births, marriages, and deaths. Most critical, the state attempted to appropriate the church's vast wealth, which was largely the result of inheritances contributed by wealthy parishioners. Under the principle of mortmain, property once bequeathed to the church could never be expropriated. With the demise of these practices toward the end of the eighteenth century the church had little of the power it once held. The struggle was to become even more perilous in the nineteenth century, as noted in inset 8b.

INSET 8B   MEXICO AND THE CHURCH

The battle between church and state especially dominated the Mexican scene. In the 1830s Liberals and Conservatives were pitted against each other in, respectively, restricting or supporting the church. As early as 1833, after the Iturbide reign, the new regime decreed transfer of patronage from church to state, partial expropriation of church property, reform of tithe collection, and secularization of schools—albeit without providing adequate substitutes for lay education. Also, use of the pulpit by the priesthood for discussion of political issues was outlawed. With the return of a conservative regime under Santa Anna, religious authorities paid little attention to the decrees of the preceding government.

By the 1850s intellectuals were still complaining of malpractices, which the prelates continued to defend. This *reforma* movement was climaxed by the presidency of Benito Juárez, adoption of the 1857 constitution and the 1859 reform laws. Since the church owned a third of the land in Mexico, redistribution or disamortization of wealth again became a primary target. Schools returned once more to lay hands, and a civil registry was adopted. The clergy was required to perform the sacraments even when the citizen could pay no fee, religious orders were restricted, and cemeteries removed from religious supervision.

With the 1863 French intervention and the brief reign of the Austrian prince Maximilian as puppet emperor, the reform measures were ignored. In the end, Juárez pushed through his secularization program. During the lengthy and corrupt Díaz period the state gradually tolerated violation of the secularization laws. Monasteries and convents survived, parochial schools flourished, morality among the clergy sagged, concubinage continued, and exorbitant fees were charged for the sacraments of baptism, matrimony, and extreme unction. It is little wonder that the Mexican Revolution was an attack against not only *porfirismo* but also the clergy. Understandably, the 1917 constitution reaffirmed and defined more specifically the Reforma.

By the 1920s Catholic Social Action supported reforms, for instance, rural cooperatives, adequate farm wages, and unionization of labor, but the decision to effect these reforms was left to the hacienda owners and other employers. More serious was the "Cristero Rebellion," which began in 1923, when the Vatican sent a representative to Quanajuato to dedicate a monument to "Christ as King of Mexico" before a crowd of forty thousand.[6] Consequently, the Calles administration beginning in 1925 enforced the 1917 measures against the church, leading to a bitter feud for nearly a decade. The *cristeros*, who were mostly bands of peasants, fought government troops, many clergy went into hiding, and the archbishop of Guadalajara fled to San Antonio. The episode was the last serious religious challenge to the government.[7] Not until 1940 did the government and religious authorities accept a kind of concordat. In view of the bitterness existing on the part of the Vatican, Mexico did not have a cardinal until 1952 and then it was the bishop of Guadalajara, not Mexico City. Still, from Miguel Alemán (1946–52) onward a presidential candidate could publicly acknowledge his Catholicism. Conversely, church leaders are free to urge social reforms, but play no political role. Surveys indicate that the mass of Mexicans would prefer the church to take a stronger social role but not enter directly into politics.[8] Vicente Fox turned in 2001 to a new approach by paying a visit to the Shrine of Guadalupe on inauguration day. His PAN party had begun expressing Catholic as well as other conservative values in the 1940s.

Division of State and Church

In most of Latin America the basic conflict between church and state is political. Liberals, as the dissidents called themselves, were particularly distressed by the special privileges (*fueros*), such as the clergy's right to be tried in ecclesiastical rather than secular courts. Also, liberals strove to expropriate, tax, or buy church properties and establish governmental schools or have parallel secular and religious schools. More and more the church had to revise its spiritual, evangelical, and educational purposes as anticlerical laws came into force by the middle of the nineteenth century in Guatemala, Colombia, Chile, Ecuador, Venezuela, as well as Mexico. Toward the end of the century the Radical Party emerged among the largely secular middle class in Argentina, Chile, and Venezuela.

The first republic to pass a law (later rescinded) separating church and state was Colombia in 1853 (the first to proclaim separatism on a permanent basis was Mexico in 1859). Disestablishment of the church was accomplished without toppling existing constitutional regimes in Brazil, Chile, and Uruguay. By the 1900s an ambivalent situation existed between the church and the state in Central America, Ecuador, and Venezuela. These countries regard the church as an autonomous institution in a free society.

Historical factors, social structure, and vagaries of different governments shape the church–state relationship. With the possible exception of Mexico, Cuba was, long before Castro, the most anticlerical country in Latin America. Several developments are relevant: (1) as Cuba had few riches, Spain gave little attention to the island other than the port of Havana. Consequently, the church had little interest in the island, particularly its rural population, with the exception of the "sugaracracy," which emerged in the late nineteenth century. (2) The church hierarchy was perceived as opposed to independence, all the more as the struggle between traditionalists and liberals grew deeper. (3) Even before the 1898 occupation by the United States, a sizeable Protestant community was at hand. Among the consequences of this secularism is a flexible divorce law—mutual consent was added to the grounds in 1939, a half-century before most of Latin America or even before much of the Western world.[9] Today the church has less freedom than in Mexico, even though the Pope's 1998 visit to Cuba slightly reduced the tension.

Crossly-Cultural Patterns

The present status of Catholicism varies greatly from country to country, from subculture to subculture. We may examine the religious situation in three representative nations.

Brazil

Catholicism was estimated to be the religion of nearly 80 percent of Brazilians in 2000. However, the term "Catholicism" covers a wide gamut in beliefs, including African cults in the northeast. Still, it is the largest

Roman Catholic nation in the world. On the whole, Brazil was spared the worst of the inquisition and the church was not as large a landholder as in Hispanic America. Yet, because of its intimate relationship with the government, the church was never permitted to develop its own autonomy. This cozy involvement was hardly conducive to a sense of commitment either to or from the people.[10] The presence of Africans, the commercial or profiteering nature of Brazilian life, *casa grande* (large estate) and *engenho* (sugar plantation) culture, easygoing ways of the people in a tropical setting are not fertile ground for the religious fanatic. The remark "even the Christ seems to hang easily from the cross" aptly describes Brazilian religious art.

After Italy and the United States, Brazil has the largest clerical hierarchy (6 cardinals, 51 archbishops, and 368 bishops) in the world; yet the scarcity of priests is a serious problem, only 15,568 priests in 1996, or 1 for roughly 10,000 inhabitants. Although several other countries have an equally unfavorable ratio of priests (Cuba and Honduras have an even lower representation), these figures mean that many Brazilians lack clerical services, and Catholicism often becomes an almost empty ritual. The individual and the community are left to their own resources.

Separation of church and state occurred peacefully in the establishment of the republic, and the Federal Republican Constitution of 1891 guaranteed religious freedom. Under this separatism the church plays no *official* role in public education, and no government jurisdiction can levy taxes on church property. The church has occasionally intervened in politics but with uneven success. During Vargas's Estado Nôvo (1937–45) Catholic leaders hoped to arouse public opinion against the 1891 constitution as "atheistic." In return, Vargas, despite his avowal of Positivism, catered to the paternalist outlook of the church as a means of bolstering his own authority and welcomed the expansion of church schools in order to lessen the hard-pressed public budget.[11] When the church opposed the election of liberal João Cafe Filho as vice president in 1950 and of Kubitschek as president in 1955, the public did not follow. The church remained adamant against communism. After the 1964 coup, opposition to totalitarianism and the desire for social action grew in segments of the clergy. The Brazilian church exhibits, with the possible exception of Chile, the strongest opposition to authoritarianism, even though the record is an uneven one.

Religious minorities have little problem in maintaining their integrity in Brazil. Positivism as based on Auguste Comte's philosophy—its influence is seen in the motto "Order and Progress" on the Brazilian flag—became an established religion before the end of the nineteenth century. Remnants still exist of the Positivists today. Moreover, Jews, numbering well over two hundred thousand, are found in the larger cities. Buddhists and Shintoists are a conspicuous minority in the state of São Paulo (the city São Paulo itself has the largest Japanese population in any urban setting outside Japan), even though many Japanese have become Christians. In addition are the African cults (Candomblé in Bahia, Xango in Recife, Macumba in Rio, Batuque in Belem, Batunque in Pará) derived from Yoruba and Dohamey beliefs.

The purest Candomblé is found in Bahia, where the chants are in the Yoruban language. Part of the origin of Candomblé was the reluctance of the Portuguese to accept the African slaves into European culture.[12] However, these religious expressions are blended in various degrees with Catholicism, becoming a means of expressing discontent toward socioeconomic distress.[13]

A widespread religious group is the Umbanda with its emphasis on healing. As its teachings are integrated with Catholicism the Umbanda cult counts at least five million followers, representing nearly the entire class spectrum of Brazil. Curiously, it first appealed to the more urbane population and spread downward in the class structure.[14] In part, Umbanda draws from a tradition of *pretos velhos* (old-slave spirits) but also has implications for self- and national-identity.[15] According to Umbandistas, if Catholics can pray to a man who rose from his grave three days after his death, they, too, can partake of a spirit world and like Christians adhere to a philosophy of social change.[16] Umbanda has a following of a million or more, and represents a mélange of seances, healing, and ingredients from African and indigenous sources. It also offers the convenience of calling on a variety of deities to solve life's problems from love relationships to finding a job or winning the lottery. Persons of various social levels accept those parts of Umbanda coherent to their own viewpoint of Catholicism, but it has made inroads disproportionately among elites. Similarly, Spiritism is in some respects a variant of Umbanda—the relationship is not too clear.[17]

For Afro-Brazilians, especially in the northeast, Candomblé draws on their ancestral rites as a means of establishing personal meaning. Although it has embodied a degree of Catholicism, Candomblé has retained its integrity since colonial times—partly a result of the ability of the original Yoruba slaves to attain their freedom and enter into marketing and a degree of social power.[18] The faith and rituals were critical in the physical and psychological survival during slavery as it gave an "alternate space" of personal identity.[19]

Probably more than in other Latin American countries, folk religions continue to have a strong following. Rural inhabitants are especially inclined to believe in magic and witchcraft, including "invocations and incarnations" among other spiritualist manifestations.[20] The news of a reputed miracle in the northeast in 1889 led to a vibrant religious movement under the leadership of a charismatic priest.[21] It is not surprising that Brazil with its cultural diversity is a favorable milieu for highly expressive religious beliefs and practices. In this context, Brazil should not be considered as unique; Caribbean areas are also inclined toward folk religions, deriving from Africa. In Cuba, where slavery began in the mid-sixteenth century, Yoruban beliefs and practices took root. Although many slaves were from the Congo, the direction of the religion is Yoruban. Still, much of the music and dance stems from the Congo. A complex system of deities—more intricate than that of the ancient Greeks—is based on *orishas* (ancestor apirits) as the foundation of the cosmology. Syncretic themes appear in the adoption of the pantheon of saints, with their African names and traditions, blending with Roman Catholicism.[22]

Most important of the minority religions in Brazil is Protestantism, the largest membership in Latin America. It has a long history beginning with the establishment of an Anglican church in Rio in 1819, followed by Methodists, Lutherans, Congregationalists, Presbyterians, and Baptists before the end of the nineteenth century. A major contribution of Protestantism is education, from primary to advanced. With the nationalist spirit of the Vargas administration, these institutions passed to Brazilian control in the 1930s.

As almost everywhere in Latin America, the appeal of Protestantism, or more precisely Pentecostalism, is to those who find no home in Catholicism. Its growth—in Brazil over ten million adherents (some estimates are much higher)—is identified mainly with the lower and lower–middle classes, who look to upward mobility through habits of sobriety and thrift. Contrary to Chile and Mexico, where *evangélicos* are also well represented, Protestantism makes inroads even among the Brazilian elites. Also, the literacy rate is higher for Protestants than for the remainder of the population. In Brazil the styles of Protestant expression range widely according to regional and class subcultures. Pentecostals, the single largest grouping of sects, is more in the U.S. tradition, with emphasis on personal evangelism and a direct, ecstatic communion with the deity, including glossolalia ("speaking with tongues").

Argentina

Argentina belongs to those countries having a Western European or what one might describe as a permissive expression of Catholicism. Still, the religion has official status—both the president and vice president must be Catholic. As the inquisition barely touched the Plata region, the history of the church is generally one of non-excess in ideology. The area became a refuge for heretics who engaged in smuggling, a principal economic endeavor of the colonial period. As leaders in the independence movement were not particularly devout Catholics, anticlerical feelings arose in the national period. Religion was not a burning issue as other tensions such as the struggle between federalists and *porteños* (residents of Buenos Aires) occupied the country.[23] After the drift to conservatism under Governor Rosas of Buenos Aires new leaders were firmly on the side of secularization, although the 1853 constitution made Catholicism a state religion. The struggle was especially acute during the 1880s as both sides reacted sharply to questions of religious education. The laws of 1884 banned religious instruction from the schools, secularized the registration of births, marriages, and deaths, and provided for public cemeteries. Whatever the drift of relationships with the state, the church functioned as a major center of ceremonial life, especially in upper-class urban circles.

As explained in chapter 4, the state–church conflict became an issue again during the Perón era, when he legalized divorce and prostitution in 1954 and suppressed Catholic education in the public schools in 1955. The end was swift, and excommunication of Perón was a factor in his fall. Although Catholicism redeemed itself by its final stand, its past surrender to the appeals of a dictatorship revived the issue of anticlericalism.

Most of the clergy remained quiet during the "dirty war" of 1976–83. Indeed, the generals who oversaw torture and executions might a few hours later attend mass and confession! As in other countries, the younger clergy urged sweeping social reforms. The country's affiliation with Fascism and Nazism in the 1930s was supported by a number of bishops who saw rightist totalitarianism as preferable to secularism and communism. This flirtation with European authoritarianism becomes even more ironic in view of the relative conspicuous Jewish community in Argentina (inset 8c).

INSET 8C   JEWS IN AN ALIEN WORLD

Jews are found in nearly every Latin American country. However, the largest community is traditionally Argentina with approximately 240,000 (down from 450,000 in 1965). Currently Brazil has 250,000 Jews (up from 139,999 in 1965, with the heaviest concentration in São Paulo). Political and economic anxieties account for the declining number of Jews in Latin America, from 804,000 in 1965 to 547,000 in 1999.

Historically, migration from Europe began in colonial times with the largest movement in the late nineteenth century, predominantly to Argentina. The two most conspicuous migrations were, first, the wave beginning in the 1880s continuing until the outbreak of World War I and renewing after 1924 when the United States restricted immigration from eastern and southern Europe. The second wave was the diaspora from Nazi Germany after 1933, especially the desperate period, 1938–39, when Jewish immigration was severely restricted in most Western countries. Tragically, except for Shanghai and a few other enclaves, only several countries welcomed Jews, most conspicuously Bolivia, Dominican Republic, and Ecuador.

In examining the nearly five centuries in which Jews found an abode in the New World a comparison is inevitable between Anglo-America and its southern neighbors. Notwithstanding the antisemitism in the United States, it has been more consistently a democracy, in which religious diversity is supported, at least officially. On the other hand, Latin America emerged out of a medieval feudalism with minimal effect of the Enlightenment and the industrial revolution. Further, Roman Catholicism was hostile to Judaism. Consequently, survival—not to mention advance—for Jews was always a struggle. In addition were internal tensions. Sephardics from Spain scarcely welcomed Ashkenazis from Central Europe, or the reverse. Fortunately, aid societies came to the rescue of new immigrants.

Although Marxism and other ideologies were current, most Jews remained orthodox, their faith being a strong support to their identity and survival. A distinct difference from the U.S. experience was the rural settlements. As pograms in Russia became more intense in the 1880s the search for refuge abroad was facilitated by the Argentine government's provision for agricultural colonies, mostly north of Buenos Aires in the provinces of Santa Fé and Entrerios. Still, migration was mostly to urban centers. Through the decades Jews settled in several *barrios* of Buenos Aires. Briefly stated, it took four generations to achieve approximately the same social status as occurred in two generations in North America.

Today, most Latin American Jews are comfortably within the middle class, partly attributable to their educational achievement, which is higher than for the general population. Individuals who might be considered as marginal receive sufficient help from the community to be included in the middle class. The correlation between income and class is positive, yet not too close. In the Western world, businessmen and industrialists often have less prestige than the professionals. This relationship is inherent in Latin America as entrepreneurs had until recently less standing than the wealthy landed elite. Despite their relatively high status, the Jews have not been able to participate in many social institutions. For instance, although they regularly vote, they seldom are invited to compete for political office.[24]

Guatemala

Possibly because of its diverse ethnicity, Guatemala is both more Catholic and less Catholic than Brazil or Argentina. The ratio of Catholics, over 70 percent of the population, is not very meaningful when one considers the attachment of most Indians to ancient practices based on Mayan and other indigenous sources. This syncretism or mixture of Catholic and indigenous is sufficiently complex so that disentangling New World from Old World influences is often impossible. Further, local variations make difficult any generalization about beliefs and rituals. In indigenous societies religion encompasses nearly all aspects of the person's world. Also, health practices (curing and prevention of disease) are intimately connected with religious beliefs and practices.

Among the features of Guatemalan Catholicism is the attachment to supernatural agencies, which represent a combination of indigenous practices and sacred images of Catholicism. The patron saint is a much more meaningful icon to Indians than are the concepts of Christ or the Deity. In all Latin America, it is not difficult to link the plurality of Roman Catholic saints with native quasi-deities and natural forces. In many communities the *cofradía*, a voluntary religious fraternity, cares for the saints and shrines connected with these deities.

Besides syncretic Catholicism practiced by Indians are the more orthodox traditions among ladinos. Comparable to the *cofradías* of Indian cultures are the *hermanidades* (sisterhoods or brotherhoods) composed of committed women who care for the church rituals. Since most villages are seldom visited by a priest, these lay organizations are major vehicles of religious life. However, since the 1960s these networks have grown thinner because of migration, fear of political violence, and Protestant expansion.

Religious activity revolves around a number of events. Most important is the fiesta for the patron saint, usually organized by the *hermanidad* under the direction of the *capitana*. The religious function is often subordinate to the carnival spirit of the event—masked dances, lotteries, fireworks, and extensive alcohol consumption. More exclusively on the religious side are pilgrimages to a shrine, which may have both Christian and Mayan significance. Other rituals such as the rain ceremony occur in those areas of Guatemala suffering from periodic drought.

Guatemala's religious institutions are not unique in failing to meet the country's ethical and physical needs. The church is caught in a struggle between the left and right, which plagued Guatemala for much of the twentieth century. The latifundista (wealthy landholders) regime, often backed by U.S interests, stifles any protest by the clergy, a trend stretching from the 1954 coup to the Reagan–Bush policies of 1981–92. Beyond these political conflicts is the liberal Catholic clergy, who protest the maldistribution of wealth and power, in contrast is the evangelical community, which often supports right-wing governments. In the late 1970s government troops and paramilitary operations began an onslaught against the peasants, mainly Mayan, which cost some two hundred thousand lives. During the presidency of evangelical Efraín Rios Montt, Catholics were under attack as U.S. evangelical missionaries entered combat zones for the purpose of making converts. The Christian Broadcast Organization, operating from the United states, was only one of such organizations supporting the Guatemalan destructive policy.[25]

Evangelical Protestantism, is estimated to include over a fourth of the population, especially among Mayans. The appeal is in part because of the call to temperance, but also the resources that Protestant missions bring from the United States. The conversion has caused considerable controversy as Mayans have abandoned their cultural traditions, for example, weaving their textiles, and have turned to the commercialism of the tourist trade.[26]

Subcultures, National Patterns, and Commitment

The comparison of these three countries in regard to their varying cultural development is an index to the religious complexity of Latin America. The contrast between folk or syncretic Catholicism and the orthodox tradition is most apparent in countries with large, unabsorbed indigenous elements as compared to mestizo or European countries, but differences are set by the national culture. For example, Inca traditions remain in the Andean version of Catholicism; the Aztec mother-goddess lingers on in the Virgin of Guadalupe.[27] Moreover, in mestizo El Salvador religious participation is less salient than in the Europe-oriented culture of Costa Rica.

As compared to other countries, Colombia is traditionally committed to the supremacy of the church. Still, the sharp division of the country into Liberals and Conservatives beginning in the 1840s was an outgrowth of anticlericalism. The threat of a conservative "theocratic government" became an acute rallying point for liberals after 1870. Until the 1980s pressure from church prelates could influence policy to a degree unknown in a nation such as Venezuela, where bishops enjoy a comfortable, low-profile relationship with the government and operate through Catholic Action and other lay organizations.[28]

In addition to the effect of national climates are the differences based on age, gender, social class, and urban or rural origin. Urban youth in particular tend to lose interest in religious dogma and ritual. In a survey of Colombian students, participation in mass and confession visibly dropped (with statistical significance) while they attended university. Those believing in the Deity as a "creative impersonal force" rather than a "personal being" increased during their university career from 15 to 25 percent.[29] A Mexican survey found greater disenchantment among university students; 28 percent never went to mass, varying from 55 percent in Mexico City to 18 percent in León—reflecting the higher rate of secularism in a national university in a large metropolitan setting.[30] In assessing these data we remind ourselves that persons tend to give socially acceptable responses that do not always correspond to reality.

In an analysis of religious behavior, gender differences usually transcend other subcultural variables except possibly social class. Apart from more traditional Catholic countries such as Colombia, where both sexes are relatively devout, it is not unusual to see four or five women to every man at mass; more often it is the younger women who attend mass and participate in religious organizations, at least in Cuernavaca, Mexico.[31] This relative indifference of the male toward religious rituals is common throughout the Latin world, whereas Protestants in Latin America show relatively less gender difference in their participation. Are "feministic" aspects of Catholicism, namely the Virgin Mary cult, a factor in encouraging greater participation by women? Possibly this gender difference is simply a cultural habit based on the separateness of sex roles. (A congregation in the United States will also show a low sex ratio; after all, women live longer than men and religious values usually become more critical in the later years.) Women are traditionally entrusted with the expressive role including the religious socialization of the young, whereas men lean to the instrumental role, which places less emphasis on religious norms.

Social class also determines religious involvement perhaps because of conformity needs, which are all too apparent in the middle class. According to my observations in cities, in worker areas attendance at mass appears to be roughly 10 percent of the adult population, whereas in middle and upper class parishes it is approximately 25 percent. The middle class is more faithful in church attendance, but men, and to a lesser extent women, of this class are critical of the church's dogma and ritualism. The lower class express less negative opinion since religion may seem more awesome to them, or they have more difficulty in verbalizing their feelings.

Roman Catholicism as a Social System

The scope and tone of Catholicism, of course, varies for a given society or subculture. A serious problem of Catholicism is its inability to serve over 80 percent of nearly five hundred million people. Its status is far from constant; in most countries a sizeable minority have reservations as to the church's functionality. This same question is asked in many countries of the world, especially in those where only one religion exists, Lutheran in Sweden, for example. In other words, what is the value of competition in sects or denominations?

The Nature of Religious Participation

A serious problem in Roman Catholicism is the small number of priests. The ratio in Latin America is the lowest of the Catholic world and a small fraction of what it is in Europe and the United States. The poorest showing is in Central America, the Caribbean, and Brazil. If anything, the situation in Latin America is barely holding its own. Table 8.1 reveals some indices of the declining religious commitment. In 1912 one priest was responsible for 4,480 persons; by 1998 the ratio varied between 4,600 in Costa Rica to 16,900 in Honduras.[32] The difficulty of recruiting clergy is a worldwide problem. Few Western nations escape the drift toward secularism; other occupations have become more attractive. In Latin America, as in most of the West, the shortage is acute because of the low saliency of the religious career for males. Until the 1960s approximately 50 percent of the priesthood were recruited abroad, principally Spain, and smaller numbers from Belgium, Canada, France, Italy, Germany, and the United States This flow has enormously slowed as seminaries are now unable to find recruits for the priesthood in most Catholic nations.

Other factors affect the declining strength of Catholicism. In rural areas it is doubtful if more than 5 percent of the population effectively participate because of distance and the lack of priests. In many countries, parishes are too large in area or population. Hence in both rural and urban areas, varied composition in educational level and social class make the role of the priest problematic.[33] Further, a perfunctory approach is also reflected in the meaning of the sacraments. Although most Latin Americans are baptized, confirmation and marriage fall far below the Catholic norm. Still, the church has a strong if declining influence on the educational and other social systems, as reflected in inset 8d.

INSET 8D   A CHURCH IN CONFLICT

A visitor can hardly fail to be impressed by the strength—and at times the weakness—of the Roman Catholic Church. Whether one sees the peasants of

Chichicastenango, Guatemala, burning their incense on the steps of an ancient pyramid, *campesinos* on a *rogación* (procession on one's knees) to the shrine of the Virgin of Guadalupe in Mexico, deference to the cardinal at state functions in Bogotá, or fiestas honoring the patron saint in any village or town, the church appears as a unifying principle of Latin American civilization. Yet, the dominance of the church today is a mere fragment of what it was in colonial times. The church hierarchy seems only partially aware of the crises that gripped Latin America in the twentieth century. The value system and ethical norms are democratic in spirit, but the church's traditionalism makes it aristocratic in style. Among the more notorious examples was the role of bishops and priests during *la violencia* (period of civil war between the two principal parties) in Colombia (1948–57), in which they most openly sided with the Conservatives.

Throughout Latin America the church is weakened by its moral ambiguity—a criticism hardly unique among world religions. Also, the church is distracted from its religious purpose by entering into alliances with the oligarchy.

As a result, educated urban society is often alienated. The church supports education and social welfare as much from expediency as from a coherent system of religious ethics. Its relationship to the population is often perceived to be passive, the greatest loyalty coming from the peasantry, that is, until the rate of rural social change is accelerated. In certain areas, as in Morelos, Mexico, the clergy is not viewed as being part of the community and is too remote from their problems of land distribution. Even when a priest becomes concerned with social problems his approach is viewed by many peasants as authoritarian and alien to the values the peasant most reveres, including local fiestas.[34] This ambivalence about the priesthood is recurrent in Mexico with its history of anticlericalism.

On the positive side, since the 1960s the church is solidifying its moral commitment to society, at least those segments it reaches. Its new roles, personal and social, can be considered as redefinitions of older roles, and indicate a less comprehensive function in society in view of the reduced power of Catholicism. Still, its educational function remains fairly intact. Religious instruction is permitted in public schools of most nations. The church operates hundreds of parochial schools in nearly all the twenty republics, and in Colombia almost half of the secondary schools. Catholic colleges and universities are found throughout Latin America; several of them have a relatively distinguished faculty and a growing student body. If anything, Latin American Catholicism expands its welfare services in order to care for what the social security system cannot. Hospitals as well as schools are maintained by over a hundred thousand nuns, but as with education this service is mainly focused on the middle class—in contrast to the work the Maryknoll order from abroad is doing. The gamut of religiously oriented youth organizations, labor unions, peasant leagues testify to a battle for the minds of the population but also indicates a growing interest in social needs.

Old and New Foci

Aware of the crisis confronting Catholicism, the clergy moves in different directions in an attempt to reorganize the church structure and functions. Ivan Vallier's analysis of the types of clergy, particularly at the elite level, still

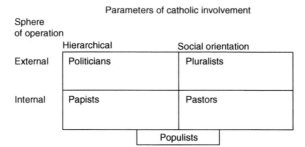

Figure 8.1 Typology of the clerical personnel.

Source: Adapted from Ivan Vallier, "The Clergy in a Changing Catholicism," in Seymour M. Lipset and Aldo Solari (eds), *Elites in Latin America* (New York: Oxford University Press, 1967), pp. 221–252.

holds, if with some revision (figure 8.1[35]):

1. *Politicians* as the traditional upper-class-oriented clergy are content with Catholicism as a set of ritualistic services available mainly to those members of society who can pay for them. By the 1960s politicians became increasingly oriented to change.

2. *Papists* are content with the church's hierarchical nature but believe that its authority should be from within the church, not from external groups. The "papists" may be included among the "newer" developments within the church since the frame of reference is the Vatican rather than the political and economic power clique in their nation. The movement involves lay activities such as retreats, youth programs, and limited evangelical activity.

3. *Pastors* are concerned with the "strong, worship-centered congregations." Key themes are "cooperation," "community," "communication," "pastoral care," and "meaning of the sacraments." Both the papist and the pastor role fit neatly into the model set by John Paul II.

4. *Pluralists* regard Catholicism as only one religion and only one approach to a complex universe. Since the church must be involved in combating poverty and other social evils, they are more or less indifferent to the other three clerical roles. We shall return to this viewpoint later in this chapter.

5. *Populist* (a role that has developed since Vallier's model) opposes the hierarchy.[36] As society becomes more pluralist, a priest may occasionally combine two roles—say, papist and pastor; clergymen are not necessarily committed to any one category.

Aside from these models, the rank-and-file of priests fulfill their careers with little change from traditional roles. The Latin American cleric can define his own degree of involvement with rituals, authoritarianism, and sectarianism. These traits became less prevalent after the Ecumenical Council, 1962–65, which prompted Latin American Catholics to reassess their goals (note inset 8e). If outnumbered by conservatives in most countries, bishops and the lower clergy who are change-oriented altered the stereotype of the professional priestly role. Inevitably, the profile of the priesthood varies among countries. As one example, the clergy is better educated and more often foreign born in Venezuela than in Colombia; yet, the church's status is greater in Colombia.[37]

The Church Popular

In the late 1960s a new movement emerged in Catholicism. The degree of poverty, political intransigence, and ineffective ritual began to trouble clergy and laity. The inspiration of the Second Vatican Council (1962–65) was an additional factor. The break with the past came most dramatically in Brazil. Dom Helder Câmara, archbishop of Recife and Olinda, led the church in a plea for social justice. With other progressive clergymen he proposed a nonviolent program for all Latin America as inspired by Gandhi and Martin Luther King.[38] Câmara also played a key role in the National Conference of Brazilian Bishops (CNBB), which organized *cursinhos* ("small courses," *cursillos* in Spanish) on rejuvenating the faith. Also, the CNBB became a mechanism for resolving problems between church and state with special significance after the 1964 coup, even though the council later lost its effectiveness in social causes.

Other leaders, inspired by what was to become known as Liberation Theology, entered the arena. Gustavo Gutiérrez, a member of Peru's secular clergy, was influenced by the theological currents in Europe and convinced of the need to establish the "Kingdom of God" on earth. Steeped in social science literature Gutiérrez spoke of the necessity of seeking "action of integral human development."[39] The focus should be on this life, not the afterlife. Similarly, the church turned to educators such as Brazil's Paulo Freire, who spoke of *conscientización* (consciousness raising) by which the oppressed—and the general public—might become aware of their situation in order that it might be changed.[40] Consequently, Freire described his work as "revolutionary." Scarcely less important was Leonardo Boff, whose book *Jesus Christ, Liberator* (1978) stressed the focus of Jesus on the poor, but extended his interest to a number of social problems, including ecology, all of which was to have effects well beyond Brazil.[41] Liberation Theology grew out of a wide range of developments. In Nicaragua as *Sandinistas* fought against the Somoza regime, local priests spoke of sin not only as the lack of individual responsibility, but as *structural* sin, that is, exploitive economic, political, and social conditions. Liberation Theology came to mean redemption of the poor from societal sin. Christians took refuge in the story of Exodus; they, too, to be delivered from slavery, with the image of Christ as the *campesino* or *Cristo Obrero* (Christ the worker).[42] These images were a far cry from the traditional fiestas surrounding Santo Domingo (the patron saint of Nicaragua). More than in most countries, Liberation Theology aroused polarization between the hierarchy, who came mainly from middle- to upper-class backgrounds, and the lower clergy, who represented the

Church *Popular*.[43] Possibly the most exciting aspect of the movement has been the emergence of basic ecclesiastical communities on CEBs (inset 8e).

INSET 8E   BASIC ECCLESIASTICAL COMMUNITIES AND THEIR IMPACT

Probably the most important event in liberation was the growth of the *comunidades ecle-siales de base* or CEBs (basic ecclesiastical communities). They began almost sponta-neously in the early 1960s, reaching a crescendo in the mid-1970s, as an outgrowth of new doctrines in the church and the hope for sociopolitical change. Paraecclesiastical organizations have a lengthy history in Roman Catholicism and bear some relationship to the populism of the nineteenth-century North American frontier. The CEB, a refuge from crisis and duress, appears especially in areas marked by suppression. In Brazil the move-ment may be a reaction against Umbanda, which tends indirectly to communicate a con-servative sociopolitical message.

Although CEBs vary by country and region, they are usually small groups of twenty–thirty people, generally from the working class. They may be activated by a bishop, priest, nun, or by lay individuals. Typical activities include help to the extremely needy, visiting the ill, sewing, and especially prayer and Bible reading.[44] Developing judgment and self-confidence was a later development, seldom evident in church teaching before the 1960s. As one Peruvian comments, the new movement is to combat the past, especially the latifundia approach to religion, and to substitute *la palabra* (the word) for *el rito* (the rites).[45] This attitude is not necessarily universal as another Peruvian observer finds an "excessive individualism" in popular theology.[46]

Women provide much of the membership and leadership of the CEBs.[47] This vigorous participation by women causes consternation on the part of males, as women—within the context of liberation theology—are questioning the masculine interpretation of the Scriptures. In many Western countries feminist scholars moved into *hermeneutics* or the search for meaning, pointing to the "patriarchalization" of the early church—a problem that remains today.[48] In an informal observation of CEB members in São Paulo at least two-thirds were rural-born, uneducated women, many of whom found a new identity in participating in the movement. They were involved in catechism, liturgy, and various ministries. Women who participated in political activity were, like most members of CEBs, inclined to the Partido dos Trabalhadores (Workers' Party). Fundamentally, the CEB experience sensitizes women to the problems they face in a man's world.

Brazil itself possibly had over seventy thousand CEBs at the movement's high water mark in the 1970s. A São Paulo study points to a variety of style and activities in regard to charity work, Bible study, and festivities. Education and class have their impact, the lower class being more emotional and politically motivated, but their goals differ little from those of the middle class.[49] No less critical, a number of lay organizations were ini-tiated or reinvigorated. The religious climate, including the fate of Liberation Theology, shifted as the spirit of Vatican II gave way to the conservatism of Pope John Paul II. Evidence of the change within Catholicism, at least on an organizational basis, is CELAM (Bishops Council of Latin America). This organization meets periodically and reflects the papal viewpoint as well as its diverse priorities.

Liberation Theology

The growth of clerical organizations is almost unending and inevitably involves overlapping personnel. Also, conflict over political issues outweighs conflict in personalities. Throughout Latin America the church has since the 1960s searched for an equilibrium between a structure "with 'nonrational' ends (faith and mission) and the church as a rational actor with institutional interests," which it must retain in order to serve its religious goals.[50] For some, liberation represents a form of postmodernism or a blend of Christian mysticism and social action.[51] A study in the city of Oaxaca, Mexico, illustrates a not too comfortable relationship between popular Catholicism and the institutional church.[52]

A review of publications on CEBs and Liberation Theology questions the ultimate contribution of the *iglesia popular* ("people's church"). Only partial data refer to the numbers reached, redirection of the church, or actual commitment to social action. Is the progressive church raising a democratic consciousness or simply perpetuating another verbal ritual? Critics of Liberation Theology point to its selective treatment of Christianity; it did not go far enough—"it contains just enough Christianity to be a threat to orthodox Christianity but not enough Marxism to be a threat to the status quo."[53] According to another observer, the Roman Catholic hierarchy is overly concerned with economic and political power. It must find a deeper moral commitment. It has lost touch with the lay public and lower clergy.[54]

Liberation Theology had special significance for areas undergoing a crisis, for instance, a natural disaster. The earthquake that destroyed Callejón de Huaylas in Peru on May 31, 1970, with twenty thousand people dead, prompted its residents to explore the meaning of religion. *Conscientización* led to probing their faith. Even a kind of *cholificación* or a convergence of Indian and mestizo took place. The concepts of sin and morality were replaced with the drive for social justice.[55] In a different arena, a group of priests working through *cursillos* and interviews activated consciousness-raising and decision making among the residents of a disorganized Panama City squatter town. As a result, clientelism among the city's leadership was redefined; political action and public services were revitalized.[56]

The Rise of Protestantism

Protestantism is rapidly growing. In some local areas 30 percent are Protestants, of whom the majority are Pentecostals. Even if entire Latin America reaches 20 percent, as some authorities estimate (it is already beyond that level in Chile, Guatemala, Brazil, and El Salvador; table 8.2), Protestantism will remain a minority religion.[57] As it is estimated that only 15 percent of the population who identify themselves as Catholic regularly attend mass or participate in their religion, the numbers of active Catholics

Table 8.2   Estimated Protestant population (1993)

	Protestant people (%)	Number of Protestant denominations	Estimated total population (1995)
Chile	27.9	60	14.3
Guatemala	24.1	215	10.6
Brazil	21.6	124	157.8
El Salvador	20.6	72	5.9
Nicaragua	17.3	79	4.4
Panama	16.7	50	2.6
Honduras	11.0	118	5.5
Costa Rica	10.7	179	3.3
Bolivia	9.3	120	7.4
Argentina	8.0	72	34.6
Peru	7.1	64	20.7
Paraguay	6.0	24	5.0
Venezuela	5.3	51	21.8
Mexico	5.1	1152	93.7
Colombia	3.8	63	37.7

Source: Anna Motley Hallum, "Taking Stock and Building Bridges: Feminism, Women's Movements and Pentecostalism in Latin America," *Latin American Research Review*, 2003, 38(1), 160–185.

and active Protestants are not too far apart.[58] However, Protestant influence extends well beyond its numbers. Along with Liberation Theology and the social crisis confronting Latin America, Protestantism prompts the Roman Catholic church to look for new directions. The fervor of Evangelism is attracting the poor, not because it offers a socioeconomic solution but because it provides an emotional escape. Unlike Catholicism, little organizational unity is apparent.

Historically, Protestantism accommodated to the scientific revolution with less friction than did Catholicism. Also, the growth of Protestantism gained from the anticlerical movement during the nineteenth century. Protestantism has its greatest appeal in areas where Catholicism is organizationally weak, as in Brazil, Chile, and Guatemala, where CEBs made inroads. It is also less attractive where the culture is overtly secular as in Venezuela and Uruguay. In Chile Protestantism flourishes in areas in which the political left is absent, moving from Santiago and Valparaiso into the towns and farms of the Central Valley.

Challenge and Response

Historically for Latin Americans, whether malcontent citizens or political leaders, from Rivadivia in Argentina to Calles in Mexico, Protestantism has been a means of expressing anticlericalism. Only the more progressive priests regard the Protestants' arrival as healthy competition that might stimulate renewal within the church. Rather, clerical and lay elites urge Catholic expansion in Protestant outposts. The relationship between Catholics and Protestants is a delicate one, neither side acts with complete discretion. The restraint of the established denominations stands in contrast to the sects constantly exhorting Catholics to leave their "religion of idolatry" for certain salvation through a particular brand of Protestantism. The proliferation of sects is a source of confusion to Latin Americans, although most members are known simply as "*evangélicos.*" The role of a given Protestant group can be caught in a political conflict. For instance, Miskitu Indians on the Caribbean coast of Nicaragua chose identification with the Moravian Church (a moderate denomination originating in Central Europe and coming to Pennsylvania in the mid-eighteenth century) as a means of expressing their ethnic identity in response to edicts from the Sandinista government during the 1980s. Hardly enamored of the Somoza government the Miskitus resented even more the bureaucratization imposed by the Sandinistas. By the late 1980s both sides came together and an accommodation was found. In other words, a Protestant sect can become an ethnic marker.[59]

Throughout the twentieth century outright persecution of Protestants was rare. The only serious encroachment occurred in Colombia. During the 1940s and 1950s the climate set by the Conservative regimes of Laureano Gómez and Ospina Pérez hampered the work of both foreign missionaries and native leaders. During the Rojas Pinilla regime (1953–57) violence flared up between Catholics and Protestants. This conflict was in part a reflection of the *violencia* that engulfed the country during the struggle between Liberals and Conservatives, but the difficulty was also attributable to the conservative nature of Colombian Catholicism, the desire of Rojas Pinilla to cater to those elements, and the intransigence of the Protestants. This intolerance was discouraged by the Liberal presidency of Alberto Lleras Camargo (1958–62), which initiated the coalition government, 1958–74. The ecumenical spirit of Pope John XXIII (1958–63) was another positive factor.

Occasional reports of persecution surfaced in other areas. In Mexico the Reforma laws against Catholicism as a "foreign church" could not but rub off on the Protestants. In this connection, major faiths became "nationalized" in certain parts of Central America. Hostility can occur because of the invasion of indigenous areas. To what degree the "apocalyptic interpretations" of evangelism may have adverse effects on ethnic traditions depends on the specific community.[60]

At present, indifference or antagonism toward Protestants is inspired more by anti-Americanism than by Catholicism. The Catholic hierarchy in Brazil has blamed ignorance and gullibility—as well as interference from North America—for the growth of Spiritism and Protestantism.[61] In reality, in several cities progressive priests join with Protestants in cooperative efforts for social betterment.

Religious Fervor and Social Service

Protestantism in Latin America may be differentiated into three types: (1) "main-line" Protestantism in areas of large colonies of northwestern Europeans—southern Brazil and southern Chile—Lutheranism being the most prevalent faith among descendants of this immigration, (2) traditional Latin American Protestantism, largely of middle-class origin, sometimes using the church facilities of the foreign Protestant colony, and (3) proselytic sects, especially Pentecostals but also Jehovah Witnesses, Mormons, Seventh Day Adventists, all of which now depend as much on domestic as foreign leadership. These fundamentalist religions show a logarithmic growth from the 1970s onward—similar to the situation in the United States (where they exert considerable political pressure in pushing conservative issues and candidates). Methodists, at least in Brazil, may feel theologically closer to Catholics than they do to certain Pentecostal variants.[62] From a historical viewpoint, Latin American Protestantism represents a mix of eighteenth-century Methodism from Britain and twentieth-century Pentecostalism from the United States In the nineteenth century the "symbols" of Protestantism—Luther, Calvin, and Wesley—were replaced by folk heroes such as Juárez, Sarmiento, and Martí, a kind of "Protestant syncretism."[63]

Not all Protestant communities show a strong discipline; adherents drift in and out of the fold. According to a report of a village in the Peruvian Andes, conversion means a choice of hard work over fiestas, social isolation over community participation, abstention over drinking. Not everyone wants to pay the cost of evangelic identity.[64] In several instances, particularly among indigenous communities, Protestantism has the same ritualized superficiality identified with syncretic Catholicism. How functional is a European type of Christianity to individuals living in a different cultural system and economic deprivation? Protestantism hardly meets these needs but it derives strength from the faith and discipline it offers the people and especially from an in-group feeling. In other words, the spread of Protestantism is fueled by cultural change. With decline of haciendas and emergence of a new economy Protestant agents are able to transform the festival complex into a new kind of revivalism. Traditional fiestas are interpreted as a subtle form of exploitation, which directly or indirectly is projected on Catholicism.[65] Not least of the Pentecostal appeals, especially for women, is its attack on machismo.[66]

The Many Faces of Pentecostalism

The growth of Pentecostal sects is remarkable. In Chile they account for 87 percent of all Protestants and they are not far behind in Brazil. The tone is not identical in the two countries but in both the *tomada de Espírito* (seizure by the "spirit") offers a dramatic escape from reality. According to observers, this *tomada* provides an emotional experience, a healing power, and a transcendence whereby the individual is able to grasp "the Truth" and become a prophetizing and convincing evangelist. Glossolalia is evidence of communication with "the Spirit." In this charismatic mood (often in a background of guitar playing) conversion is for many persons irresistible.

Besides Pentecostalism, other sects compete for allegiance. In most of Latin America, Seventh Day Adventists account for more converts than do Mormons and Jehovah Witnesses—with the exception of Mexico, where Mormons number 783,000.[67] Another eighteen sects are recorded in Peru alone.[68] A number of them, especially the Unification Church or "Moonies," appeal mainly to youth. A Peruvian sociologist ascribes the attraction to social dislocations: family breakdown, *tráfago* (depressiveness) of the cities, and school dropout.[69]

If Pentecostalism and related faiths are the most dramatic aspect of the Protestant cause, the more traditional churches—Methodists, Lutherans, Presbyterians, and to a lesser extent, Baptists—appeal to the middle class, principally since 1950. Although most Pentecostals are working class, the religion also attracts the middle class. The striking multiplicity of Protestantism is shown in table 8.2.

The 1980s saw the rise of televangelism in Latin America. I recall my shock on turning on my TV in a Rio hotel room and finding three different channels offering North American evangelists dubbed in Portuguese. A principal theme was the seeming unimportance of the individual's material well-being as compared to personal salvation and the rewards of the next world—hardly a plea for social reform. In Brazil the Universal Church of the Kingdom of God and other Pentecostal churches turn to the medium not only for conversion but for supporting right-wing politicians—with intriguing financial overtones.[70] In Guatemala, Pentecostal presidents Efraín Ríos-Montt and Jorge Serrano Elías, however different, encouraged the military–oligarchical rule of the country.[71] In appeasing the landed elite Ríos-Montt authorized a scorched-earth program against the Mayans in order to end the guerrilla insurgency.[72] Finally, the arrival of the later-discredited Jimmy Swaggart in Managua in 1988 offered something of a new perspective as he ascribed Nicaragua's problems neither to the *Sandinistas* nor to the Contras; they were simply "due to the devil." The *Sandinistas* could only hope that if North Americans had seen the evangelist on their TV screens, they might have "run out of arguments for the Contra war!"[73]

The Protestant Contribution

A major achievement of Protestantism is in education, medicine, sanitation, and other remedial programs. As one example, Seventh Day Adventists reportedly maintain a mobile hospital service on the Amazon River with a fleet of nine well-equipped launches, dispensing medical treatment for over twenty-five thousand patients in a year. Bible translations, book publishing, literacy programs, and radio broadcasting are all extensions of the concept of service along with conversion. Protestants would seem to be the inheritors of the Jesuits of the colonial Plata. In recent decades Catholics are also driven to social service.

The Chilean human rights organization Committee for Peace was founded in 1973 by Catholics, Protestants, Jews, Eastern Orthodox. Catholics and Protestants in that country also collaborate in "radical" programs. On the other side, Pentecostals lean toward conservatism. In Brazil they opted for the military coup of 1964, and tacitly supported Pinochet in Chile. Generally the evangelical movement has been apolitical, but with their increase of adherents they are now more critical of the status quo.

The Church and Social Protest

In the 1960s lay Catholics placed much of their eagerness for reform in the Christian Democratic parties in Chile, Peru, and Venezuela. For them the answer to Communism lay in moderate reforms within the system. However, the risks to this viewpoint are unmistakable. The party is not always able to fulfill its promises of social transformation, as the fate of the Belaúnde and Frei regimes demonstrates. Christian Democrats do not necessarily subscribe to the "Christian-virtue-is-the-road-to-social-reform" theme voiced in sermons throughout Latin America.

Setting the stage for the development of social action and the fate of Liberation Theology has been the series of meetings of the General Conference of Latin American bishops (CELAM). The Medellín conference in 1968 was inspired by stirrings in the church following Vatican II and the social doctrine of Pope John XXIII, notably his encyclical *Pacem in Terris* and its stress on human rights, which was to have a major impact on both clerics and lay members. Inset 8f tells of the transition in the life history of a Colombian priest. Liberation Theology was approved not only by the Conference but also by the Vatican. Paul VI passively assented to the proposals. However, by the time of the 1979 conference in Puebla, Mexico, the tide was shifting. Many bishops became nervous about the extent of CEBs, particularly the threat to the hierarchy and to traditional Catholic doctrine.

INSET 8F   EVOLUTION OF A PRIEST–MARTYR: CAMILO TORRES

Every one, including myself as a visiting Fulbright professor, in the School of Sociology at the National University of Colombia was impressed by a young Dominican priest Camilo Torres Restrepo (1929–66), who came from a prominent Bogotá family. After an M.A. degree in sociology from the University of Minnesota, the bishop encouraged him to pursue theology at the University of Louvain in Belgium. After Camilo's return to Bogotá it was clear that he had a deep affection for people and was appalled by poverty and violence in the world, notably in his own country. In 1962 he was appointed university chaplain. Soon after, a strike was called by student groups because ten students were expelled for reasons that Camilo considered unjust. The cardinal called for his dismissal, writing to him "I have been absolutely convinced that the pontifical directives forbid a priest to intervene in political activities and in purely technical and practical questions in the matter of social action, properly speaking."[74]

The following year Camilo left the church and organized a movement, Frente Uniido, among workers and peasants, finally deciding that peaceful, nonviolent action could not change the status quo. In 1966 Camilo as a guerrilla was killed on a lonely, frigid *páramo* by government troops.

An analysis of Camilo's psychosocial development is given by priest Germán Guzmán, an adjunct at the Facultad. Guzmán had served as a parish priest in Tolima, one of the areas most affected by *la violenica* (1948–57), which strengthened his social conscience. He writes of Camilo as utterly dedicated to the Christian doctrines of service, brotherhood, and especially justice. In Camilo's early career science was a commitment as he demanded factual knowledge. As he perceived increased misery in his land, science gave way to service, and eventually militancy. He was frustrated by the church's intransigence. Liberation Theology had not yet come to conservative Colombia. Drawing from Pope John XXIII's *Mater et Magistra* and its plea for human betterment, in the early days of his movement (United Front), Camilo tried to bring together discordant elements of the left. At first, Marxists rebuked him because of his Christian pacifism, but in the end he came to believe that Christianity had to accommodate to Marxism and as peaceful methods were not effective, he accepted what for him was the inevitability of violence.

Essentially the clergy could be divided into three groups—conservatives, centrists, and liberationists. The conservatives were dominant, with Bishop López Trujillo of Colombia as a principal voice and supported by the new pope John Paul II. As it turned out, the Puebla meeting might have been a disaster for the liberals, had not the Mexican publication *Uno más uno* leaked a letter from López Trujillo to another conservative bishop. The letter contained directions about keeping the conference in the hands of the conservatives as well as making derogatory remarks about several of the liberal clergy.[75] Still, López gained the upper hand in CELAM in securing Vatican support for a conservative orientation in the Latin American hierarchy. In CELAM's next meeting in 1992 in Santo Domingo lip service was given to CEBs but the thrust was toward more missionary work and discouraging the clergy's political activities. Pressure for social reforms was reduced by the decline of dictatorships in the 1980s. Indeed, by 1982 Brazilian conservative bishops replaced liberals, the rest of Latin America followed suit.[76] The future seems unclear as John Paul II urged more conferences to deal with the North–South conflicts, yet the Vatican continued to discipline dissonant bishops. With the election of Pope Benedict XVI in 2005 it would appear that the Vatican will hold on to a restrained stance toward social change.

The Struggle in Central America

Until the 1960s Catholicism in Central America was institutionally indifferent to social injustice. However, the arrival of new priests and nuns, both of national and foreign background, began to reflect the spirit of Vatican II and eventually of the Medellín spirit. Nowhere was the conscience of the church placed in a more difficult position than in revolutionary struggles in Nicaragua and El Salvador in the 1970s and 1980s. In Nicaragua the church was highly critical of Somoza, but the Sandinistas' drift to the left produced a crisis. The new regime placed several liberal clergymen in the cabinet— minister of culture, of education, and of national planning. However, Cardinal Obando y Bravo and other bishops reacted sharply when the FSLN (Frente Sandinista de Liberación Nacional) adopted slogans and programs that smacked to them of Marxism. Somehow the cardinal ignored excesses of the Contras but objected to the Sandinistas' actions, reportedly dismissing over sixty priests and nuns who supported the government.[77] After the 1992 election of a more conservative government, church spokesmen were declaring a return to "traditional values."

El Salvador offers an especially tragic history of persecution of the church. As in Nicaragua, CEBs began in the 1970s, and a number of orders— Franciscans, Jesuits, Maryknoll, among others—were involved in this commitment to the "social gospel." Still, several bishops remained cautious in criticizing the government. Disappearance and death were meted out to a number of priests by the oligarchy, operating through quasi-government channels. From 1977 onward, Archbishop Oscar Romero from his pulpit in the capital condemned violations of human rights. This stance resulted in his murder in 1980 by death squads of the right-wing Nationalist Republican Alliance (ARENA). As Romero's successor, Msgr. Arturo Rivera attempted to carry on his policy in spite of mistrust by the elites, the army, and the U.S. embassy. On the whole, Romero's murder had the effect of somewhat subduing protest by the church. The CEDES (Episcopal Conference) strove to bring the two sides together. Yet, possibly more than any other Latin American country, El Salvador shows the clearest example of strong and continuing bonds between religious organizations and peasant revolt.

Reformism in Brazil

The development of reform thinking in Brazil is not necessarily typical but represents the climate of social change in Latin America. Although Brazilian Catholicism is marked by moderation, since the 1960s this relative tranquility has given way to defiance vis-à-vis the political scene. For instance, in rural areas the church played a role in organizing peasant cooperatives, agricultural labor unions, and in the Northeast, peasant leagues and Catholic Action, a movement imported from Europe, recruited working youth along with secondary-school and university students.

These organized efforts (perhaps more accurately described as coping attempts) led to social strain as opinion became polarized in Brazil between those leaning to radical solutions and those preferring a restrained approach. Most priests who belonged to dissident reform movement preferred to work within the system. During the fluid atmosphere of the Goulart regime in the early 1960s, conservatives, moderates, and radicals within the church could move in almost any direction they wished, albeit with friction between upper and lower levels of the clergy.

In short, the church became more liberal, even radical, notably during the repressive atmosphere from the 1964 coup into the 1970s. Internal conflict increased during the 1980s and 1990s because of the conservative thrust of the Vatican. An explanation of the relative liberalism of Brazilian Catholicism lies in a cluster of factors: Protestant pressure, *conscientización*, CEBs, and greater decision making by local bishops, priests, and laity rather than the upper hierarchy. Militancy in the working class and removal of ties with conservative political parties contributed to this autonomy of the church. As one sociologist explains, radicalization in the Brazilian church arose because of interaction between the "organizational value set" and the elite reaction to changes in the social order.[78] That is, the state was no longer underwriting the church as it did the first four hundred years. Catholic schools ceased to be the chief socializing agency for middle- and upper-class children. To a lesser extent this transformation occurred in Chile in contrast to the regressive stance of Argentina and Colombia. Leftist political movements placed more pressure on the church in Brazil and Chile than in the other two countries.

Chile: Opposition on the Left

Traditionally the Chilean clergy is socially conscious, partly because of a history of political liberalism as compared to its neighbors. The ultimate test for Chile's church came in the 1970s. The Allende period (1970–73) inevitably stirred the church in different directions. In 1970 segments of the Catholic clergy and laity along with a minority of Protestants organized Christians for Socialism. Yet, nearly all bishops were negative toward the regime's Marxist tone. Even for Santiago's humanitarian cardinal Raúl Silva, political involvement of priests violated Catholic tradition. The hierarchy could support socialism but not what they perceived as communism.

However burdensome the nearly three years of Popular Unity—especially the chaos of its last months—the 1973 coup devastated liberal Catholics, Greek Orthodox, Jews, and Protestants in Chile as probably no other event in that nation's history. The church can hardly be criticized for recognizing the Pinochet regime, as several Western nations (including the United States, which was involved with Allende's demise) did so within ten days of the coup. As torture, murders, and disappearances rose, members of the clergy were arrested for actively helping victims. With rising unemployment, Catholics, Protestants, and Jews worked with international aid agencies in organizing soup kitchens and development projects in various parishes. In spite of the horror of the dictatorship, responses to the crisis probably made the Chilean church more flexible and creative. For instance, the Vicaria de

Solidaridad administered legal and material aid to victims and their families. Not least, the crisis brought bishops to realize that deeper problems exist beyond doctrine and ritual. Still, the 1990s saw minimum consensus between progressive and traditional clergy, who were caught in the "consolidation process"—the bringing together of civilian and military. As the present Vatican appoints conservative bishops, conflict with local priests increases. To what degree accommodation can reduce tensions in Chilean society remains uncertain.[79]

Conclusions: Activism and Social Change

Today minorities within the church are convinced that a radical restructuring of Latin American society is the only solution to Christian humanitarianism. Throughout the region many neo-Marxists avoid offending the church, but many left-wing politicians link religion with imperialism, including a liaison between Protestantism and the CIA. The church itself has an ambivalent and generally hostile stance toward Marxism. The statistics for those clergy who followed their conscience is grim. According to one census, 488 priests were arrested in Brazil between 1964 and 1978, and few countries were without some harassment from the authorities.[80] Although it would be difficult to catalog developments in each country, a few examples can be cited not only in Brazil and Chile, but also in Peru, Argentina, Paraguay, and elsewhere.

In Argentina, for example, the military dictatorship in 1976 brought oppression eventually of the entire population; reportedly eighteen priests (including one bishop) were murdered, eleven tortured, ninety-three arrested, and nine kidnapped or disappeared. Yet, for the most part the church power structure remained mute spectators to the dirty war, as the reign of terror was called, the clergy losing much of its credibility once the Falklands/Malvinas War brought down the government. For all Latin America between 1964 and 1978 forty-one priests were murdered (six as guerrillas), eleven disappeared, four hundred and eighty-five arrested, forty-six tortured, and two hundred and fifty-three expelled from the country.

The case of Cuba illustrates the church's accommodation to a changing universe. Although a scant majority of bishops opposed the Batista regime, in the lower clergy antipathy was nearly universal. Most opposition came from the Spanish-born clergy. After Castro's drift toward communism, tension appeared between church and state. But by the late 1960s a curious parallelism developed. The church was relatively free to do as it wished as long as no counterrevolutionary activity was involved. Also, developments in Nicaragua had a positive effect as a sizeable minority of priests played conspicuous roles in the Sandinista government, despite opposition of most bishops. This lesson further moderated Castro's attitude toward the church.

Given the evolution of Latin American society, can one hypothesize that religion represents conflict more than harmony? First, witness the shift by

Vatican decree to a more restrained social action at the time of the harsh economic realities of the 1980s. Although the church gave lip service to the need for reducing poverty, Liberation Theology was abandoned.[81] The Catholic presence is more evident and has improved its political tactics. It is as yet impossible to know how the church will interact with other institutions such as the state and the economy, or what role it may play in given social processes. Moreover, evidence in much of the Western world points to a decline in religious authority.[82]

In their study of the Latin American church, particularly Chile and Peru, Michael Fleet and Brian Smith conclude that the church will continue to react to external pressures, as it has increasingly over the last century—rise of secularism, separation of church and state, and pluralist ideologies. Three possible scenarios include: (1) strengthening the traditional religious stance, especially in view of the Vatican's selection of conservative bishops, (2) bifurcation of the church into "retreatists" and "actionists," and (3) stress on spirituality but attention to social justice. Of the three, the third is the most likely. Even though the church will hold to its mission of spirituality, it will be sensitive to humanitarian issues, conscious of justice, and devoted to the rights of the suppressed, notably women.[83] All this depends on leadership, which leans toward traditional evangelization rather than the combative world of politics.[84]

In summary, religion remains as a complex institution attempting to provide integration to conspicuous social structures and competing ideologies. According to a Mexican study, segments of the public are demanding new approaches to their spiritual and emotional needs. Esoteric and mythical themes associated with the "New Age" appear in new cults, more than occasionally inspired by cyberspace.[85] In other words, in view of inevitable conflicts social institutions seldom function with consummate efficiency or predictability, but significant social change appears periodically, and the church will likely attempt to provide a special meaning to the society.

Chapter Nine

Education: A Continuing Challenge

Throughout the world, people are asking, "what is the quantity and quality of the educational system? Latin Americans find their schools and universities lagging behind those regions of the world with which they would like to be compared, Schools, although continually expanding, do not meet the needs of society. Among other problems, school construction and teacher training fail to keep pace with population growth—despite a falling birthrate since the 1970s.

Schools inevitably reflect the value structure of the culture in all of its complexity of social, political, and economic movements. Only recently has a mandate for educational reform appeared in the emerging urban society of Latin America. Rigidities in the culture—abrasiveness of its class system, maldistribution of resources, vagaries of the political structure—are all relevant. Until the economy becomes healthier, more diversified, and better distributed, educational expansion will not lead to satisfying outlets for its graduates. Education must be considered as both a capital and a consumer good, but neither can function unless articulated with the society.

It would be unfair to think of Latin American education as a unitary structure or process. Besides the three levels (primary, secondary, and higher), a distinction exists between public and private, national and regional, urban and rural, formal and nonformal education. Schools in several countries, Brazil for instance, are generally under federal supervision; in others local hegemony prevails. In several capitals a few schools are rated among the finest in the world, whereas in all Latin American nations, and most of the world, the rural school falls far short of what might be called a basic education.

Education and Society

In any society, education mediates the societal norms and prepares a new generation to acquire given skills in order to assume the statuses and roles the particular culture needs. Usually the power elite determines what these statuses are to be. In many societies the citizenry are helpless to interfere. In most developing nations the educational system remains in the colonial–capitalist mode as the dominant economic classes—both national and multinational—attempt to create a disposable workforce. In one respect, the educational institution embodies the crux of the functionalist–structural model as the school is one more integrating force in society. The school serves a number of functions: among the manifest ones are employability, personal and social development, upward mobility, and civic and interpersonal identity. Latent functions include custodial care, removal of the young from the workforce, creation of networks and youth subcultures. Education is also related to various power systems providing: one, the individual with knowledge of manipulative skills, and two, from the societal viewpoint as the kind and level of education the state requires for its functioning. As in other cultures, education in Latin America is designed to promote status allocation according to the needs of the upper and middle classes. In this respect, education offers potential conflict as children and adolescents have unequal access to the limited assets society offers or to the evaluation their mentors place upon them.

Likewise, modernization—or globalization, as it is more currently known—seems to hinge on an improved educational system. Evidence abounds to the close relationship between years of schooling and the modernist attitudes in the population.[1] According to Marxist and conflict theories, education favors modernism but generally fashions workers into units of production for the ruling economic interests. Formal education is a critical means for creating a competitive society. Usually education simply carries out the norms of given social order; it is seldom an agent of micro or macro social change. Yet, a new social order, as with Castro in Cuba or the Sandinistas in Nicaragua, may provide an innovative orientation.

A number of specific questions are raised about the status of education in the Ibero-American world: what historical events led to the present impasse? What is the nature of the primary and secondary school? How effectively do they carry out their functions? Can adult programs fulfill the task of providing literacy, vocational training, and orientation to change? Why is higher education moving at a faster rate than basic education? How do students and teachers view both their careers and societal problems? What is the relationship between education and social change? In other words, what does the educational institution offer the total society? This chapter will focus on these issues.

The Past and the Present

In early colonial times, school was the privilege of many but not all whites. In fact, most of society remained illiterate. Conducted almost entirely by the clergy, the few hours of instruction each week stressed religious dogma. In the eighteenth century, conditions improved somewhat with the establishment of the *colegios* (private schools) under secular control. The indigenous and mestizos were denied education; even their acquisition of Spanish might disturb the caste society the colonials desired.[2]

After independence, constitutions of almost every nation gave lip service to universal education, but this goal remained meaningful only for the upper classes. However, after introducing new subjects into the curriculum, education was broadened from its narrow Iberian mold. Eventually, authorities pledged to follow the hopes of revolutionary leaders, for example, Bolivar and Nariño, to bring schooling to segments other than the aristocracy.[3] In this direction, Nueva Granada (Venezuela, Colombia, and Ecuador) borrowed the Lancaster system from England. Under this scheme a master taught over a hundred children in a single classroom with advanced students monitoring the pupils. The Lancaster philosophy insured morality and regimentation in its emphasis on standardization, hierarchy, and order, which leaders thought would be useful in the rural scene and for the industrial world that was to come much later.[4] For most of the nineteenth century Latin America intermittently felt the shadow of educational philosophers running the gamut from Jeremy Bentham to Horace Mann.

Schools existed primarily for children of the *hacendado*, town officials, and merchants, with little input by either teacher or pupil. Both lay and clerical teachers stressed goals such as meticulous penmanship at the expense of a functional education. Corporal punishment was the official method of assuring indoctrination. Worse, the school was plagued by inherent societal problems—poverty, disease, corruption, and indifference to intellectual values. Only the late nineteenth century saw any serious effort to impose academic standards. By this time most national constitutions had decreed free and compulsory education under secular auspices.

On the whole, the nineteenth century brought a gradual improvement. Among other countries, Mexico attempted to resolve the recurring struggle between secular and religious control. This conflict appeared as early as 1830, and the Juárez regime made valiant efforts to secularize what education there was. Illiteracy was nearly universal, and little change occurred during the Díaz period. On paper the number of primary schools more than doubled, with a quintupling of students in all public schools. However, state officials undoubtedly reported more progress than actually took place.

In Brazil, where education particularly lagged in the colonial period, Dom Pedro I as emperor tried to make education available to lower-class boys by opening a public secondary school in 1837, long before most Hispanic American countries. Another advance was teacher training, including standardization of teacher's examinations and the opening of four normal schools by 1846. These efforts were further spurred by the establishment of the Republic. In 1886 Brazil had over six thousand public primary schools, in addition to religious facilities. Still, over 80 percent of the population remained illiterate. The 1891 constitution vested responsibility for instruction in the states, but after the 1930 Vargas "revolution" the federal government became the principal agent for overseeing the school system. Despite a fivefold growth per capita in school attendance since 1900, Brazil in the 1990s still trailed thirteen other Latin American countries in the ratio of children in primary school.

Among the smaller nations, Ecuador made considerable strides under President García Moreno, who dominated the political scene from 1860 until his assassination in 1875. He strove to adhere to the decrees of 1821 and 1833 specifying primary education as compulsory. School enrollment more than doubled during his dictatorship, and progress in economic affairs, science, transportation, and culture was unmistakable. Under Moreno's rigorous religious outlook the educational system was dominated by Jesuits. For all their parochialism, Ecuadorian schools proved to be a model for other Andean nations.

Uruguay's high level of education dates from the leadership of José Pedro Varela during the late nineteenth century. He reflected the influence of the Argentine Sarmiento as well as U.S. educational exponents, issuing the Law of Common Education in 1877. Varela insisted on free schooling under secular control, and today education is virtually without tuition at all levels in Uruguay.

The twentieth century ushered in a commitment to universal education, whatever its actual accomplishments. Growth of trade and incipient industrialization permitted most countries to implement their educational program during the first half of the twentieth century.[5] As one instance, Mexico pursued the aims of the constitutions of 1857 and 1917, which pledged to bring literacy to the masses. The appointment of José Vasconcelos as minister of education in 1921 led to an expansion of primary schools and teacher training. Classical European education was infused with a revolutionary ardor for social action and indigenous influences in art, folklore, and literature. As one innovation the House of the People (La Casa del Pueblo) stressed the "three R's," yet the "curriculum" included music, art, sports, and fragments of the sciences of sanitation and agriculture. Vasconcelos recruited learner–teachers who themselves were barely able to read and write. The Cárdenas (1934–40) and Calles (1940–46) administrations continued this theme of educating the total population.

Other countries similarly increased their educational plant and to some extent the intellectual climate in the schools. Argentina made remarkable strides in education, largely because of its European background. Still, the nation had its own special problems. With the advent of Perón, education became a propagandistic tool of the state. Sarmiento and other democratic leaders were downplayed, whereas the military dictator Rosas became a national hero. After Perón's downfall in 1955 schools generally recovered their academic standards. Even so, the quota of students permitted to enter high school and especially the university, if larger than in most of Latin America, leaves at least a sizeable minority of young people frustrated. Like Perón in Argentina, Getúlio Vargas in Brazil used the schools to promote his nationalism, but expansion was illusory. Most reforms reached only the two major metropoli, Rio and São Paulo, despite Vargas's pledge to reduce illiteracy, which stood at 83 percent in rural areas. Among the false promises, lists of films were sent to rural schools in Goiás, in which there was no electricity.[6] But all was not lost, reforms did occur. The break with the past promoted by Vargas indirectly found its way into the classroom—formality was giving way in the 1930s to stress on motivation as well as innovations.[7] The most dramatic increase in funding for Brazilian education followed redemocratization after 1984 when governmental spending stressed education especially primary schools but an increase at all levels.[8]

Educational progress in post–World War II Latin America corresponds to what occurred in Western nations more than a generation earlier. Indeed, most countries in the world underwent a mass educational expansion after 1950. When we compare the Latin American profile with other developing areas, Latin America stands quite well. In 1960 approximately 42 percent of its children aged six–eleven were without any schooling, whereas in 1980 only 18 percentage were so handicapped. Comparable percentages for Africa are 70 and 40, for Arab countries 60 and 31.[9]

The School System

Notwithstanding a number of advances, Latin American schools still fall short of the dreams of its nineteenth-century spokesmen. Despite considerable improvement in the twentieth century; illiteracy rates vary from 71 percent in Haiti to less than 4 percent in Cuba.[10] In rural areas of several countries half the adults have never attended school. Valiant attempts notwithstanding, reduction of illiteracy and provision of vocational training for adults are unable to realize the national constitution's objectives. In assessing a school's function it must be remembered that socialization, especially in developing countries, is carried out primarily by the home. Not only rudimentary skills but also cognitive abilities are acquired even in lower socioeconomic households.[11] This aspect is all the more important in rural areas, where schooling is marginal.

Children and adolescents in the school setting operate in a structured climate by which they acquire a "tolerance for education" as based on a cost/rewards estimate, or "punishment–reward ratio."[12] Students derive a motivational system through interpersonal stimulation, largely parents and peers, and are conditioned by what they perceive as realizable goals. The school environment also shapes, however marginally, the students' perception of the anticipated payoff from their learning. What passes for educational evaluation is often psychological punishment as their performance is compared with peers. The academic institution developed more slowly in Latin America than in Europe or Anglo-America, and students have fewer, if improving, options for self-development and curricular choice.

Structural Problems

The challenges confronting education in Latin America are to a large extent economic. Also, with some exceptions schools suffer from traditionalism and a number of specific rigidities.

One, the financial underpinning is inadequate. Over a generation ago an inter-American conference set the goal of 4 percent of the gross national product to be spent on education (the United States and Western Europe spend roughly 5–6 percent). In 1979 a regional meeting of ministers of education agreed to an accelerated program for the advance of education with three principal objectives: elimination of illiteracy, expansion of instruction from eight to ten years, and improved quality in basic education. These reforms were to be financed by raising the educational outlay to at least 7 percent of the GDP. In the 1990s few nations moved significantly. Only two (Bolivia and Brazil) reached the 5 percent level and three (Dominican Republic, Guatemala, and Haiti) were under 2.5 percent.[13]

Because of the economic crisis most countries reduced their outlay for education in the early 1990s.[14] In all but four countries (Paraguay, Colombia, Costa Rica, and Paraguay), minimum salaries for teachers decreased in purchasing power between 1980 and 1989 by 25 percent.[15] Rural areas, of course, fare less well than urban, particularly the national capital. Despite economic stagnation the percentage of children attending elementary school rose from 83 in 1960 to 96 in 1991, but again the urban middle-class sectors had the advantage.[16] The percentage of students attending at all three educational levels vary markedly between countries, as seen in table 9.1. Further, allocations differ among nations in respect to the proportion given to primary, secondary, and higher education, and to

Table 9.1   Students enrolled in levels of primary and secondary education as reported for years 1991–98 (in %)

Country	First level	Second level	Third level
Argentina	99	62	40
Bolivia	77	34	12
Brazil	84	45	14
Chile	86	40	27
Colombia	89	53	17
Costa Rica	94	52	25
Cuba	99	62	30
Dominican Rep.	63	26	19
Ecuador	97	27	19
El Salvador	80	25	8
Guatemala	59	35	8
Haiti	26	21	2
Honduras	89	22	10
Mexico	99	33	17
Nicaragua	83	28	14
Panama	92	53	30
Paraguay	91	39	11
Peru	91	53	31
Uruguay	96	57	36
Venezuela	91	24	19

Source: Adapted from Latin American Center, *Statistical Abstract of Latin America*, vol. 38 (Los Angeles: University of California, 2002), pp. 259–269.

the ratio of the budget absorbed by administrative costs. In addition to national differences are class divisions. As their children attend private schools, the upper and upper–middle classes seldom concern themselves with public schools. Parents prefer private because of better facilities, smaller classes, and the students' social status. In order to extend choice Chile experimented with a voucher system—the results are as yet undetermined.[17] With the declining birthrate, Latin American budgetary problems may become less severe.

Two, largely as a result of inadequate financing, facilities remain continually on a low level. Textbooks and other instructional aids are also deficient. Libraries and laboratories are nonexistent in many schools. As school construction lags, most students are housed in dilapidated or unfinished structures. In this respect, Cuba is the most depressed, partly because of the U.S. embargo. On the other hand, Cuba has brought literacy to a population ignored by previous regimes.

Three, educational quality suffers from a lack of teacher training institutions. Few teachers in rural schools and possibly half in many urban schools conform to minimal requirements. As for the teacher shortage, major factors are low salaries, lack of tenure, and inability of secondary and normal schools to keep pace with the need, not to mention maintaining standards. In a Brazilian survey, teachers' salaries were less than 60 percent of the minimal wage. The limited salaries are rather striking in view of the duties expected of a teacher: keeping detailed records, performing custodial tasks, and acting as counselor to forty or fifty pupils. It is no wonder that teacher absenteeism is a problem. In addition, teachers' unions, often dominated by corrupt leadership, seldom succeed in furthering the interests of teachers or pupils.

Four, administrative appointments are usually made ad hoc and with cronyistic overtones. Not only the minister of education but his entire staff can change with any political gyration. As most administrators are political appointments few are trained as professional educators. With the politicization of education—a problem in many countries of the world—schools seldom promote ideological or social diversity.

Five, the curriculum suffers from a middle-class bias in its predilection for the traditional academic program. Further, whatever the formal requirements, they are not too predictive of what is actually taught; rather the focus depends on the administrator's preferences and the teacher's intellectual skills. Professional teacher training is of a hit-or-miss variety, with neither practice teaching nor cross-fertilization from other educational environments. Procedures prescribed by the state are seldom questioned. One may ask: what is the significance of the school for pupils who are exposed to few books, magazines, or newspapers during their childhood—or adulthood? Nor is the curriculum intended for an urban student body meaningful for the child of a *campesino*. Over a third of the rural population will someday live in the city; yet, little that they acquire in school is relevant, even if retained—a problem hardly unique to Latin America.

Table 9.2   Educational norms on age and duration of schooling in years

Country	Average age range	"Compulsory" education	Elementary	Secondary
Argentina	5–14	10	7	3 + 2
Bolivia	8–14	8	8	4
Brazil	7–14	8	8	3
Chile	6–13	8	8	2 + 2
Colombia	5	8	6	4 + 2
Costa Rica	6–18	10	6	3 + 2
Cuba	6–16	9	6	2 + 3
Dom. Rep.	5–14	10	6	4
Ecuador	5–14	9	9	3
El Salvador	7–15	9	6	2 + 3
Guatemala	7–14	6	6	3 + 3
Haiti	6–15	6	6	3 + 2
Honduras	7–12	8	6	3 + 2
Mexico	6–14	6	6	3 + 3
Nicaragua	6–12	6	6	3 + 2
Panama	6–15	6	6	3 + 3
Paraguay	6–12	6	6	3 + 2
Peru	6–12	6	6	2 + 3
Uruguay	6–14	6	6	3 + 3
Venezuela	6–15	10	9	2
USA	6–16	10	6	3 + 3

Source: Adapted from Latin American Center, *Statistical Abstract of Latin America* (Los Angeles: University of California, 2002), 271.

Six, pupil absenteeism and a large dropout rate signal the school's inability to meet either individual or social needs. The average duration of primary school education in Latin America is 4.2 years. This average falls short of the ideal norm (table 9.2). This problem is related to a variety of causes: distance from the school, indifference of peasants and workers to symbolic behavior, language conflicts, lack of incentives in the home, need for remunerative work, and resistance of the *patrón* (boss) to mass education. The school appears as an alien world thrust upon peasant and landowner by a remote government. Because class differences are accentuated by dress and other means, the bulk of children feel ostracized and many parents are reluctant to encourage attendance. Rural schools have their special problems. For instance, in Peru attention must be given to bilingualism, Aymara and Quechua are incorporated into the curriculum.[18] Yet, whether rural or urban, questions of migration, working mothers, health, and nutrition are constant problems.[19]

Additional problems complicate the process. With gender, cross-pressures occur. The family is more concerned with the boys' education, which he will presumably find useful in his future employment. Consequently, the dropout rate is higher for girls. Until the 1980s continuation into secondary school, for the minority who complete the elementary program, was predominantly a male prerogative. However, in rural areas, girls more likely stay in school, as boys are confined to work in the fields. Reducing sexism in the school has implications for democratization of the entire society.[20]

The failure to promote is both cause and effect in the lack of attendance. Repetition of the first year is reported to vary among countries between 6 and 23 percent, of the second year were only slightly less.[21] This situation results from negative factors surrounding the school setting—malnutrition, inadequate instruction, parents unsympathetic with educational values, and teachers drifting in and out of their careers because of political favoritism, irregular pay, or personal whim. Nearly half the teachers in rural areas work only 75 percent of the school year. Interpersonal tensions in the staff and administration are a factor in this high rate of absenteeism.

Finally, the most difficult aspect of the school situation—other than the lack of financial means—is the method of instruction. From the first grade to the university, many students are taught according to a traditional rote system, a heritage of Lancasterism. Materials are memorized with little desire on the part of the teacher or student to use a logical approach to the subject matter. Whether quantitative or verbal skills, the child and adolescent learns by rote memory.[22] Motivational psychology and attention to individual personality needs may be lost once the teacher candidate leaves the normal school or a national university. Even there, "theories" are often learned verbatim rather than as a guiding philosophy for the teacher–student relationship. Still, exceptions are found. For example, over the last generation in certain areas of Colombia the Escuela Nueva, as fostered by UNESCO, offers a lower student–teacher ratio, better prepared and paid teachers, improved physical facilities, and—for financial reasons—in a multigrade classroom. Among the outcomes is a lower dropout rate, even though the instructional cost is only 5–10 percent higher than in traditional schools.[23] In this connection, monitoring of teaching methods, curricular options, and evaluation of student achievement. Chile may well lead in its examination of curricular effectiveness and cross-national comparisons of the educational process.[24]

Educational processes are compounded by several barriers. One is centralization of decision making. Venezuela represents a classic case of rigidity in curricular decisions and personnel deployment, whereas in neighboring Colombia a tradition of regionalism allows more local autonomy. Argentina decentralized educational administration in 1978; the provinces are increasingly responsible for curriculum and personnel. Yet, within the provinces, centralization in Argentina generally holds sway, rural schools still lag urban ones. Attempts to decentralize meet difficulties because of conflicting interests in teacher unions as well of a host of national, regional, and local officials, as found in Mexico. In several countries political elites turn to the usual clientele networks for their choice of superintendents and directors. A related problem is the confusion engendered by ideological and interpersonal conflicts as political parties succeed each other with little consensus about priorities. A Mexican critic points to the lack of making firm decisions about specific projects. Even when priorities are established they are not implemented.[25]

Criticism of educational systems runs throughout the Americas. A fundamental charge is curricular irrelevancy and lack of serious evaluation, compounded by the failure to find a balance between centralization and decentralization.[26] Complaints also focus on the neglect of various areas from science to sex education—not unrelated to the persistence of religious influences in the public school. Another appraisal points to the isolation of education from other social agencies, that is, competing governmental departments only marginally articulated with the educational process. Moreover, parents should be more involved in the school.[27] Fundamentally, the problem is one of the government establishing priorities among its various demands—military, welfare, crime, drugs, AIDS. Even when the government invests funds it often ignores crucial services—among others, improving teaching personnel.[28]

Elementary Education—Urban and Rural

For at least half of Latin Americans, education occurs only at the primary level. With a large young population (average age of twenty-five for Latin America as compared to thirty-four for the United States), the idea of free and compulsory education remains little more than a dream outside urban areas. In theory, if not in reality, the school system begins with a prepri-mary program. Once again, social stratification determines what the child will receive. Wealthier urban areas offer an extended nursery school or kindergarten for two years. Yet, in the 1990s 14 percent of elementary school-age children and 84 percent of secondary school-age were denied an education.

About age seven the child enters elementary school, usually a six-year program. In rural areas the school can be a three- or four-year affair, with many students completing only the second grade. Ironically, in nearly all Latin America education is compulsory by law to age fourteen or fifteen. Reality continues to fall short of the dream. Even in Chile, which represents a more advanced educational system than does most of Latin America, no more than 70 percent complete the first cycle of four years, only 50 percent finish the full cycle of eight years. Again, parental absence, poverty, health problems, inadequate facilities, curricular dullness, and teaching methods take their toll.[29]

Parents in the most economically distressed areas are often ambivalent about school attendance. They may send their children to school if they per-ceive the income potential sufficiently enhanced to make up for the loss of immediate earning power. On that point, the Brazilian federal government over the twentieth century adopted thirty-three measures to prevent child labor.[30] A marked gradient appears in the percentage of children attending school in more prosperous areas, for instance, the states of northwestern Mexico as opposed to the marginal economic areas of the south and the north central highlands. School attendance is correlated with a number of positive and negative variables ranging from migratory patterns to the tendency to go barefoot.

A wide range of educational norms is found in the hemisphere. At one extreme, Uruguay reportedly has nearly 90 percent of children ages six through eight years in primary schools. Yet, with 102 children on the average for each schoolroom, double and triple sessions are the norm. The majority of urban children stay in school through the sixth grade, but retention in the five-year rural school is less dependable. A survey of Uruguay's schools, both urban and rural, ended with some fifty specific suggestions for improvement in dealing with class size, course failures, individual differences, material resources, and other inadequacies.[31] At the other extreme is Haiti, where families reckon marginal gains their children will make in life by having a primary education (only the most affluent can consider secondary education). Public and private lotteries become the only hope of winning the chance for their children to escape from "nearly intolerable conditions," even though the cost of a lottery ticket can mean food deprivation.[32]

We return to the lack of services of rural schools, from school buses to nutrition—many children arrive at school without breakfast. Also, ethnic and language problems affect rural schooling. For example, as the Mexican government launched industrialization projects in the modernization push of the 1950s, rural areas were no longer a priority. Consequently, formal education suffered from lack of financial support. (The economic disaster of the 1980s had similar effects.) According to a study of Huejutla in the state of Hidalgo, Mexico, children of wealthy ranchers are sent to urban schools, whereas parents in certain villages have to pay for private tuition to assure their children's education. Cultural differences further complicate the situation. As few *rancheros* understand the Nahautl language and culture they increasingly move in urban circles, where they are detached from the indigenous, who in turn became even more isolated. Debates arise as to the desirability of bilingual education: will children appreciate the national goals and values if they become too absorbed in the indigenous heritage? In 1984 several bilingual teachers organized a protest against the local school inspector. On this point, national and local policies are not always in harmony. The question of pluriculturism is complicated by the division between a federal system and a parallel system of indigenous education.[33]

Several problems and possible solutions are highlighted in a study of Mexican primary schools. Even though the thirty-one state governments are responsible for education, the federal government initiated research because of its concern with high repetition rates during the first two years. Findings reveal the importance of several factors: social setting, presence of a public library, family's socioeconomic status, and parents' level of education. Other factors include the negative effects of single-parent households, the need for preschool education, and innovations in instruction.[34]

Secondary Education

Latin America's secondary schools (*educación media*) have more affinity to Europe than to the United States, although fewer young people attend. Basically, schools are for five, six, or occasionally seven years beyond the primary, often divided into two "cycles," (1) *básico* and (2) *superior*, which on completion leads to the *bachillerato* (secondary school diploma). Each country has its own variation. The first three or four years of secondary are a continuation of general education, and the last two-year program is the *preparatoria* or university-oriented; vocational and technical programs exist as well. The second cycle is oriented toward the career chosen, whether university preparatory, teaching, commercial, industrial, or agricultural. In Brazil, most students—that minority who enter the second cycle—are usually enrolled in the academic track. The *normal* (teacher preparatory) and commercial tracks vie for second place, with less in the business and fewest students in agriculture (despite its importance in the national economy).[35] Some secondary schools are vocational for both the lower and upper cycle and the student may complete one or both cycles. Employability in these programs is not as successful as are for graduates of nationally financed occupational training entities, such as SENAI in Brazil or SENA in Colombia.

With the awareness of a narrowly middle- and upper-class bias in secondary education, the curriculum has become more diverse. In Brazil, after four years of primary school, students may enter the *ginásio*, which approximates the North American middle school, for four years, followed by three years of *colégio*, where students have the choice of two types, the *clássico*, emphasizing humanities (including classical languages), or the *científico* (scientific) concentrating on sciences, mathematics, and modern languages. Or students may attend a commercial or an industrial school. In all countries, students are to decide by the early teens whether they wish, or can afford, to attend secondary school. After another three years they face another crossroad, whether or not to prepare for university, for which completion of the *bachillerato* is necessary.

Since the national government does not consider secondary as essential as primary education (nor, ironically, as important as higher education), responsibility for developing this level falls disproportionately on state and local governments and on private resources. Although *educación media* is free in several countries, the middle class is favored because of admission standards, type of curriculum, and the family's economic situation. Secondary schools become one more mechanism insuring social inequalities in university access. A survey of ten countries demonstrates the superiority of private over public schools, both primary and secondary, but some of this attainment is due to selectivity.[36]

In certain regions only the city offers secondary education, whether public or private. In Brazil's impoverished northeast a few *ginásios* or *colégios* may serve an entire state. However, for students applying for entrance at the secondary level the greatest frustration occurs in the city, where multiple applicants appear for every vacancy. Not surprising, the number who qualify in rural area is minimal. Small towns usually offer no facilities; those few who complete *primario* apply for a scholarship in order to enter the *colégio* in a nearby town. In addition to public schools are a large number of private *colégios* or *institutos*. Generally they include both primary and secondary programs, and are often church-related, largely Catholic but occasionally Protestant. At present, private *colégios* assume the major role for secondary education in several capitals. Elite institutions may have a tuition higher than the per capita income of the respective nation. In addition, in almost every capital, British, French, Germans, North Americans, or some other national group, establish schools catering to the upper–middle class as well as to relevant foreign enclaves. Historically, importation of educational systems from abroad can be troublesome to national governments (inset 9a).

Inset 9A Nationalism and the School: A Nicaraguan Episode

Education must meet the cultural, not least the political, needs of a nation. An interesting case is Nicaragua at the end of the nineteenth century. The basic problem was a minority population on the Caribbean Coast, complicated by foreign powers. In the colonial period Britain had occupied certain areas of the coast, with the Creoles, largely Afro-Caribbeans, moving into Bluefields and other centers. Especially troubling was the United States' Manifest Destiny, as dramatized by the William Walker expedition, economic domination, and actual occupation intermittingly during the first third of the twentieth century.

Principal occupants of the coast were various indigenous groups, particularly the Miskitos, whose kingdom was recognized by the British as early as 1687. The Miskitos and *costeños* in general were largely ignored after independence in 1821. But all this was to change. After General José Santos Zelaya came to power in 1893, war with Honduras ensued, Zelaya demanding integration of the eastern territory into the national orbit. Not the least troubling was the English influence, notably the Anglican Church. Even more threatening was the dedicated Moravian missionaries from the United States, who established their mission in 1849. By the 1890s they had established several schools, teaching children the Bible in the Miskito language, a direct challenge to Catholicism. English was also incorporated into the curriculum. Consequently the crisis was one of allegiance—citizenship, religion, and language.

For Zelaya, a self-assured dictator, the situation called for aggressive action. Not only outraged by the spread of the Moravian influence but also by the Anglican Church providing instruction in English, he ordered the minister of education to move rapidly on the construction of schools and the training of teachers over the next decade. By 1905 thirty-two schools were established accommodating nearly eighteen hundred students (25 percent mestizo or "Spanish," 38 percent Creoles, and 37 percent Miskitos.) This scholastic explosion was historically the largest educational concentration outside the capital Managua. On that point, not until the 1979 *sandinista* regime did the proportion of elementary school-age children actually enrolled in classrooms reach above 35 percent.[37]

The type of school, public or private, determines the curriculum, which tends to be highly academic with as many as ten or more subjects being "covered" each year. Nonetheless, because of state supervision both courses and examinations are rigidly structured, and failure to promote is more frequent than in Europe or the United States. Particularly in Brazil a *cursinho* (private lessons) industry flourishes in order to prepare students for entry into the university. All in all, problems haunting the primary school are found (fortunately diminishing somewhat over the last generation) at the secondary level: absenteeism, dropout, inadequate facilities, low professional commitment, and rote memorization. Whatever the pitfalls, secondary education is on the rise. Public *ginásios* and *colégios* have increased three times as fast as population growth in Brazil since the 1960s, if with little change in the curriculum. This increase occurs more in private schools because of a growing urban middle class, yet newer types of vocational schools constantly appear. For instance, private schools are on the increase in Paraguay, where recent decentralization permits more experimentation in curricula in both public and private sphere.[38]

Another question is the diversity of secondary schools. In a sense, programs are an index of the nation's socioeconomic development. At the bottom of the scale are countries with a highly classical type of *educación media* for the small number of middle- and upper-class persons, for instance, Guatemala, Haiti, and Honduras. At the other extreme is Argentina, where almost no status differential appears between programs since the vocational school permits access to the university. According to this paradigm, countries where less than 20 percent of the age cohort attend secondary school favor a traditional education. On the other hand, if the percentage increases to over 35, as in Argentina, secondary education can become an end in itself with a broadening of programs. These relationships between a practical and an academic program may be measured in the relative importance a given subject occupies in the country's total curriculum. The percentage of time committed to mathematics and the sciences in given secondary schools varies from 43 in Uruguay to 19 in Brazil. This difference is significant in shaping university access, vocational skills, and their role in national development. Beyond these questions are those of role conflicts, effectiveness, and morale of the educators (inset 9b).

INSET 9B   AN IMPOSSIBLE ROLE

In all societies, it seems, teachers have relatively low income and uncertain status. Role conflict and strain are universal for this career. In Latin America underdevelopment of teachers' organizations and the comparative absence of tenure are understandable in view of the small number of certified teachers. In reality, not many institutions can afford fully accredited teachers.

As agents of socioeconomic development, neither teachers nor students are truly effective. Teachers who have insight and commitment to change realize that they are reaching too few qualified students. Because of academic traditionalism, secondary school students are inadequately trained for a global, society. Even so, the small minority of fourteen-year-olds who have the rare privilege of studying two foreign languages and Euclidean geometry along with seven or eight other academic subjects are more fortunate than are most of their cohorts, who rarely receive any kind of high school education. Planners are wrestling with these priorities. For example, Mexico is monitoring its secondary system and urges more emphasis on social and psychological skills including citizenship training for the purpose of building a more effective democracy.[39] Educational leaders are asking about the reality of implanting attitudes of integrity and good citizenship when political and economic elites are so visibly entrenched in corruption.[40]

Lack of self-assessment by many school systems perpetuates the status quo. This reluctance to innovate is not confined to educational circles. For all their revolutionary façade, Latin Americans—like others—can approach change with considerable rigidity. Also, research plays a very limited role in policy planning. As one instance, in a survey conducted by the Mexican Secretariat of Public Education, only a fifth of the regional directors review research in the process of arriving at a policy decision. Even for societies undergoing change the problem can be staggering. In the 1980s Nicaragua curriculum and teaching methods could not keep pace with the educational aims.[41] Exceptions occur, as when a charismatic leader appears on the scene—Paulo Freire directed educational policy in the city of São Paulo from 1989 to 1992. Among the achievements during this period was a 10 percent rise in the retention rate of primary school pupils. A critical factor in this reform was higher pay for teachers.[42]

In most countries, teachers' morale is low. Since they represent a restless and disgruntled portion of the population, teachers are not incapable of being oriented to change, but they are seldom able to translate this desire for change into action. In recent decades teachers have shown a militant attitude toward their economic situation. In 1961 Lima teachers walked out of the classroom, closed ranks with labor, and a general strike followed. In Uruguay teachers were repeatedly on strike during the 1960s and 1970s. In most Latin American countries teachers' strikes continue into the present. Over the world, teachers acknowledge that their middle-class position is frequently illusory. However, not all unrest is due to economic pressures. In 1998 Colombian teachers threatened a strike because of the risk of violence to them and their pupils.

National Goals and Nonformal Education

Educational attainment in Latin America must be viewed in the reality that many young people are denied an education. In addition, traditional values still have priority over modernization. This lag is finally leading to a restructuring of secondary education. One question is often raised: where should the money be invested—primary or secondary education? Further, if the need for technical personnel is to be solved, it will be by expanding the number of high schools and colleges. Fortunately, a falling birthrate partially helps solve the school crisis.

What role does education play in modernization as well as in the fulfillment of individual goals? Debate continues as to whether the stress should focus on the nation or the individual. One Brazilian educator asks whether individualism has any meaning in a materialistic world, whether Marxist or capitalist.[43] Brazilian educators are becoming aware of the school's need to consider questions of community belongingness and ethnic identity. Because of their rough similarity in size and heterogeneity, both Brazil and the United States can learn from each other in respect to identity development and the meaning of a multiracial society.[44]

Nor do all countries give the school equal priority. One index of this commitment is the proportion of the national budget going to education. Reportedly Mexico and Costa Rica spend 26 percent of the governmental budget on education, Argentina 12.6 percent. As another index, Cuba invests 6.2 percent and the United States 5.4 percent of the GNP on education.[45] In contrast, Brazil allocates less than half of what it spends for national defense.

A nation's educational commitment has significant consequences. Achievement based on education remains the entrée to middle and upper–middle class (which in the end merges with an upper class largely based on ascription). Moreover, in less advanced countries, probably at most a third of those who attain a secondary or university education during the present decade can realistically expect to find technical or professional work. In this connection, the lack of technical resources is critical.[46]

Society, Nonformal Education, and Literacy

Education can be *informal* as with socialization in the home or the range of personal experiences by which we are socialized and resocialized. *Formal* education refers to schooling from primary school to the university. *Nonformal* education includes a broad spectrum of instructional programs generally aimed at adults. Any one of these three types of learning experiences represents aspects of the other two: informal may encompass the formal as in so-called bush schools in rural areas, nonformal education involves certificates, and in formal schools the peer group can be important.[47] The "intergenerational" passing on of copybooks is only one instance of how informal formal education can be.

Nonformal or "popular" education is recognized as a critical link to social change. Statistics on illiteracy have far-reaching implications for democratic stability and economic development. The range of literacy both between and within nations is striking. However, literacy is differentially defined as the ability: (1) to handle a few written words and to write one's own name, (2) to understand and to answer a simple letter, and (3) to function as an vital participant in an urbanized society. The difference between official and functional literacy for selected areas of Brazil is found in table 9.3. Since most statistics about literacy in Latin America focus on the first two of these meanings, implications for a democratic social order are often ignored.

Table 9.3 Percentage of illiteracy for persons of 15 years or older for given states of Brazil

	Illiteracy	Functional illiteracy
Federal district	5.1	14.9
São Paulo	6.2	19.3
Inas Geraís	12.2	29.4
Bahía	24.7	48.3
Pauí	31.6	53.1

Source: *Anuário Estatístico do Brasil* (Rio de Janeiro: IBGE, 2000), 140.

The question is, again, whether the focus should be on the elite by expanding secondary and higher education, or on teacher training and primary schools, or whether the focus should be on the over eighty million Latin Americans fifteen years or older who cannot read or write. Authorities are not in agreement as to their priorities. As over half the nations have reached an 85 percent "literacy level," emphasis could shift toward adult vocational training, notably improved agricultural methods. In that context, in all but five countries (Argentina, Chile, Costa Rica, Cuba, and Uruguay) illiteracy remains a problem. The earliest serious attack on the problem occurred in Mexico, where José Vasconcelos initiated a literacy campaign as early as 1922. In 1936 President Cárdenas accelerated the process by appointing a staff to organize a movement of *educación popular* (education of the people) by which 225,000 adults acquired some level of literacy. An intriguing impetus occurred under the Avila Camacho administration: in 1944 each literate person between the age of eighteen and sixty was to teach at least one illiterate for a period of one year. As classrooms could accomplish the same objective with less teachers, the system was phased out.

Research studies find a positive correlation of literacy with a number of indices, especially economic growth and life expectancy. In a classic study of five Colombian villages the correlates of literacy included mass media exposure, innovativeness, achievement motivation, travel to urban centers, political knowledge, and empathy (the ability to suggest actions the respondent would take in given secular or urbanistic roles).[48] It appears that literacy classes in a village attract *campesinos* who are achievement-oriented, have a relatively higher standard of living, and practice their literacy training by reading and letter writing. Also, the movement focuses not only on increasing literacy but also on ideological formation. Latin American educators point to the importance of adult literacy for parents as a means of encouraging school performance of their children.[49] A Chilean report reveals the importance of informal education to women in their attempts to compensate for the grossly unequal role they meet in the employment market. In this empowerment process they acquire work and social skills, coping techniques, and a higher self-image.[50]

What role can nonformal education perform in political socialization? The relation between urbanization, governmental stability, and mass participation tends to be positive, even though conditioned by the cultural climate within a given country. Several advances are in Central America, where varied delivery systems of adult education emerged. In El Salvador and Guatemala, programs, often under military auspices, are more than occasionally funded by the government with support of U.S. Agency for International Development (AID) programs. Notably in Guatemala the tractability of agricultural workers is assured, as indigenous education is regarded somewhat ambivalently. Teaching Indians agricultural techniques under the Educación Extra-Escolar (extracurricular schooling) has increased crop yields, but has in no way aided the *campesinos'* integration into the national social order. Even these programs virtually disappeared under the military excursions beginning in 1979.[51] After twelve years of civil war El Salvador presents a similarly mixed history of adult education. Because of the U.S. AID stress on modernization, programs were concentrated on technical and professional training instead of the bottlenecks in the implementation of general education. Yet, between 1977 and 1987 no less than 1,246 literacy centers were established, largely through grassroots organizations. Despite its position as the least advanced country in Central America, Honduras through various organizations—Radio Suyapa, National Peasants League (UNC), and Council for Coordination for Development (CONCORDE) with its Training Centers (*Centros de Capitación*)—provides both verbal and vocational skills to rural areas. It is significant that in the mid-1980s Honduras and Mexico devoted over a fourth of their public radio programs to education, almost unique for both advanced and developing nations. Costa Rica has for over a half-century taken the high road. The Party of National Liberation in 1948 placed a premium on expanding education. Still, as the rural population was largely neglected until the 1970s, the Open University of Costa Rica initiated twenty-six evening schools. As in all Central America, both Catholic and Protestant groups play a role in this educational campaign.

Educación Popular

Latin America's most significant innovation in education are grassroots movements. A unifying principle in Latin America, but especially in Brazil, is, again, the work of Paulo Freire.[52] Conscious of the socioeconomic, regional, and ethnic asymmetries in the nation, he was determined to break through the bureaucratic barriers.[53] Also inspired by Liberation Theology, this movement became aware of the importance of *conscientización*. Along with education in general, popular education reflects three viewpoints: (1) *structural functionalism*—education as the path to national development, (2) *radical functionalism*—placing emphasis on the underclass or underprivileged minorities with an aggressive attack against traditional curricula, and (3) *interpretive functionalism*—encouraging individual expression.[54] The models are not mutually exclusive, they simply suggest different goals and methods. Approaching adult education in a different context, a given political regime shapes the philosophy underlying the program: one is the view of military governments as "social engineering" or "corporatism," the state being the directive agent, as in Peru under Velasco Alvarado (1968–75). In contrast is the "modernization-human capital" outlook, with the expectation of bringing an achievement-universalistic orientation through acquisition of knowledge, skills, and incentives.[55]

Among the goals of *educación popular* (the people's education), consciousness-raising becomes a means of changing one's perception of reality and acquiring skills, but its fundamental aim is transforming society. The movement is designed to bring various skills to the poor and primarily to the rural sector. In Chile the momentum is from the Ministry of Education and even more so from nongovernmental agencies (NGOs), including the Roman Catholic Church and various grassroots organizations. Essentially these agencies work with teachers, parents, and children, along with civic groups committed to assessing values, goals, and self-growth. However, *educación popular* in the Chilean scene goes well beyond the schoolroom, three features being salient: (1) alleviating deficiencies of poorer school districts, (2) strategies for "survival and satisfaction of basic needs" among people living in poverty, and (3) educating a wider public, especially the dominant sectors, toward a just social ideology.[56]

The most conspicuous examples of innovative education appear in societies undergoing vast social change, that is, socialist states, Castro's Cuba, Allende's Chile, and *sandinista* Nicaragua. We may describe the most recent one: under the Somoza regime, few beyond the privileged received a truly formal education. As early as 1969 the *sandinistas* proposed "a massive campaign to wipe out illiteracy immediately." On assuming power a decade later they launched what became known as the Literacy Crusade. The Sandinista Youth Organization was in large part responsible for preparing sixty thousand secondary students to become *brigadistas* (brigadiers) in the People's Literacy Army (EPA). Stress was placed on training, including role playing, debate, simulation, the arts, and pilot projects before the *brigadistas* took to the field. Their work was not confined to literacy; distributing antimalaria pills, conducing agricultural surveys, and concern for popular culture were also on the agenda.[57] In other words, *brigadistas* themselves were on a learning mission for at least six months sharing the life of rural people. According to reports, within nine months of the beginning of the campaign in 1980 the illiteracy rate for people over ten years of age declined from 50 to 15 percent. The literacy campaign became a "model of social change with decision making shared at the grassroots level."[58]

Nonformal learning in these transitional societies takes different directions. In several countries, literacy is regarded as the reading proficiency of a second-grade student, which falls short of functional literacy. Even a moderate degree of literacy can give the individual a sense of identity and accomplishment. On that point, findings from a survey of the indigenous population in Guatemala and Bolivia stress the payoff from education. Although earnings did not equal those of *ladinos* (Indians who have become somewhat Europeanized) the return was better than for the noneducated. In Bolivia, with somewhat less ethnic discrimination, education brought perceptually better earnings.[59] A different perspective took root in Peru, where both indigenous and *cholo* find that once they leave the classroom, promises of upward mobility are not realized. Also, economic stagnation, poverty, and massive urbanization promote the idea of restoring the greatness of their ancient civilization, replacing *criollo* heroes with indigenous ones such as Tupac Amaru, the eighteenth-century revolutionary martyr. In part, inspiration comes from neo-Marxism, Maoism, and the Sendero, but revitalization of the Quechua past is the main theme.[60] In respect to socioeconomic change, revolutionary regimes are less concerned than capitalist ones in preparing the population for participation in an individualist, competitive economy. Socialist governments are generally inspired to create community values and goals. The end result differs greatly. For example, Cuba remains a totalitarian state. Nicaragua in 1990 made the transition from socialism to a quasi-pluralistic, multiclass, multiparty national community.

Nonformal education, Thomas La Belle concludes, falls into three strategies:

1. improving life chances of the underclass, ranging in a broad spectrum from agricultural instruction to psychological techniques of raising self-worth.
2. "human capital" or developmental approaches as part of a modernization process, usually promoted by NGOs, U.S. AID, and similar programs.
3. "*educación popular*," whether consciousness-raising techniques of the *Educación de Base*, largely the educational efforts of revolutionary movements in Cuba and Nicaragua.

The potential of nonformal education is more limited in laissez-faire, democratic societies than in the revolutionary ones. Fundamentally, popular or basic education is committed to three purposes: to make decision making more participatory, to "forge linkages" across and within social classes, and to reduce the concentration of power in the oligarchy. The task is not easy and the costs are high.[61]

Along with the rest of the West, Latin America is caught in a "peoples education" based on commercially driven mass media, notably movies and television, along with the cybernetic. Electronic media, usually modeled after North American image-making, is dysfunctional in regard to the goals and values of many Latin American citizens. The "culture industry" discourages intellectual and artistic production of spontaneous and folk origins. Happily, a popular theater—amateur and professional, public and private, traditional and revolutionary—remains active in most countries.[62]

The Challenge of Higher Education

Postsecondary education is responsible for the training of a professional and semiprofessional corps in any society. Besides being a major avenue of upward mobility for the individual, higher education can be a channel of innovation. Also, in most developing nations university students play a special role as change agents.

Institutions of higher learning in Latin America number over three hundred, although well less than a hundred of these can be considered to be of true university rank. Some six million students attended postsecondary institutions in 2000, twice the number of two decades earlier, but barely half of the U.S. ratio with its nearly 13 million students in a population much smaller than Latin America. Brazil leads with 1,470,000 students, followed by Mexico with 1,252,000, and Argentina with 755,000.[63] What is remarkable is the growth in the number of students from 1960 to 1997—for example, eightfold for Honduras, twelvefold for Mexico, and fifteenfold for Brazil. As compared to several areas of the developed world, the ratio of university students is high. For example, Argentina has proportionately more students than do several European nations. With a teaching staff is roughly 70,000, the student–teacher ratio is not unfavorable, ten to one in several nations, but this ratio is misleading in view of the part-time status of many professors. Academic salaries are marginal, but vary greatly both within and between countries. Tenure provides a degree of security but is less institutionalized than in Anglo-America or Europe.

The Latin American university has a lengthy past. Both the University of Mexico and the University of San Marcos in Lima were established in the 1550s. Bogotá, Morelia, Quito, and Cuzco all had universities before 1600. The program was largely theological and classical, although fragmentary instruction in the inductive sciences appeared in the late eighteenth century. During the nineteenth century universities languished in their restricted curricula of medicine, law, and theology. In several countries no viable university was established until well into the twentieth century. For instance, the University of Costa Rica, Central America's foremost institution, opened its doors in 1941. (The nation had been without a university since 1888.) For most nations the second quarter of the twentieth century marked the true beginning of a system of higher education. Brazil did not have an accredited university until 1920. Indeed, until recent decades secondary schools were the principal breeding ground in shaping future social change agents (inset 9c).

INSET 9C   EDUCATION AND SOCIAL CHANGE: FORMATION
OF A REVOLUTIONARY

In 1950 the fourteen-year-old Carlos Fonseca, later to be a major leader of the Nicaraguan *sandisinista* revolution, entered Matagalpa's only public secondary school, the Instituto Nacional del Norte (INN). The institution was largely male, a few students obtaining government scholarships, including a modest stipend enabling boarding students to survive, "he learned fast how to steal food." Private schools, catering to the middle and upper classes, provided a quite different *ambiente.*

Despite—or maybe because of—the repressive Somoza regime, *the school* had a reputation of student dissent. In 1946 the student publication *Vanguarda Juvenil* (young vanguard) surfaced briefly; several students spent a few days in jail for their subversive writing.

Fonseca soon found his way, first into a conservative organization and later an anti-Somoza underground group, finally choosing the communist party, Partido Sandinista Nacional (named after Augusto César Sandino, who in the 1930s opposed Somoza). As no Marxist literature was available in the school library, one of his teachers came to the rescue with various left-oriented writings. With two other young intellectuals Fonseca organized a Centro Cultural at the INN. The group published a modest Journal *Segovia,* Fonseca writing a dozen articles on literary and political issues. A major purpose of the journal was to bring learning to the rural population, of whom 80 percent were illiterate. By the late 1950s Fonseca and his coterie graduated. His idealism was to continue into his participation in the Cuban Revolution and founding of the Sandinista Front of National Liberation (FSNL).[64]

The university pattern has more in common with Europe than the United States, Yet, in many respects it is like neither. Traditionally the university is a combination of separate schools or colleges (*facultades*) occupying scattered buildings throughout the central city, with little of the North American campus atmosphere either physically or socially. However, during the past half century a number of universities adopted the idea of the *ciudad universitaria* (university city) built in a special terrain, often on the edge of the city. This movement to a kind of self-contained island emphasizes the university's autonomy. With this new unified setting the separation of *facultades* became somewhat less rigid.

The Basic Structure

A distinguishing mark of the Latin American university is its *autonomy,* a trend going back to the University of Salamanca, Spain's first institution of higher learning. Supported financially either by the federal or state government, the public university appears as a sovereign institution, in theory immune to government intervention, as concerns the rights of both students and professors. Consequently, university buildings are de jure, if not de facto, off-limits to law enforcement officers. Rioting and student strikes have few restrictions as long as they remain in university quarters. However, governments intervene in university affairs when they are convinced that this action is necessary for national security. For example, in Bogotá in the 1960s civilian police entered the National University in search of rioting students. Brazil never conceded a great deal of autonomy, but universities have made substantial gains, interrupted by the 1964–85 dictatorship. Whatever the excesses of students in exploiting university autonomy, the system gives higher education a remarkable degree of academic freedom. Otherwise the university might be one more government fiefdom. As stable democratic governments become institutionalized, the autonomy principle is less crucial.

A second aspect of higher education is the *co-government (co-gobierno)* principle by which students contribute to the university's policy making. Students, who are elected by their peers, sit with deans and professors on the academic council, and to a lesser extent on the superior council. The specific pattern of numerical representation varies with the university; in several institutions the students constitute as much as a third of the academic council.

A third principle, related to the principles of autonomy and co-government, is the reform movement (*la reforma universitaria*), which began in Argentina at the University of Córdoba in 1918. The fundamental issue was academic democratization, including access by all social classes to higher learning. Implicit in this demand was admission to the university of any secondary school graduate and the right to repeat a course as many times as students wished until they passed the examination. Only minor fees were to be charged for higher education. Among other aims were selection of professors by open competition, periodic review of the teaching staff, broader communication within the university, modernization of the curriculum, and university involvement in the national and local agenda. The movement spread to neighboring universities, La Plata being the first. In the following two decades most universities in Hispanic America adopted at least a few of the principles, notably the lowest possible tuition. Success of the *reforma* movement was a function of the strength of at least quasi-democratic institutions and specific historical developments. In a relatively permissive country such as Chile, students' rights began early in the century. Although the Students Federation of Chile (FECH) was founded in 1906, its status was further enhanced by the Córdoba developments.

The principles of reform, co-governance, and autonomy are important means of the students' self-development. Also, the students' right to contribute to policy making, as American educational leaders discovered in the 1960s, can be justified since educational institutions presumably exist for the students in their search for decision-making skills as well as abstract and technical knowledge.

Discontinuities in the University

To what extent do principles of autonomy and governance actually add to the university's success? Several problems prevent most universities from providing high-level instruction and contributing to socioeconomic development. In part, these defects are carryovers from the inadequacies of primary and secondary education. They are aggravated during periods of

Table 9.4 Percentages of expenditures on education for selected countries

Country	Primary	Secondary	Tertiary
Argentina	46	35	20
Bolivia	44	36	20
Brazil	54	20	28
Ecuador	59	36	21
Peru	49	31	18
USA	59	36	2

Source: Rebecca Marlow-Ferguson (ed.), *World Education Encyclopedia* (Detroit: Gale Group, 2002).

political unrest and rightist political regimes:

1. As throughout the educational system, the university suffers from a faulty financial structure, as seen in the budgets for libraries, laboratories, dormitories, and scholarships. Countries vary in the degree of financial commitment to universities as compared to what they allocate to the lower and middle levels of education (table 9.4). As with all budgetary commitments, the government establishes priorities, particularly when an economic crisis occurs. For instance, during the years of the oil bonanza Venezuela sent thousands abroad. With the collapse of oil revenues this privilege vanished and by 1990 students applying to the Central University in Caracas might wait two years to begin their studies. Moreover, professors in 1986 went on strike to collect their back pay.

2. Student admissions vary from overly lax to fairly stringent, depending on the university's capacity and the number and quality of secondary school graduates. In a competitive environment, say Santiago or São Paulo, candidates are turned away, whereas in less privileged areas the university tends to accept substandard students. Another problem is the high dropout rate. Largely because of financial pressures, only a fifth to half—depending on the institution and the country—of those who begin a university education will graduate.

3. As implied earlier, structural disunity in the university makes for an uneconomical usage of funds and facilities as well as the functioning of teachers and students. Most universities are composed of autonomous *facultades* held together only by the *rectoría* (president's office) and the council. Most universities have from five to ten *facultades* but some have fifteen or more, often with the difficulty of students' taking courses outside their own *facultad*. The problem is more acute in older universities, notably in the provinces. Newer and revitalized institutions provide for more integration among *facultades*.

4. The teaching body and the administrative apparatus are only partially committed to their tasks. While full-time professorships (*catedráticos*)

are found in larger universities, a greater number of professors are part-time (the present drift in the United States toward adjunct professors is a similar trend). Since they leave their law office or a ministry to teach a class in their specialty (hopefully), they are known as "taxi professors," more than occasionally arriving ten or fifteen minutes late. Despite the abuses, the system has the potential advantage of obtaining a specialist who has a continuing professional experience. With increased professionalization in the university since the 1980s, commitment has improved.

Except for superior institutions, student performance frequently follows in the mode of preuniversity education, along with the institution's reputation of sterile lectures and verbatim memorization. Until recently, up-to-date textbooks, libraries, and laboratories were deficient. In other words, sciences lack an empirical framework. As students enroll in as many as eight or nine courses, a superficial regurgitation of the subject matter follows. As a related problem, students are only marginally committed to the academic process. In a public university, particularly those offering an evening program, the student has fifteen–twenty hours of lectures in addition to preparing for examinations and often part-time or full-time employment. Consciousness of the need to improve standards is growing; however, rapid growth of higher education has not improved the quality of the product.

Political involvement of students was at one time a major distraction to their studies. In the 1960s and 1970s students were trapped in the polarization of Latin American society. The 1980s economic crunch, a harder line by governments, growth of private universities, and the struggle of students to find a career are all factors in this relative depoliticization of students. Vagaries of the political process as in Mexico influence this tug-of-war. Co-option can occur on both sides, as the Echeverría regime in the 1970s sought to buy off the students in the hope of discouraging their interference in national decision making.[65] Absorption in politics can be counterproductive to students because of the neglect of career training, but the nation may gain in their participation in social change.

Finally, a serious problem is the lack of research activity in most universities, as reflected in both the number of publications and doctoral programs. Until the 1970s only several institutions such as the University of Buenos Aires, National Autonomous University of Mexico (UNAM), University of Chile, and University of São Paulo had a serious research commitment. Fortunately, natural sciences are no longer subordinate to medicine, law, or the humanities. And even the social sciences are finding increased recognition.

The University in Transition

On the whole, the university occupies an increasingly conspicuous role in society. Even more than in North America the university degree represents a symbol of political or economic power. Diversity among universities remains most striking; they can be viewed as being local, regional, national, or international in scope. Others may be classified according to the nature of policy making, administrative style, and its own historical development. The status of the university and the degree of faculty and student involvement in societal goals and decisions vary according to the permissiveness the national culture allows.

Private and Public Education

Possibly the most remarkable change in Latin American universities over the last few decades is their growth. In reality, four types of institutions are found: (1) federal or national universities; (2) provincial or state; (3) private religious institutions, largely Catholic; and (4) private and secular. In smaller countries one national institution dominates, although others exist on a marginal basis. In larger nations such as Chile all four types are found. In regard to public institutions the federal government, as in Brazil, plays a greater role in the university than in lower levels of education. This aspect may prove to be an advantage because of the financial base. Harassment of professors became a serious problem during military rule.

Private institutions have a lengthy history. Some were converted from religious to secular institutions, others are of recent origin. Except for occasional course requirements in morality and theology, functioning of the two kinds of institutions is little different in curricula. However, the administrative structure is more hierarchical, the concept of *co-gobierno* (co-government) ushered in by the Reform had little effect on private schools. The Catholic University of Chile, which is the equal of the University of Chile in certain disciplines, is among the exceptions, as students in the late 1960s obtained considerable power, even though, like public institutions, it was shattered in the Pinochet dictatorship

Many private universities receive residual aid from the government, otherwise their tuition would be higher than it already is. Since public universities often have a graduated tuition based on parental income, the cost for students from affluent families may be little more in private than public universities. With increasing entry of the middle class (and occasionally upper–lower class) into higher education, private universities are no longer the exclusive domain of the upper class. Still, because of its size, diversity of *facultades*, and a broad geographic base in its student recruitment, a national university usually has a more heterogeneous student body than a private one. A country's commitment to higher education as compared to lower and middle levels varies as shown in table 9.4. Further, a sort of shadow boxing for funds, facilities, and students occurs between large and small, public and private universities. Chile resolved this problem by providing federal support to its eight public institutions in the country, at the same time encouraging them to find private funding. The expansion of private universities in the 1970s is attributable in large part to the elite's disenchantment with the direction of national, state, and local government. Mexico is only one instance of the importance of private schools in fashioning an intellectual elite. With rising enrollments in state institutions throughout Latin America the oligarchy feels threatened. For most nations, these bulging enrollments disturb the "boundary maintenance" of the upper classes against middle-class invasions. Hence, new private venues become mandatory in order to insure a higher social status. As implied earlier, other questions are relevant. Politicization threatens public institutions from delivering to society the economic and technical roles necessary for modernization.

Whereas private universities, for all their inadequacies, are meeting needs in the humanities and sciences, public institutions usually carry on the more expensive *facultades* of engineering, medicine, and dentistry. As compared to most of Latin America, in Brazil private generally surpass public universities in quality. Again, either because of outside employment most students in public institutions lack sufficient time to give to their studies, or an inferior secondary education does not equip them to perform according to the desired norms.

Restructuring the University and Its Curricula

Universities founded in the postwar period lean toward innovation, examples being Los Andes in Colombia and the Oriente in Venezuela. In these institutions reform means the provision of a general studies curriculum for all students, advisement services, coordination of liberal arts programs with professional schools in order for students to postpone their choice of major of *facultad*, and emphasis on research as an accompaniment to career training. The more progressive universities, both old and new, have formed liaisons with European and North American institutions. This departure is particularly valuable to science and technology. In some instances this kind of exchange may be less than functional. For example, business schools patterned after North American models assume a rationalistic approach that hardly harmonizes with the patriarchal management style of Brazil, among other countries.[66]

Whether new or old, private or public, universities suffer from, as one Brazilian educator reports, financial, administrative, political, professional, social, and ethnic handicaps. As compared to traditional institutions, newer ones strive to meet commercial and technical needs but are criticized for the neglect of cultural values that should infuse higher education.[67] Another factor is size. Many institutions are too small to be effective and others are unwieldy because of their size. The University of Buenos Aires with two hundred and fifteen thousand students could serve better if divided into several universities. Seven Argentine institutions have from thirty to two hundred thousand students, eight others have less than five thousand.

At both extremes is a question of functionality—mass industry in one case, lack of scientific and technical resources in the other.[68]

In highly politicized settings the university becomes subject to pressures that may be less than favorable to reform. Or reform may mean only further indoctrination, as the 1975 *perfeccionamiento* ("purification") reforms in Cuban universities, which attempted to secure ideological orthodoxy along with financial efficiency. An interesting case of politicization is Mexico, where student movements have been active since 1929, with a strong integration with workers' and teachers' unions and a commitment to improving wages and related aims.[69] In reference to the university setting, successive administrations moved both for and against establishing selected goals and standards. Basically, the conflict appears to be between one, the *modernist* viewpoint emphasizing individualism and two, the *populist* focusing on working-class needs and the redirection of political power. The populists stress an open admission procedure with little or no tuition. They were particularly successful in several states as an indirect result of the crisis following the October 1968 tragedy when federal troops fired on an estimated three hundred students in Mexico City. As public opinion and student resentment over this event ran high, President Díaz Ordaz and especially his successor Luis Echeverría accepted populist arguments in order to reduce hostility on the part of students. An analysis of graduation rates may show the results of populist control. For instance, at the Autonomous University of Puebla (UAP) less than 30 percent of entering students graduated at the end of four years as compared to nearly 60 percent for the nation as a whole during the early 1980s. Financial questions are also relevant—Puebla students come disproportionately from low-income families. With the modernism-oriented administration of Salinas de Gotorí (1988–94) the theme was one of providing the nation with sufficiently skilled personnel, tightening the norms of four-year programs, and establishing new two-year technical programs modeled after U.S. community colleges. Moreover, for the better part of a generation, government policy leans to developing regional universities that would not have the political strength enjoyed by UNAM with its some one hundred and forty thousand students (over three hundred thousand students according to some estimates based on extension courses and other programs).

Profile of the Student

Latin American students show as much heterogeneity as do their counterparts elsewhere in the world. They represent a variety of backgrounds, life goals, and attitudes as they confront the university milieu, a specific *facultad*, a given career cult, and especially the peers they encounter. In addition, vagaries of the national scene and confrontation with political ideology and action have varying effects. In view of this spectrum, the student's background and values are relevant (inset 9d).

INSET 9D   THE PERSONAL EQUATION AND THE SOCIAL SETTING

No typical student exists, but studies suggest a portrait of those attending large public universities in major cities. The average age is twenty, generally a ratio of 1.5 men to 1 woman. Social class composition differs with the particular country and institution. Personal adjustment unquestionably varies. According to my survey, the Colombian student's social life appears to be somewhat restricted as judged by North American standards. Only a fourth reported any kind of interpersonal encounter (dates, visits to a friend's home, party, or "going to a cafe or bar") during the previous week. In cities such as Buenos Aires, Santiago, and São Paulo, where university life is more urbane *centros* within the *facultad* provide an opportunity to socialize.

A student trapped in economic frustration, unsatisfactory living conditions, and few extracurricular activities is a likely candidate for anomie or other disorientation. Even so, students appear to have more social life than their parents and are primarily friendship-oriented, whereas the older generation is kin-oriented. These various pressures produce a cynical outlook toward the world. To the question, "how many of the people you know are really happy?," 52 percent of Bogotá students replied less than half. Approximately a third (32 percent) of the students thought that their future would be less happy than the present one. Again, responses were less pessimistic than among the general adult sample, but were more foreboding than findings from student and adult samples in other western countries I investigated.[70] Public opinion polls are not strong on validity and reliability, but they are not meaningless. Students usually have a more critical outlook toward the status quo than do the general public. According to sociologist Gabriel Careaga, consciousness of the "phoniness" of the middle-class's love affair with consumerism on one side and misery of the lower classes on the other side are a key to Mexican students' leftist leanings.[71]

Careers, Student Identity, and Socioeconomic Development

Again, we are reminded that only a minority of young people (less than a fifth in most countries) have a realistic chance of participating in postsecondary education. In nearly all Brazil and certainly in Manaus, federal and state funding does not keep pace with population growth, severely affecting the ability of secondary school graduates to pass the *pre-vestibular* in order to be admitted to the university.[72]

Latin American students enter the university disproportionately from private elementary and secondary schools. They come to the university for a number of reasons (and may leave for still others). They are motivated by their image of a career, whether school teacher, accountant, doctor, or cabinet minister. Others are primarily interested in maintaining their upper-class respectability. As often as not they attend a private university or go abroad, originally to Europe, but after World War II more often to North America. Students are also vaguely intrigued by the idea of social service within their country. Finally, a few, probably less than 1 percent, are professional agitators. Commitment to societal betterment is secondary to career mobility.

Students' images are structured by events within the university, their expectancies of what their career will be after graduation. In any instance, students might be considered members of the marginal elite. Until the 1970s they probably had, or expressed, more power than students in any other university system in the Western world. Also, because of the age pyramid, the person of seventeen–twenty-three represents a large segment of the population in developing nations. Further, as in developed societies, the problem of self-identity has to be viewed in the perspective of the student's function in society. Moreover, the relation of career and social service is critical for students in developing areas. National universities in Latin America often prescribe an internship for its graduates in medicine, dentistry, teaching, or other schools for service in underprivileged rural areas or occasionally in the urban slum. The history of student activism in countries as different as Nicaragua and Chile document the commitment of at least a conspicuous minority of students to help the less fortunate members of their society and to correct injustices in the social system.

This idealism is still tempered by the necessity of completing their *carrera* and the anticipation of placement difficulties. A study of employment in Bogotá sheds some light on this problem. In this sample of some four thousand workers the score on the university entrance examination along with the prestige of the university had a positive impact on the individual's occupational status and earnings. The father's educational attainment was another favorable variable. Curiously, course repetition at any level had little prediction for success in employment.[73] Generally, graduates of private universities have a better choice of jobs than do those from public ones.

University recruitment of students has to consider both personal and social motives. In recent decades, postgraduate foreign study has become increasingly attractive. Escape through the "brain drain" appeals markedly to advanced students of engineering, biology, and physical sciences. This phenomenon reached its peak in Latin America in the 1970s; a sagging economy in the Western democracies became a deterrent in the 1980s (in addition to the Latin American fiscal crisis), with a revival of the brain drain in the 1990s. Despite a rough market, students look forward to something more than a routine job. Society together with the university may have to work out a more effective reward system if students are to be of maximum service to their nation. However, students occasionally become impatient with the status quo (inset 9e).

INSET 9E   STUDENTS AND REBELLION

The most important criterion by which students judge the university is the probability of securing employment on graduation. Yet the university has to be judged by various criteria and obviously changes in space and time. In addition to national and regional variations, an institution may undergo cyclic patterns. For example, the National University of Colombia was for a generation the apex of higher education in the country. But as regional universities gained momentum in the 1960s the prestige of degrees awarded by the National was less certain. Another problem was the rising involvement in political activism. As a result of the lengthy strikes in the 1970s, few public institutions in Bogotá, Cali, and Medellín completed the academic year. By the dire 1980s public universities became as quiescent as the private ones.

For casual observers the stereotype of the Latin American student was, at least until the 1970s, one of constant rebellion and strike activity—an image standard for the developing world as well as for Japan and Western Europe. Even the United States had its moments of student revolt during the Vietnam War. Latin American students strike for or against various issues within or outside the university. Demonstrations and strikes are occasionally directed *against* an intolerable social condition such as a rise in bus fares or *for* new legislation, whether in housing, labor codes, or social security. Increasingly, students protest social injustice rather than the university bureaucracy.

The role of students in ending dictatorial regimes is well documented. Students at the University of São Paulo were active in denouncing the Vargas dictatorship. Student revolt was also a major factor in bringing Perón under scrutiny, although opposition of the armed forces was decisive in removing him from office in 1955. As early as 1952, students applied intermittent pressure on Batista in Cuba and paved the way for Castro's entry. Students were also instrumental in the fall of Pérez Jiménez in Venezuela and Rojas Pinilla in Colombia in the late 1950s, the military junta in Ecuador in 1966, and dictatorships in Brazil and Chile in the 1980s. In these events disaffection among students channeled their discontent toward the power elite—industrialists, the clergy, the military, rival political groups—making possible an overthrow of the ruling clique. Student restlessness reappears periodically, from 1999 in Caracas, Lima, and Mexico City to 2005 in Quito and La Paz.

University intellectuals are occasionally uncertain about the role that students should play in politics. Jorge Millas, a University of Chile philosophy professor, was very critical of the political role of students in the turbulent 1960s. Yet, after the 1973 military coup, he came to defend the policy he once upbraided. In the end he decided that military rule of the university was one more form of politicization, but possibly the least desirable.[74]

The University and the Sociocultural Milieu

As students traditionally come from the upper and middle classes, most are not exposed to dissent prior to their arrival at the university. The students' background and aspirations are a factor in attitude formation. A survey of graduating secondary school students in Cuzco, Peru, surveyed in regard to their attitude toward the Sendero Luminoso movement, reveals a comparison of Marxian and Weberian concepts. Students were asked how they perceive their career and future earnings as well as their views as to what is the best solution for the country's dilemma—"social revolution," "more education," "change of government," "eliminating corruption," and "other." As ideological formation about a given society can be "the result of asymmetric relationships between power elites and young people," these questions seem relevant. Adolescents who come from both low- and high-income groups choose almost equally "more education" and "social revolution," but those who expect low rather than high income and anticipate a lengthy search for employment disproportionately choose revolution. These responses must be seen in the context of how different classes perceive income, and the degree of legitimacy students attach to their society and its educational process. In view of what might be labeled rationalistic decision making, the study supports Weberian functionalism more than neo-Marxism.[75]

Beyond student input is the debate about the role of the university in contemporary society. With the economic crisis of the 1980s came a hiatus in university development in several countries. Especially in once wealthy Venezuela university budgets were severely cut.[76] One critic points to the bureaucratic nature of Latin American institutions, which produces a lack of articulation between university and industry.[77] On the positive side, many universities make significant contributions to social change, as the role of Brazilian universities in research and extension courses in developing the Amazon basin.[78]

The needs of the state and its economy can determine the shape of university offerings. Nowhere is this more apparent than in Mexico. As the ruling party, the Institutional Revolution Party (PRI) traditionally gave the university flexibility in the training of future political chieftains. That is, specific curricula are to prepare professionals and managers in the economic and political arena. Because of bulging enrollments from the late 1950s to the 1970s the government evaluated its needs and goals and demanded various reforms. The 1980s economic debacle placed additional stress on the chances of attending university. Even after the introduction of free tradition the offspring of the lowest quartile in household income were ten times less likely to attend a public university as compared to the wealthiest quartile. From 1992 to 1996 public universities were actually reinforcing socioeconomic inequality—a factor in student unrest and the 1999 strike.[79] Moreover, the gloomy job market and neoliberalism set limits as technicians became more of a priority than professionals. With Salinas' neoliberalism (1988–94) private education was promoted over public. Not surprising, privatization began to affect state institutions. Moreover, academic concentrations shifted. The triumvirate of medicine, engineering, and business replaced law as a major professional commitment.[80]

Ideology and Social Change

Since most nations are in divergent levels of socioeconomic development, no stable set of intellectual values is available to students. Whether in an authoritarian or a democratic setting, a barometer of the country's revolutionary fervor might be the sensitivity governments have to the stirring of students. Among other examples, as the Argentine dictatorship (1976–83) intervened in public and private universities, both staff and students disappeared. Courses in "moral education" were substituted for programs oriented to social problems. When students in El Salvador, Guatemala, and Nicaragua became more active in the 1970s than in the 1960s, oppression of students increased. Moreover, under Pinochet, student organizations were banned, activists expelled or arrested, social science courses came under extreme scrutiny, and military men or similar functionaries were chosen as rectors and deans. As a result, the mood varied between a sullen resentment and apathy. In a 1987 survey only 2.4 percent of Chileans eighteen–twenty years old said they discussed politics "often." Many of the generation who had grown up during the Allende and Pinochet regimes expressed a distaste for both socialism and democracy.[81]

At one time nationalism was a rallying cry for student action. It still assumes various forms—political, economic, cultural. Almost any action students initiate are brought under the broad umbrella of national development. More than with any other issue, the theme of liberation from foreign domination, usually the United States, is a stimulus to a strike. Furthermore, since most of Latin America—and most of the Western world—is perhaps only midway in its transition to a truly functional social order, as an elite group, students may become torchbearers of this process.

Any model for classifying student movements would have to accommodate those societies undergoing social change such as a revolution. In Cuba the Federation of University Students (FEU) was completely reconstituted in 1960 with a new brand of stable *fidelistas* supplanting the older ones who had rebelled against the corrupt Batista regime. When the revolution was bureaucratized, student protest could only be directed to specific flaws in the system, not to the system itself. The need for protest was negatively affected by the shift after the revolution in curricular offerings—from the humanities and social sciences to the practical areas of engineering and education.[82]

Conflicts in Latin America revolve about the distribution of power in major societal groups. The socialization process at the university includes the exercise of power. Students who learn to wield power as leftists at age twenty, turn conservative before thirty, and may become a governor at age forty! Students often acquire an analytical orientation to the political structure. In undemocratic and military regimes, alignment with secret and occasionally terrorist movements, as with the Sendero in Peru, becomes an appealing outlet for student frustration.

A Final Note

In reviewing a number of monographs on education and social change in regard to underlying themes, in most societies education is a matter of public policy. In capitalist societies, education, especially university and adult, serves to legitimize a market economy or similar system.[83] These generalizations apply to formal as well as nonformal education and raise a number of questions. For one, in view of limited resources how should a nation determine whether to improve quality or expand quantitatively? Or whether to establish fewer schools of high achievement or build more schools in order that everyone may have at least an elementary education? Further, why does higher education receive a disproportionate amount of the public educational budget? A developing society needs a technical and professional work-force, yet humanistic values cannot be ignored. Why does the military traditionally receive more of the national budget than does education in all but one nation, Costa Rica? (The power elites fear uprisings on the part of the underprivileged in addition to the feeling of insecurity about neighboring countries. Costa Rica decided in 1948 that international aggression was not to be feared; if it did, the United States would come to the rescue.)

Despite reform efforts, structural deficiencies still haunt Latin American education. A number of rigidities must be solved if the region is to join the advanced nations of the world, however far ahead it may be above the remainder of the developing world. Education depends on the entire school sequence from the elementary through university. Further, it reflects the health of other social institutions, especially the economic and the political.

Chapter Ten

Communication, Science, and Technology

When Latin America reached the mid-twentieth century, wider means of communication were indispensable. Urbanization and mass education became significant factors in the second half of the twentieth century. By the twenty-first century new breakthroughs in science and technology reached a crescendo. Clearly, these developments have enormous sociopolitical implications. For one, mass media are critical in the road to democracy. Also, they set the agenda in national goals and identities. Inevitably these issues are inseparable from those of previous chapters—economic and political forces, national legitimacy, and religious behavior, among others.

Communication and the Mass Media

A principal role of mass media—the printed page, radio, and television—is to extend the individual's and the group's cognitions through a myriad of symbols and images. Mass media are highly diverse in content and form. In attempting to adopt a given version of reality, a difficult task is to change the culture and society. The media not only affect public opinion, but values and norms determine levels of feeling and emotion—degrees of optimism or pessimism, acceptance or rejection of political frustration, aggression, violence, and attitudes toward changing patterns of authority. Further, it is critical for the mass media to recognize actors in the social scene, the differences between, for instance, elites and masses, in order that the community can cope with conflict in various sectors—locality, region, nation, and the international scheme as imposed by globalization. Latin Americans fear that NAFTA and similar organizations may condemn them not only to standardization of their economy but also to the obliteration of their cultural institutions.[1]

The output of the TV culture industry through the *telenovela* (TV serial) brings both a Cinderella fantasy and idealization of the nation with a passive acceptance of reality.[2] What one reads in the press or views on the screen defines one's mental set toward either a narrow or an expanded universe. Like most of the West, Latin America increasingly looks to a vaster world, sped on by cyberspace.

The media, whether press, radio, or TV, are fraught with the basic problems of financing, commercialism, governmental meddling, and a struggle for professional autonomy. These barriers are noticeably acute in authoritarian regimes, but even in the democracies, discontinuities arise with governmental shifts. Moreover, the quest for national autonomy is not an easy road. British, Spanish, and French influences were predominant in the nineteenth century, but by the twentieth century the United States achieved a leading role in news and programming in radio and especially TV. For several countries it is cheaper to import programs than to produce their own. Even as late as 1990, 75 percent of the content was from Anglo-America. Today, giants such as Mexico's Azteca and Brazil's TV *O Globo* dominate the media. However, each country has its own mix of domestic and imported offerings.

Varied Profiles

Countries differ considerably in governmental style, resources, literacy levels, and informational needs—urbanism, international involvement, and the repertory of mass media, as portraits of several nations illustrate:

Chile: A Changing Profile
In view of its democratic tradition, censorship was seldom a serious problem in Chile; yet in the early 1960s the Jorge Alesandri conservative government issued an "Abuse of Publicity Act," providing for libel suits. As the atmosphere eased during the Frei presidency (1964–70), radio, which had been entirely in private hands, was open to educational and other types of public sponsorship. TV broadcasting fell primarily to the universities. However, throughout the 1960s party polarization found its way into all mass media. The Christian Democrats set up the National Television Channel, which riled both the right and the left. Still, in practice, the three parties, left-center Christian Democrat, the socialist Popular Unity, and the ultraconservative Nationalist, had access to one channel or another.[3]

Although the Allende socialist regime did not have the votes to change the mass media structure, it did increase public ownership of certain media. Moreover, an opposition press continued. Most significant, U.S. authorities appropriated over three million dollars to radio and newspapers, notably the prestigious *El Mercurio*, in order to provide advertising space for the opposition. In the 1973 putsch the media were muted. While foreign correspondents could report home on the reign of terror, self-censorship by domestic mass media became the rule, later institutionalized in the National Security Decree of December 1975. By 1979 two antigovernment newspapers *Hoy* and *Mensaje* were allowed to print limited criticism of the regime. The following decade was an uneasy one in censorship, but in the 1989 return to democracy with Aylwin, toleration of dissent was restored. At the same time, in a country where fifteen families "own most of the country"—a phenomenon hardly unique to Chile—self-censorship becomes the code of many journalists.[4] In 2001 Congress with the signature of President Ricardo Lagos passed a law removing the more severe restrictions on freedom of expression. Even so, the nonestablishment press offers more probing news reporting than the well-entrenched *Mercurio* and *Ercila*.

Mexico: Reality and the Verbal Façade

Except for the Juárez period, censorship reigned in Mexico until the 1910 Revolution. However, even during the late *porfirato* underground newspapers—the "penny press"—circulated among the working class and fragments of the intellectual elite. These satiric weeklies documented opinion about events of the day—strikes, a typhus outbreak, scandals, temperance movements, and so on.[5] Although the cry for land, bread, and freedom was uppermost in the revolution an uncensored press remained a remote goal. As the Institutional Revolutionary Party (PRI) orchestrated control of the country, at least from the 1940s onward, a cozy relationship arose between government officials and publishers—payments (*mordidas, iguales, gacetillas*) varying between outright bribes and "consultancy" fees from authorities to the press in order to assure a favorable chronicle of events.[6] Still, the question of press control goes beyond the prowess of the PRI or any other political entity. It is rather the conflict between rival economic interest groups and reform-minded elites, both public and private.

Insensitivity of the press to the 1968 student massacre at Tlatelolco provoked a lengthy national debate. Congress held lengthy public hearings and debates on the media's function in national life. Both presidents, Echeverría and López Portillo, questioned whether the media were giving sufficient attention to all social sectors. Yet, Echeverría reportedly sanctioned the massacre of a dozen rebellious students in 1971, suppressing disclosure of the evident until revelations shocked the nation in 2004. As articulated by a government spokesman, communication was a public commodity indispensable to a democracy. Yet, ideals do not always suit individual or state demands. The media treatment of the 1988 Salinas and the 1994 Zedillo presidential campaigns was far from balanced, and the failure of *Ecselsior* and the *Atzeca* TV to report the realities of the Chiapas revolt was a reminder of the limits placed on mass media autonomy.

Peru: A Continuing Rise and Fall of Authoritarianism

Like its neighbors Peru has a jagged history of media censorship. Perhaps for the first time in this country's history President Manuel Prado in 1945 lifted controls in order to insure an open election. The freedom was short lived. General Manuel Odría's coup (1948) brought back censorship and outlawed the leftist APRA (American People's Revolutionary Alliance) party. But with a new mandate in 1958 Prado reasserted press freedom, which was soon to be tested. Peru's oldest newspaper *El Comercio*, founded in 1839, chose to ignore the verbal and physical attacks (mostly stones and vegetables thrown at his car) during the 1958 visit of Vice President Richard

Nixon. Although the disorder embarrassed the government it did not deter the editor from reporting the event. Similarly, an aging Prado did not meddle with the press when it ridiculed his attempt to secure a church annulment in order to marry his young companion. On more substantive issues, such as the Belaúnde presidency maneuvering in national and international economic affairs, for example, the sale of oil reserves to a subsidiary of Standard Oil of New Jersey, the opposition press's sharp attacks were a factor in Belaúnde's downfall.

Despite its promises of media permissiveness, the 1968 Velasco military "revolutionary" takeover led to a redefinition of national culture, including issues of business and industry sharing their profits with workers, land reform, and expanding education. New laws restricted the press's right to criticize the government. Accordingly, several newspapers and radio stations were expropriated, both TV and radio came under supervision by the Ministry of Education, followed by the initiation of ENTEL-Peru, a national telecommunications company. In the administration of General Francisco Morales Bermúdez (1975–80) these structures were dissolved. In retrospect, Velasco's attempt to integrate the poor and the indigenous into national life was well intentioned, but opposition to his plan was widespread, journalists being caught in the maelstrom. The weekly *Caretas* revealed corruption of politicians and the military, as for instance, the action of death squads in Barrios Altos, a Lima impoverished neighborhood.[7]

The waves and counter-waves of Peruvian mass media assumed a new intensity in the Fujimori era. Most radio, TV, and newspapers, tempered by the Odría and Velasco regimes, held to self-censorship. However, fearing harsh attacks from the press because of his police controls and obliteration of civil liberties in his campaign against the Sendero Luminoso, Fujimori ordered a military occupation of mass media offices and the closing down of several newspapers and magazines. In their anxiety about basic needs of food, shelter, and personal security, most Peruvians find the question of integrity in mass media of secondary concern. Yet, the desire for "access to resources" played a role in Fujimori's 1995 bid for reelection.[8] Even more insensitive was his 2000 campaign, the mass media calling for his resignation before the end of that year.

Brazil: Challenges to Corruption

From the 1930s to the 1960s literacy levels and economics made radio the principal source of information. In Brazil, as in most of Latin America, a newspaper equals nearly a third of the price of a loaf of bread. Only the middle and upper classes, who traditionally determine the course of political and economic development, can afford a paper.

In his Estado Novo, Getulio Vargas organized a pattern of corruption and censorship that was to haunt the country decades later in the military dictatorship (1964–85). With Vargas's resignation in 1945 the rejuvenated media permitted even the Brazilian Communist Party access. This freedom had a limited life span; 1950 brought Vargas again to power with further harassment of recalcitrant critics, in particular Carlos Lacerda, publisher of the Rio daily *Tribuna de Imprensa*. Still, the press had its effect in reporting economic slowdown and governmental corruption. Finally on August 24, 1954 Vargas reported to his cabinet that public opinion had turned against him, then went to his office and shot himself.

The mass media enjoyed full rights from Juscelino Kubitschek to João Goulart. Indeed, the press, especially *O Globo* and the *Jornal do Brasil*, made much of Goulart's leftist "*clientelismo*." Several papers did support Goulart, for instance the *Ultima Hora* and *O Diario*, which were conveniently allowed to import newsprint without paying interest on government loans. Influential in Goulart's forced resignation was the São Paulo *O Estado*, as it had rumors of his plans for further collectivizing the economy. Also, *O Globo* inspired dissent as it had assembled sufficient wealth in order to bypass government support. Pro-Goulart radio stations floundered. In 1964 the stage was set; Castel Branco and his fellow generals marched in.

The military regime or, as some prefer to label it, a "bureaucratic authoritarian regime," held itself in power through the mass media. In reality, TV more than radio shaped the contours of control. TV *Globo* became the world's fourth largest system, covering over three thousand of Brazil's municipalities, all neatly timed for the generals, as the regime came into being in 1964. The state-owned EMBRATEL (Brazilian Corporation of Telecommunications) assembled the technical infrastructure for *O Globo*, permitting the government to "enlighten" thirty-seven million households about the new regime's achievements—the trans-Amazon Highway, the country's victory in the 1970 World Cup, and the perceived need of a terrorist campaign against pockets of resistance in the countryside. As newspapers reported a violent, chaotic world outside Brazil, TV *Globo* delivered a soothing message that the nation was in good order.

As the economy weakened in the 1970s and 1980s the mass media shifted gears. However, censorship continued, the media found ingenious means of protest—Radio *Jornal do Brasil* in Rio suspended news broadcasts from Brasilia. The press and TV devised imaginative tactics in tandem with political parties to bring about the return to democracy. Journalists became still more active in the 1990s, notably in the Collargate. Whereas *O Globo* supported Collar for the presidency, *A Folha* lost little time in exploring his corruption. Collar responded with lawsuits against the press, including arrest of at least three newspaper executives. After Collar's fall, *A Folha* pursued corruption in Congress. Further, regional and local papers often encounter problems when they report ecological destruction or military brutality.

Uruguay: Disarray in a Classic Democracy

Perhaps no country in the Western Hemisphere has a more socially constructive past than has Uruguay. The working relationship of the Blanco and Colorado parties, and especially the remarkable leadership of President José Batllé y Ordóñez (1903–07 and 1911–15), who left a heritage of democracy and social welfare that was to endure until the economic debacle and runaway inflation of the 1960s. Because of workers' strikes in 1967, including newspaper employees, radio and TV became principal communicators of the worsening crisis. In 1970 the daily *Acción* published the Tupamara "Manifesto to Public Opinion," leading to the newspaper's suspension, followed by a similar fate for other papers. By 1972, kidnapping of public officials, a walkout by health workers including doctors, and intolerable inflation led to a shackled press and restriction of civil liberties. Montevideo's eleven dailies were reduced to six. Still, several radio stations could outmaneuver the government. Broadcasters were forbidden to mention strikes, but in order to inform the public about transportation options they would say "service will be normal tomorrow."

The military dictatorship (1973–84) closed down more newspapers and magazines, censoring broadcasting, popular music, even a number of tangos. In view of public outrage, the military somewhat loosened its reins after 1980. Broadcasting gradually found freedom; political parties (although carefully monitored) returned.[9] In 1985 a civilian government was again in place. Private media once more replaced public sources. By 1990 Uruguay enjoyed the most extensive per capita radio facilities, with 99 AM stations and 227 TV sets per 1,000 inhabitants in 1993, the highest in Latin America.

Areas of the Mass Media and Their Problems

Beyond the political dimensions are other questions of the effectiveness of mass media. First is the type of ownership and its implications. In Brazil, Colombia, and several other nations, concentration of ownership makes for limited programming and lack of originality. The press and TV often belong to a capitalist oligarchy, discouraging competition. Paradoxically, competition can also create monotony as each channel or station fears deviating from the norm.[10] The entanglement of public and private in transitions from democracy to dictatorship and the reverse may create a monstrosity. In Argentina the mass media that President Raúl Alfonsin inherited from the military regime was a Kafkaesque, corrupt bureaucracy. Without a budget to operate the media Alfonsin invited private entrepreneurs to enter the field. Most countries represent a mélange of private and public, only Cuba enjoys total—and tight—control.

Second, considerable discussion and controversy surround the degree of cross-national or foreign influence and control. This issue is hardly new. An early Anglo-French–U.S. rivalry surrounded the trans-Atlantic cable, which was to connect Europe with Latin America. Despite German competition, the United States dominated the early days of radio in supplying the technology, all to the satisfaction of North American entrepreneurial and military interests.[11] As TV came to Latin America in the late 1950s, programs remained North American well into the 1980s. However, the major countries increasingly produce and export their own fare, *telenovelas* being an inexhaustible source of revenue for Brazil, Mexico, and Venezuela. As a Brazilian observer explains, formal as opposed to informal communication can be of European, North American, or domestic sources. By 1980, 73 percent of urban Brazilian homes had TV with dominance passing to the major networks—*O Globo* and later *Tupí*. Domestic programming, especially *telenovelas*, was offered at prime time.[12] Still, as another form of dependency, North Americanization of mass media and other aspects of culture remains a source of friction. Accordingly, as early as the 1960s, comic characters such as Donald Duck connoted subtle forms of a capitalist culture.[13] As *Sesame Street* inspired materialist consumption patterns, the Velasco Alvarado regime removed it from the Peruvian screen.

On the whole, the Old World still counts. A content analysis of newspapers and magazines suggests more European than Anglo-American influence. Between 1949 and 1982 in issues of some twenty prestigious newspapers and magazines, European articles and comments appeared nearly twice as often than those of the United States, Yet, domestic merged with Latin American material was greater than that of Europe and United States combined. Moreover, the flow is not always a one-way phenomenon. As an instance, *The Kiss of the Spider Women*, a popular play and film in the United States, was the work of the noted Argentine dramatist Manuel Puig, the director turning mainly to Latin American stars—Hector Balenco, Sonia Braga, and Raúl Julia.[14]

A positive development of recent decades is the growing number of women in journalism, radio, and TV. They still have to fight the stereotypes emerging from a male world. Democracies are more receptive to women than are authoritarian regimes. Societies undergoing socioeconomic change such as *sandinista* Nicaragua are reported to welcome women journalists. Generally, women from the middle and upper classes are more able to break the male barriers. María Jimena Duzán of Colombia represents the more daring of these new professionals. Daughter of a prominent Bogotá journalist, she plunged into analyzing what she calls the "disarticulation" of her nation, interviewing narcotraffickers and guerrilla leaders in the countryside.[15] Similar vignettes appear of women in the media from Mexico to Chile.

The Press

Historically, the press was the sole means of informing the public of national, regional, and local events. The first regularly published newspaper *La Aurora de Chile* appeared in Santiago in 1812, followed a year later by the nonofficial *El Seminario Republicano*. The first paper with political cartoons was Mexico's *El Iris* in 1826. The first daily newspaper was *El Mercurio de Valparaíso*, which remains the oldest continually published daily of the Spanish-speaking world. Still another first was the journalism program at the University of La Plata in Argentina in 1834. Then as now, because of poverty and functional illiteracy, newspapers had a limited circulation. Even today, prominent newspapers have a following well below most of the West. Even in inordinately large cities such as Buenos Aires, São Paulo, or Mexico City a prominent newspaper may have a local circulation of less than three hundred thousand. The number of newspapers varies not only to the country's size and wealth, but also to the climate of freedom. In 1996 Mexico had two hundred and ninety-five daily newspapers, Haiti and Nicaragua four (plus some ephemeral publications—inset 10a), and Cuba only one.[16] Of course, as elsewhere, alternative media are now available—radio, television, and Internet.

INSET 10A   THE PRESS, NATIONAL CRISIS, AND
"HUMOR ERÓTICO"

During the *sandinista* regime in Nicaragua (1979–90) Róger Sánchez Flores was determined to offer humor as both escape and social criticism. His cartoons first appeared in *Barricada*, with an international edition expressing a liberalism familiar to the United States and Western Europe. His cartoon character Polidecto offered a satire of U.S. military intervention. In 1980 he inaugurated a sixteen-page tabloid *Semana Crónica* (weekly chronicle), directing a savage humor with acute sexual overtones to political and clerical figures. He especially targeted reactionary priests (notably Bismarck Carvallo), hinting at their sexual misconduct. There were other villains, for instance, Somoza-like characters and CIA agents.

Not surprising, circulation grew briskly, reaching sixty thousand (comparable to five million in the United States). Its reputation spread to other countries, particularly those sympathetic to the *sandinista* regime—Mexico and Cuba. Readership in Nicaragua was largely male, and ironically oriented to the political right. In the 1986 plebiscite a majority of those who supported the *sandinistas* were predominantly the less educated masses, who could hardly afford purchase of a newspaper.

Criticism of the *Semana* emerged from various sectors. The government was not enthusiastic about what some critics thought of as pornography. Especially the Communist Party (representing a distinct minority) felt that concentration on sex was a distraction from revolutionary goals. Women were less than rhapsodic about erotic humor. They also felt that Sánchez could have done more in supporting women's rights. However, he did favor family planning, which elicited the ire of President Daniel Ortega (a father of seven children) who insisted that Nicaragua needed a larger population. Sánchez responded by drawing a cartoon with five decidedly pregnant nude women sitting by what appeared as a reproductive machine. This and other affronts to the government led to closure of *Semana Crónica*.

In summary, Sánchez's cartoons lampoon conservatism as well as political bureaucracy. His treatment of sexuality is satirical and existential with parodies on both prudery and machismo. As compared to glossy *Playboy* portrayals, his salty, witty images on cheap newsprint seem quite neutral. Sadly, stomach cancer brought him (at the age of thirty) and the *Semana Crónica* to an abrupt end.[17]

The Challenges of Publishing

As suggested earlier, several characteristics are apparent in the Latin American press:

1. Problems of finance, materiel, and technologies. Newsprint is at a premium, especially during economic crises, which for some countries are nearly perpetual. A government may control the newsprint allotted to a given publisher, cutting off the supply to recalcitrant newspapers. Even sizeable countries search for means of manufacturing their own paper. Venezuela is such a case as circulation remains brisk, particularly in secondary cities.

2. Division between private and public. Ownership of newspapers, like ratio and TV, can be commercial, governmental, educational, or occasionally a combination. Financing the private can be expensive as in giants such as Brazil's *O Globo* (which grew in nationwide circulation from 220,000 in 1985 to 620,000 in 1995) and also for dailies in smaller nations. Whether public or private, commercialism, all too often inherent in the press and other mass media, are trapped in clientelism, bureaucracy, and parochialism. The pressure from—and for—advertisers inevitably compromises editorial policies and autonomy of reporters.

3. Press freedom or state surveillance. A sliding scale operates in the press's degree of freedom as nations move from democracy to dictatorship or the reverse. Even in a stable democracy, interference is not unknown. In Costa Rica in 1993 a journalist was sentenced to ten days in jail and heavily fined for "libel" in reporting the shady dealings of a former congressman. TV reporters are especially vulnerable.[18] As is apparent in national profiles, a free press is essential to a healthy democracy. "Watchdog journalism" is most visible in a climate of well-organized political parties. Indeed, many nineteenth-century newspapers were identified with a given party, and a score of editors or publishers became presidents. This tradition lingers on into the present, especially in Argentina and Colombia. Also, among the dailies with a pungent investigatory bent are *El Clarín* of Buenos Aires, *El Tiempo*, and *El Espectador* of Bogotá, and *El Comercio* and *La República* of Lima, to mention several. In the reporters' zeal to expose corruption, they take on not only the government but other powerful entities. For instance, the Medellín cartel declared war on the press in bombing editorial offices and murdering reporters. As another example *Pagina 12* of Buenos Aires revealed corruption of President Menem's relatives in laundering drug money as well as corrupt financial dealings within the government itself. In the 1990s several Brazilian newspapers took a strong stand on environmental degradation by emphasizing technological and industrial growth and its impact.[19] These exposés have a healthy influence on other mass

media. However, journalists may pay a heavy price. Throughout
South America 116 journalists were killed between 1965 and 1985.

4. Ambivalence, insecurity, and the status of journalists. Both the state
and publishers debate requirements for reporters. Cuba in 1942 introduced the *colegio*, which mandates a university degree for reporters.
At present, half the countries maintain a collegium. On the surface,
the system offers professionalization, but the norms are unevenly
applied and have political overtones. In 1994 both the Declaration of
Chapultepec and the Declaration of Santiago stipulated that membership in the *colegio* is a violation of freedom of expression and should
be considered as voluntary. It is questionable how much the rituals of
professionalism improve the press. Can one compare the recent graduate with a well-seasoned reporter who has fought in the trenches
over the better part of a lifetime?[20]

A serious problem for the journalist is the risk of job loss, arrest, and
especially violence. Because of dictatorship, drug traffic, and guerrilla
warfare Argentina, Chile, Colombia, Mexico, and Peru have an unsavory record of detention, kidnapping, disappearance, and death. In
Peru alone no less than twenty-four reporters were killed by the military or the Sendero Luminoso.[21] On the other hand, in some countries
skirmishes are mere annoyances, but nothing came of the suit.

5. The search for objectivity and prestige of the press. The journalist is
obligated to check his or her sources, documenting the data with
visual accuracy, credibility of eyewitnesses, and thoroughly weigh the
"facts." Truth is relative, but the prestige of a newspaper or journal
depends on its factual reporting. Latin American reporters have an
uneven—but an ever more positive record—not always with acknowledgment. So often proof in the press of fraud or other delinquency is
met with a light sentence or a non-guilty judgment. On the whole,
Latin American journalists have performed impressively in recent
years.[22] One may hypothesize that globalization in its broadest sense
(at least in Brazil) tends to produce a relatively open system in news
reporting. Even though corruption still prevails, the telecom infrastructure forces both private and public media to widen their horizons
and reduce national interests and conflicts.[23]

Radio and Television

Radio has since the 1920s been the major avenue of public communication.
In my 1960 survey of San Salvador shantytowns I recall how two or three
families would listen in front of the *choza* (hut) of a resident who had a
radio. Today in the impoverished shantytowns nearly every hut has a radio
and possibly every second, a TV. Consequently for rural areas radio is the
major communication from the world outside.[24] Yet, it would be difficult to
overestimate the impact of TV. For instance, a study of its effects of an

urban and often multinational world on the Yucatec Maya community reveals a transformed approach the individual feels toward interpersonal relations, self-identity, and the meaning of the community and its economy.[25] Beyond the ubiquitous TV—90 percent of urban Brazilians have access to TV sets (33 percent of rural and semirural homes with TV in Mexico, 30 percent in Argentina, 26 percent in Brazil)—is the growing use of video and video cassettes. Religious, human rights, and other organizations are able to document their programs. Also, indigenous groups can record their ceremonies. Local communities move to establish their own TV stations, however modest, to counteract the overwhelming influence of national, public and private systems.[26] Most viewers are fixed on the *telenovela*, which is contextualized for nearly every phase of Latin American culture—class belongingness and mobility, ethnicity, gender relations, economic stress, all portraying both reality and escape.

Besides *telenovelas* are films, mostly imported, whether Mexican or Argentine, European or Anglo-American. But the political dialogue is also available. The presentation can be biased, or it may be balanced. For example, the 1989 presidential contest between Alfonsin and Menem offered an open discussion of Argentine realities. As in other parts of the world, the TV medium often focuses on neopopulism including novelty in personalities as in the 1989 election of Fernando Collar in Brazil and the 2001 election of Alberto Fujimori in Peru.[27] The public, weary of routine politicians, found a new genre in the visually appealing Collar. His appearance fit in well with the photogenic vigor set by a blond, sexy female bombshell entertainer "Lula," who has dominated the Brazilian TV screen and its marketing of products. Generally, political partisanship in TV follows in the tradition of the press.

Whereas radio has been, with some exceptions, decentralized, TV is centralized and commercialized. Advertising is the driving force, and in Colombia accounts for over 1 percent of the GNP. TV provides in transnational empires like *O Globo* a homogenizing of culture. Images on the screen offer a variety of consumer items, which feed the economy. These images convey powerful imagery of people's lives. Also, a sociopolitical ideology is woven into the content, usually supporting the establishment.

Personal and Societal Effects of TV

No less than Europeans and North Americans, Latin Americans are concerned about the positive and negative aspects of TV. Especially, they are troubled about the loss of their cultural values. With the growth of cross-national communication giants and their cable systems, regional and national traditions and identities are in danger. In the words of one critic, superficiality, confusion, and chaos are inevitable.[28] Most appraisals are impressionistic, but methodologically sophisticated surveys appear. One focused on the medium's impact on the values and behavior of Argentine adolescents during the transition from authoritarianism to democracy in the 1980s. Questionnaire and interview data mirror the importance of family members in sustaining meaning and identity during this period. At the same time, TV exposure was greater in the lower than in the middle class. The findings reveal few resounding relationships, for instance, frequent TV viewing, especially on Saturday evenings, is associated with lower educational goals. Exposure to violence on the screen has less inimical effects than what is reported in U.S. samples. These like other findings must be judged according to cultural variables such as age, gender, and social class.[29]

Another research project, specifically oriented toward societal change, was the weaving of social themes into soap operas. The Mexican Institute of Communication Studies together with Televisa (the leading TV network, later to be overshadowed by the private Azteca system) investigated communication and social learning models underlying soap operas. Six *telenovelas* with characters expressing role conflicts in the family and other interpersonal situations were chosen in order to study attitudinal shifts in a variety of themes—sex education, family planning, women's rights, and child abuse, among others. Audiences processed archetypes and stereotypes along with innovative themes. Feedback indicated the payoff from this entertainment–education strategy with explicit and implicit messages. Viewers as opposed to nonviewers were predisposed toward new attitudes and personal growth.[30] Again, the potential for social reform through mass media, notably TV, is structured by the individual's perception of her/his stimulus world and receptivity to new images and ideas.[31]

Science: Delay, Awakening, and Societal Response

The late development of science in Latin America stems from the mystical bent of Spanish culture. At the moment when early astronomers were examining the universe, the Tribunal of the Holy Office of the Inquisition was installed in Spain in 1478. This hostility toward science characterized Roman Catholicism from the Middle Ages to at least the nineteenth century. In Latin America this doctrinal outlook persisted throughout the colonial period and beyond. By comparison, New World ancient civilizations had a pantheistic concept of nature. Their science merged with a natural religion in a cyclic form. The syncretism of Catholic and indigenous beliefs remained as part of a "collective memory" into the twentieth century.[32]

Historical Trends

Indigenous cultures made a striking contribution in science and technology. Especially intriguing was—and is—the use of plants in a holistic approach to medicine. Also, the Aztecs prescribed steam baths, massages, flagellation, and most intriguing, performed dental and orthopedic surgery. Much of indigenous medical care was inspired by magic, such as imbibing bird eyes for correcting visual problems or using testicles of animals to treat impotence—practices no more ludicrous than those of Western medicine until the late nineteenth century.[33] The dedication of pre-Colombian civilizations to mathematics, astronomy, and agricultural technology is well known. Perhaps the most visible contribution to the West was the discovery of the potato, *solanum tuberosum*, which made its way to Spain in 1570, as did tobacco about the same time, and brought to England by Walter Raleigh in 1586.

The early colonial period was not without its achievements. In New Spain, Francisco Hernández led an expedition (1570–77) in order to trace the area's natural history, with a focus on medicinal plants. In the explorations during the seventeenth century other scientists examined geography, mineralogy, and astronomy. For one, Gabriel Soares de Sousa wrote *Tratado Descriptivo do Brasil* in 1587, followed over the next half-century by more complete treatises on the vegetation of that vast country.[34] A number of Mexican scientists tested the findings of Galileo and other Europeans for validation in the New World.[35] In the seventeenth and eighteenth centuries the inquisition made war against adherents of Copernicus, Newton, and Harvey. Because of entrenched scholasticism, Newtonian physics was not taught until the late eighteenth century in the universities of Lima and Bogotá, still later in Mexico. Harvey's treatise on the circulation of blood was similarly delayed.

Most remarkable in the eighteenth century were the expeditions documenting fauna and especially flora. Other surveys demarcated regional and national boundaries as between Spanish and Portuguese territory. One expedition (1778–88) in Peru produced 589 drawings, approximately 3,000 plant descriptions, and a herbarium of 1,980 specimens. Another team carried out a far-reaching expedition of Nueva Granada, succeeded by the establishment of a scientific observatory in Bogotá directed by geographer Francisco José de Caldas. Although botany was the principal concern throughout Ibero-America, astronomic and other observations followed, for instance, examining the transit of Venus across the solar disc in Baja California and Montevideo, respectively.[36] Most influential was German scientist Alexander von Humboldt's expedition (1789–1804), providing a wealth of geographic, biological, and cultural descriptions. He was also impressed by advances in Peru and Mexico, in as disparate areas as mining techniques and teaching of mathematics in the university.[37]

The National Period

A growing sociopolitical ferment drew from the Enlightenment. Also, Spanish American intellectuals were aware of experimentation, including the work of Benjamin Franklin. Independence itself provided an impetus to scientific inquiry.[38] One instance is the founding in Mexico City of an Institute for Science, Literature, and Arts in 1826 and a national museum in 1828. Similar developments occurred in Caracas, Bogotá, and Lima. Nineteenth-century Peru brought forth a diversity of scientific projects—climatic accounts of the dry coastal areas, ethnology of several tribes, a lexicon of place names, and astronomic pursuits, including records of shooting stars at Ayucucho.[39] Thanks to Dom Pedro II's interest in science, Brazil moved in various directions: physics, metallurgy, mineralogy, geology, paleontology, and zoology from the nineteenth century onward. Because of its application to industry, chemistry received special attention, notably in late-nineteenth-century São Paulo, which increasingly assumed predominance as a scientific center after 1870. However, both Mexico and Brazil—as well as other nations—were caught up with differing ideological currents (inset 10b).

INSET 10B   SCIENTIFIC FADS AND NATIONAL IDENTITY

By the middle of the nineteenth century, intellectual currents began arriving in Latin America. Medical doctors, philosophers, and politicians were speaking of Charles Darwin's contributions and other scientific breakthroughs. Ideas about material progress were also in the air. In this context, the philosophy of Positivism, outlined by French "sociologist" Auguste Comte, gained attention. Lip service was given to pragmatic theories as opposed to abstract principles or vague utopian plans. At the same time, universities were leaning to the French model as opposed to the more religiously oriented Iberian model. However, instruction took precedence over research.

An intriguing aspect of scientific development was the conflict between evolutionary theory and positivism, that is, Charles Darwin and Auguste Comte. Each had his fervid followers. In part the controversy stemmed from differing national attachments. In Mexico enthusiasm for France and positivism did not offer an inviting milieu for Darwinism. The devotion to Comte was somewhat diminished by the French military intervention in the 1960s. In any event, in 1877 at least one scientific society discussing evolutionary theory declared that it did not conform to the scientific method. On the other hand, pro-Anglo sentiment in Argentina and Uruguay offered a more favorable environment for Charles Darwin and Herbert Spencer, a theorist imbued with the idea of progress through evolution of social institutions. Despite the hostility of the church, by the end of the century, medical schools and most scientists accepted Darwin. Blends of positivism and Darwinism became the norm. Even so, Brazil still leans to Comte, who even Spiritists espouse, positivism indirectly finding its way into health practices, notably homeopathy.[40]

As another case study, scientific development came late to Chile. The University of Chile was founded in 1842, inspired by educator Andrés Bello, who dreamed of intellectual and scientific advance for his country. However, the conflict between conservatives and liberals complicated a commitment to science. Also, by the 1880s a scientific community was emerging in both the Universidad Católica (founded 1888) as well as the Universidad de Chile, further documented by professional journals (for one, the *Revista de Medicina Chilena*, appearing as early as 1872).[41]

Concerning human survival, the most significant research through the century was in health and medicine. The stimulus came mostly from abroad as biologists and physicians arrived principally from France and Germany, with a return migration of young Latin Americans for study in Europe. The contributions of Pasteur, Lister, Koch, and Bernard inspired the development of laboratories in several cities, culminating in the integration of laboratories into hospitals, notably in 1865 in Colombia. Because of several epidemics, for instance cholera in Guadalajara, medical efforts focused on public health. Moreover, sanitation was necessary in order to attract foreign investment. After 1900 the Rockefeller Foundation operated in several countries in eradication programs against malaria, yellow fever, and other tropical diseases. Nor was mental health ignored in the late nineteenth century. A novel approach occurred in Lima, where a mental hospital initiated a treatment based on rest, a balanced diet, isolation from the family— and hypnosis. Unfortunately fame of the experiment led to growth and overcrowding, the institution becoming all too similar to other asylums.[42]

The Twentieth Century and Beyond

As Argentina and Brazil led the way for modernization, universities strengthened their offerings in physical and biological sciences. For example, in the 1920s the University of La Plata imported several German physicists, who established a well-equipped laboratory. Refugees from the Spanish Civil War (1936–39) and National Socialist Germany (1933–45) enriched the scientific community. On that point, the Argentine Psychoanalytic Association was founded by a Spanish exile, a German Jew, and two Argentines. European refugees led the modernization of medical research. Similar scenarios occurred throughout the hemisphere.

In the middle of the twentieth century scientific endeavor moved from applied fields—engineering, natural history, medicine, and public health— to theoretical research.[43] After World War II scientific research reached a higher level, especially in biology. Argentine Bernardo Houssay's endocrinological research led to a Nobel Prize for medicine in 1947. Similarly, another Argentine, Frederico Leloir, received the 1970 Nobel Prize in chemistry for his work on carbohydrate metabolism. A third Argentine Cézar Millstein won the Nobel Prize for medicine and physiology for his work in immunology, specifically advances in cancer treatment. However, in a reverse direction, the Onganía 1966 dictatorship drove dozens of scholars and scientists into exile; consequently, science in Argentina did not recover its former status until the mid-1980s.[44]

Following the heritage of Carlos Ribeiro Chagas and Oswaldo Cruz, Brazil led in research in tropical disease. Physics continued to enjoy more prestige than other sciences, at least in the academic community, but it was not until the 1960s that physicists published widely in their own country; previously they chose foreign journals for reporting their research.[45] The sociopolitical setting can be relevant. After the 1959 revolution, despite financial barriers, Cuba stressed agricultural biotechnology and medical-pharmaceutical production.[46]

Beyond biology and health, advances are occurring in physics. Nuclear physics leads in Argentina, Brazil, and Mexico. *Informatica* or the microelectronic, digital, and software industry is especially strong in Brazil, but growing in most nations. The 1990s saw a renaissance of scientific activity, following the 1980s economic setback. The present and future also depend on the socialization process (inset 10c).

INSET 10C   THE CRAFTING OF A SCIENTIST

The relatively late development of science—a few inquiries in ancient Greece, further breakthroughs in the late Middle Ages, and the flowering over the last four hundred years—suggests an unusual set of capacities for human beings to arrive at a level of "cognitive control."[47] Even today, relatively few of us understand the intricacies of nuclear reaction, cerebral circuits, or the mystery of the cosmos.

To be successful, a scientist must have a number of psychological traits—curiosity, critical detachment, a determination to investigate, weigh evidence, test, and retest. Basic to the mental set of a scientist is a composite of personality characteristics, including motivational stimuli, emotional control and release, imagination, creativity, and, most important, objectivity. Not all cultures deliberately cultivate these traits in the socialization process. Undoubtedly in most societies a certain degree of self-selection occurs as the scientist role is partly a matter of genetic traits and primary socialization in the home, in addition to secondary socialization by groups and institutions. A study of a sample of the National Autonomous University of Mexico (UNAM) students reveals how "scientific identity" is acquired from undergraduate to graduate studies through face-to-face interaction with staff and fellow students along with self- and group-evaluation. A charismatic teacher or a committed in-group can nourish a scientific mystique. In view of the limited resources available in most Latin American academic settings a basic problem is an overidealized ambition in research aspirations.[48]

The importance of the academic setting can hardly be exaggerated. Chapter 9 described the growth of higher education in the late twentieth century—university enrollment in Costa Rica rose from one per thousand to thirty-three per thousand population between 1941 and 1980, continuing to grow, with the ratio of scientists rising geometrically.[49] In order to strengthen its program, beginning in 1971 UNAM established research centers for instrumentation, applied mathematics, atmospheric sciences, cellular physiology, nitrogen fixation, and genetic engineering. Also, the curriculum at UNAM was restructured in the 1980s and 1990s, becoming increasingly intensive, at least for medical students, who combine study and research with emphasis on problem solving. Although students complain of the workload, they feel that they gain skepticism and curiosity.[50] Argentina always prides itself on the quality of its institutions of higher education. However, the Perón era was less than favorable; still worse were the 1966 and 1976 military takeovers. Today scientific programs are available in some twenty national universities and eighteen private ones. The University of Buenos Aires has resumed its prominence with a number of centers. Coincidentally, a third of the national research budget goes to academic centers.

Finally, socialization has to include some degree of theory and practice. Too often the academic is separated from the laboratory and workshop. The European tradition of pure science as detached from the practical and technological sphere persisted throughout the twentieth century. The problem is partly moral and epistemological, the investigator must distinguish between the professional role and the applied or technological.[51]

Research and National Development

As implied earlier, science, both pure and applied, had its stirrings in the mid-nineteenth century. Decades passed before the establishment of a serious laboratory in most countries. Through the second half of the twentieth century scientists struggled for their status, identity, and for governmental recognition of the need for tangible support. Progress was evident during the 1970s but slowed in the fiscal crisis of the 1980s. As the 1990s moved toward neoliberalism, globalization provided a more productive arena— exchanges with scientists abroad, aid from cross-national research entities, and increased finances from international agencies such as the Inter-American Development Bank. Still, as a group of Brazilian scientists reflect, a twin problem in the global free market universe is sustainability and competitiveness for which the region is in an inferior position. A more intimate relationship between the pure and the applied, public and private, is mandatory. In this context other issues emerge.[52] First of all, less than 1 percent of the Latin American GNP goes into research and development. Consequently, relatively less personnel is involved in science and technology as compared to the West, with marked variations between Latin American countries (table 10.1). Funding has low priority and is subject to cronyism and bureaucratic decision making, often concentrated in a few urban areas, usually the capital. Second, as stated in chapter 9, with several exceptions, educational institutions traditionally offer a questionable setting. Academic salaries tend to discourage qualified scientists to enter teaching or research. Limited course offerings reflect this hiatus, and laboratory facilities are lacking. One Brazilian scientist speaks of doing his research in the basement of his home. Moreover, critics complain of insufficient evaluation of research,

Table 10.1 Scientists, engineers, and technicians employed in research and development for years 1992–98

Country	Researchers		Technicians
	Total	Female	
Argentina	22,897	10,655	5,092
Bolivia	1,300		
Brazil	2,654		2,940
Chile	4,630		2,940
Colombia	1,732		1,024
Costa Rica	1,866		
Cuba	17,667		12,288
El Salvador	102		1,812
Mexico	16,434		6,676
Peru	607		34
Uruguay	2,093	720	
Venezuela	4,258	1,490	650

Source: Adapted from Latin American Center, *Statistical Abstract of Latin America* (Los Angeles: University of California, 2002), p. 294.

qualitative unevenness between given science departments, and relative lack of interdisciplinary interests.[53] In Brazil a Reforma Universitaria was adopted in 1975 to resolve these problems, but the only significant improvement was an increase in funding.

An Uruguayan scientist outlines several indispensable priorities: (1) the precise relevance of the innovation, (2) competence of the investigators, (3) monitoring of the societal impact, and (4) examining the efficient use of resources.[54] Fundamental to this process is commitment. For instance, in a comparison of four countries between 1970 and 2000, investment in research and development as a ratio of contribution of governmental and corporate funds rose by a factor of 2 in Mexico, 4.5 in Brazil, 5 in Spain, and 9 in Korea. The private sector accounted for 24 percent of the investment in Mexico, 40 in Brazil, 50 in Spain, and 73 in Korea. A significant variable in Mexico is the industrial leaders' stereotype of scientists as primarily academicians who seldom move into the applied. In 1998 the rector of UNAM finally established a liaison office to establish links between the university and the entrepreneurial world.

The Social Sciences

Behavioral or policy sciences suffer even more than the natural sciences in both funding and lack of coordination with national needs.[55] Economics developed largely after World War II with the advent of national and international agencies such as CEPAL (Economic Commission for Latin America). Universities increasingly offered courses in macro and microeconomics, as business schools appeared in the 1950s, either as part of Facultad de Economía or independently, as determined by the degree of private or state funding. Neoliberalism and cyberspace aid the growth of business schools. Interdisciplinary programs emerged as *facultad* and departmental walls became less rigid. Consequently, research and teaching broadened with consequent interest in business history, particularly in Argentina, Brazil, and Venezuela.[56]

Until World War II, sociology was predominantly social philosophy. However, an empiricist trend, partly a result of Latin Americans turning to graduate studies in the United States, took hold in the 1960s and beyond. Political science, as one Colombian reminds us, was a late entry, largely because it too easily threatens the status quo.[57] Social scientists have become more professional and supported by a number of organizations, both national and regional, and the Latin American Studies Association (LASA or ALAS) has over five thousand the members from both North and Latin America presenting at their meetings every eighteen months with approximately fifteen hundred presentations in both social sciences and humanities.

Under the UN umbrella several important groups are found—FLACSO (Schools of Social Sciences in Latin America), CLACSO (Council of Latin American Social Sciences), among others. In addition are national organizations for each discipline. Despite these groupings of scientists and scholars, national planning elites fail to implement the findings of anthropologists, economists, political scientists, and sociologists.[58] All these disciplines are engaged in research. In particular reference to sociology, Mexico's Pablo González Casanova points to the lack of a holistic approach. Research studies tend to focus on a particular problem in a specific setting, whether crime, family relations, labor-management conflict. It is mandatory to integrate these studies into a conceptual system and implementation in order to facilitate societal change.[59] As compared to sociologists, anthropologists have been more successful in persuading the government to adopt given recommendations, notably regarding the indigenous population.[60]

Problem Areas

When compared with other developing areas, Latin America is far advanced in science relative to Africa, but lags Asia. Likewise, the area represents a greater ratio of women (15 percent) in the sciences than Africa (9 percent), but lesser as compared to Asia (23 percent).[61] The region also suffers from a scarcity of technicians in its science laboratories. Sciences enjoy differential prestige. Many universities lack a strong program beyond a few disciplines; budgets extend to only a few significant offerings. Physics, for instance, is ranked higher than chemistry, yet receives limited support. Further, as implied in chapter 9, relatively few cross-disciplinary programs emerge.

These problems, including the balance between pure and applied research, affect even leading countries—Brazil, Argentina, Mexico, Chile, and Venezuela (in descending order of the annual number of scientific publications).[62] An idea of the barriers and challenges can be seen in the discussion of a few representative nations.

Brazil

Perhaps more than its neighbors, Brazil presents a dualism in its background, values, and goals affecting science along with related aspects of development. One is struck by the contrast in effects of educational levels, rural and urban, indigenous and African versus European, traditional versus innovative, the tropical north versus the industrial south. In this context is the unresolved dilemma of scientific and technological development of the Amazon.[63] In scientific inquiry the nation drew mainly on Europeans fleeing economic or political pressures. These immigrants created the first laboratories and taught young Brazilians the scientific method and its potential. By the 1930s Brazil was on the threshold of serious scientific endeavors, largely because a generation of Europeans formed the nucleus of research. This trend continued into the postwar period. The University of São Paulo was *the* center of these activities, but since the 1980s probably a half-dozen universities with expanded graduate studies account for scientific advance.

Expansion or contraction of experimental science depends, of course, on the sociopolitical climate. The 1964 military intransigence curtailed the possibilities of pure research. Only with industrialization and university expansion of the late 1960s and the 1970s were adequate funds available—even a limited amount for pure research, posing an interesting contradiction about academic priorities in the thinking of the nation's intelligencia.[64] During the 1978–79 fuel crisis a different mood emerged. NGOs and foundations continued their role, the Rockefeller Foundation delivering breakthroughs in health and agriculture. However, since the 1990s national centers for given research areas [the most important being the Conselho Nacional de Pesquisas (National Council for Research)] moved in gear for the nation's surge toward development. Nuclear development is the centerpiece of the Centro Brasiliera de Pesquisas Físicas (Brazilian Institute of Physical Research).

The 1990s brought new funding sources, restrictions of the military era were gone, and scientists increasingly joined international networks. Growth of the democratic process—removal of Collar and election of social scientist Cardoso to the presidency—fostered a commitment to serious research. Federal and state agencies along with corporate investments increased the financial underpinning. A striking aspect in this expansion was the strengthening of the space program in 1999 by establishing the National Space Research Institute and launching a microsatellite. Emphasis is on microgravity research in cooperation with similar programs in the United States— always with tenuous budgeting questions.[65] Problems remain as to ecological costs and the outcome for the hundred million Brazilians who live in poverty.[66] Brazilian scientists also complain of the centralized decision making, non-implementation of planning, and inability to diffuse research facilities to lesser cities and institutes, among other problems.

Peru

Even more than in Brazil, scientists in Peru depend on the government for support. Historically, precolonial societies had an impressive knowledge of certain areas of science and technology. From the colonial period through the nineteenth century the country remained detached from European intellectual traditions. However, major universities emerged. The University of San Marcos, dating from 1551, offers respectable departments in sciences, complemented by the National University of Engineering, in addition to nearly a score of other centers of higher education.

The Velasco military regime provided the impetus for coordinating science and technology in the Consejo Nacional de Investigación; still, for political reasons little integration occurred.[67] Other advisory councils directing scientific investigation toward national resource development supported applied over pure science. As the economy flounders, inflation rises, and investment in science sinks to a new low, as occurred in the Fujimori era. The amount of the GNP allotted to science and technology fell in Peru from 0.32 percent in 1981 to 0.23 in 1998, as compared to 0.80 percent in

Argentina, 0.66 in Brazil, 0.60 in Colombia, 0.55 in Chile, but 2.8 in the United States and Japan.[68]

Costa Rica

As in other Latin American nations, Europeans, mainly Germans, were conspicuous in raising a scientific consciousness during the nineteenth century. One impressive figure was the Swiss Henri Pittier, who arrived in Costa Rica in 1887. After considerable prodding by Pittier the government founded the Instituto Físico-Geográfico in the 1890s. The institute assumed a number of tasks, but its outstanding achievement was the geological and botanic mapping of this volcanic and ecologically diverse country. After 1900 North Americans gradually replaced European scientists. Still, Pittier's heritage remained, bolstered by the political elite, particularly the Liberal Party, who recognized the importance of scientific inquiry.[69]

A critical concern is the advance in scientific medicine. Ignorance, charlatanism, and patent medicines dominated health practices into the first decades of the twentieth century, as illustrated by a Cuban immigrant Carlos Caballo Romero, who remained a prominent *curandero* (healer) until 1932. However, the most serious health problem was the scourge of tropical diseases. Prompted by the United Fruit Company, the Rockefeller Foundation established a program in 1912, attacking malaria, yellow fever, and intestinal diseases. Between 1915 and 1931 the Department of Health, often working with educational authorities, aggressively pursued public health measures. In the 1920s the number of MDs, all trained abroad, increased approximately 70 percent, permitting care not only to the middle sectors but also the *clase popular*.[70]

Unfortunately, economic turbulence often prevents the ideal. Of the less than 1 percent of the GNP for science and technology, only a quarter comes from the private sector. Moreover, budgets forestall the acquisition and maintenance of equipment. As one instance, in 1978 the University of Costa Rica applied to Spain for high-level equipment for its chemistry department. When the item finally arrived in 1985 the government failed to come up with the customs duty of approximately five thousand dollars, and the item remained at the airport for years. The dilemma continues. Of Costa Rica's export earnings 25 percent goes to importing oil—a factor in the nation's four billion dollar debt. Yet, in 1974, industry overtook agriculture, and by 2000 tourism (including ecotourism) exceeded both. As compared to most developing countries, Costa Rica offers a favorable university environment, all the more remarkable for a nation that prior to the 1920s had only a law school.[71] As Central America's most advanced country, Costa Rica leads in scientific research, which is primarily focused on public health, marine studies (notably fisheries), veterinary medicine, botany, and agriculture, which focuses on food cultivation and processing.[72] Concentration on education and the social sciences is also evident. Finally, Costa Rica suffers from the constraints that affect much of the Third World (inset 10d).

Despite a marked improvement in the potential of scientific and technological research, inhibiting factors persist. One is the isolation of developing countries from major centers in Europe, East Asia, and North America. Two, most research projects in Latin America have a severely limited number of scientists and technicians assigned to them. Three, because of health needs, mathematics, physics, and chemistry are generally, but not always, subordinate to the biological sciences in funding and personnel—hardly propitious for the research potential in national development. Four, only a limited linkage exists between science and production planning, including implementation. Five, rewards for scientists are limited, many turn to administration or management as the only way of advancing their careers. The "brain drain" is another aspect of this problem.[73]

Technology: Goals and Realization

The difference between science and technology is arbitrary; it is a question of the applied versus the abstract or theoretical. Technology may be defined as a combination of knowledge, skills, and organization in order to produce and coordinate a system of goods and services. It is the means by which individuals manipulate their environment for meeting their needs. In addition to fashioning new products, technology may lead to savings in capital and labor.[74] All this must happen in an interrelationship that favors creativity (figure 10.1).

As with science, technological advance increased at a geometric ratio from the seventeenth century to the present. Technological breakthroughs were especially decisive in socioeconomic development during the last third of the nineteenth century. In this context, steam power made possible an

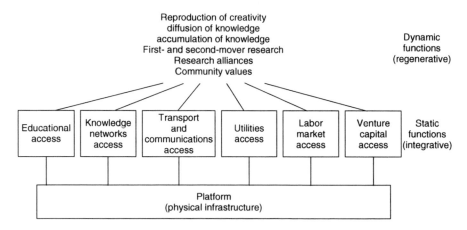

Figure 10.1   Relationship of infrastructural functions to invention.

Source: Luis Saure-Villa, *Invention and the Rise of Technocapitalism* (Boulder, CO: Rowman & Littlefield, 2000), 190.

expanded agriculture in Latin America. Further, the societal significance of technology, notably transport, is illustrated in the effect of railroad expansion under Díaz. Without this technological diffusion, mobilization of peasants in the Mexican Revolution would not have been possible. Although railroad workers were not unified, many aligned themselves with the *zapatistas* (followers of the charismatic Emilio Zapata), supplying the revolutionary movement with personnel and supplies.[75]

The impact of change raises questions as to the meaning and validity of our social institutions. Technology involves "multilayered meanings."[76] We may welcome vaccines but question nuclear power plants. We now live longer, but our atmosphere is polluted for both north and south. In many technological breakthroughs, society cannot adapt to the social and personal dislocation that change involves.

Latin Americans feel bewildered by the rapidity and compulsion of change, not least because it is usually imposed from abroad. Technological transfer, mediated by corporate structures, comes disproportionately from Anglo-America, although Japan and the European Union are ever more visible in this process. Communication satellites bring the region into the Western orbit with an array of economic, political, and scientific effects—all to an area historically ambivalent about scientific and engineering feats. Yet, it would be misleading to think of Latin America as indigenously deficient in technology. The Mayans had a calendar superior to the Gregorian one. The Incas devised masonry construction never to be surpassed. A number of cultures created ingenuous farming systems despite climatic, topographical, and hydrographic barriers. However, innovation was apparently not encouraged. Moreover, after the *conquista* the cultural milieu was not favorable to technological inquiry. The contemplative nature of Roman Catholicism was more oriented to "being" rather than to "doing." The work ethic remains even today a questionable variable in some quarters. Discovery and invention have only recently been rewarded. Still, managerial acumen might be accepted if it did not disturb the secular–clerical hierarchy. The role of worker is also relevant (inset 10e).

INSET 10E MACHINE, THE WORKER, AND INNOVATION

Through the twentieth century Latin America industrialized, if at an irregular tempo. However, because of globalization armed with cyberspace, a new—and questionable— tempo emerged. For example, the *maquila* represents a new level of efficient production in combining microelectronics with the assembly line bringing Detroit to Ciudad Juárez.[77] This technology, now a commonplace in the Third World, not only supplies the First World with cheap consumer items, but a mark-up of likely 800 percent assures the owners a comfortable profit. In addition, the *maquiladora* offers a new style of bureaucratic management as another instance of hierarchical control of women (men are also exploited). Still, counter to this marginalization, a social solidarity arises among informal groups, who agitate—if none too successfully—for autonomy, wage increases, and fringe benefits. By no means are workers completely passive agents. For instance, should production flounder and engineering personnel be unavailable, a repair technician more than occasionally intervenes by redesigning circuit boards.[78] Generally, exploitation at the workbench is the rule, as in the clothing industry, where new technology places severe demands on female workers, who are also subject to the arbitrary demands of a male manager.[79]

Questions of management, technology, and workers' rights appear throughout Latin America, indeed in most of the world. A Brazilian sociologist analyzing the automobile industry asks why workers are not given more initiative. Incentives and innovations are a departure from the conventional. Nevertheless, in the 1980s group decision making replaced the bureaucratic, mechanistic approach.[80] Moreover, growth of the metropolis tests the limits of technology and labor. Bureaucratized management instead of decentralization creates a new proletariat.[81] At the same time, this type of marginalization leads to self-protective measures. An instance is the collectors' (*pepenadores*) recycling of Mexico City's rubbish and garbage. As rural migrants, forced to sort, select, and transport trash as well as dealing with urban realities, including ruthless businessmen, they organized a cooperative for improvement of wages, housing, and other means of raising their marginal status.[82]

Technological innovation is often a response to a discontinuity or crisis. The 1973 oil crisis led Brazil to search for a fossil fuel substitute. Using sugar surpluses converted into anhydrous alcohol and mixing it with gasoline or proalcohol became an ambitious program in 1975. Several developing nations adopted similar programs, but Brazil was the major actor through the 1980s. During the external debt crisis Brazil, along with most of Latin America, sought self-sufficiency. However, energy demands, including automotive transport continues to grow. The negative effects of proalcohol on the environment, including agricultural balance and pollution are evident. Consequently, Brazil built massive hydroelectric plants, and looks to other frontiers such as solar energy.[83]

Technological, like scientific advance, hinges on ideology and the climate of decision making. Implementation rests with industrial, governmental, and professional elites. In view of limited resources a nation's leaders establish priorities, as revealed in a report of Argentina and Brazil in their development of nuclear energy and degree of computerization. International competition can influence the options in processes of production. For instance, "pragmatic anti-dependency 'guerrilla' groups" urged governmental elites on the necessity to establish autonomy. Advance in the computer industry in Brazil and a commitment to nuclear energy in Argentina arose from the convergence of ideology and varied actors in the national and world economic scene.[84] Bewildering many Latin Americans in their hope for industrial development is the hiatus between government planning and implementation, the division between private and public sectors, and the coordination of the academic and the industrial–technological.[85]

Technology, Agriculture, and Resource Management

The rural scene amply displays the conflict between traditional and modern. The Aztecs, Inca, and other indigenous cultures had a formidable system of sustainable agriculture with terracing, *chinampas* (water agriculture as in Zochimilco near Mexico City), elevated fields, crop rotation, and other forms of diversification. The colonial era modified these practices with the introduction of European crops. However, the twentieth century was caught in two cross-pressures: one, the needs of a larger population, and two, commitment to export crops, both with the result of overgrazing, soil degradation, erosion, and deforestation. *Campesinos* are forced into marginal lands and staggering poverty—a scenario for most of the Third World. Predictably, millions leave the countryside for the city.

To the degree that improvement is possible (the situation in Latin America is fortunately less drastic than in Africa and Asia) new agricultural technologies are relevant. Indigenous practices can be combined with novel methods of pest control, selection of disease- and pest-resistant types of crops, applicable fertilizers, mixed farming systems of crop cultivation and livestock, minimum tillage, and agroforestry.[86] Counter-effects are also possible. According to a research study of a Maya area in Campeche, Mexico, *ejido* lands are merged into NAFTA-inspired export agriculture with drastic effects on traditional farming. For the elders the situation is reminiscent of the oppressive *porfirato* haciendas. The present short-term, hierarchical, private ownership style violates the twin approach of "collective ownership and individual responsibility" characterizing indigenous agriculture of the tropical forest. In addition, rapid Western technological advance violates the Mayan cyclic view of the universe.[87] A balance must be maintained between local control and the demands of national and transnational agencies. Various organizations, both private and public, experiment in farming technologies. Yet, when *campesinos* have the opportunity to participate in decision making, success is more likely than failure. In several Andean countries extension programs based on governmental agencies (NGOs), or local grassroots organizations are changing both levels of productivity and natural resource management. However, these agencies often work at cross-purposes. The risks are clear when modernization and land intensification are geared to commercialization as opposed to sustenance of the local population. The degree of "modern" versus indigenous technology differs for each locality.

Similarly, the history of substituting European for indigenous models and bureaucratic for community standards is illustrated in irrigation systems. A study of Cabanaconde, a valley in the Peruvian Andes, points to problems of the Velasco and post-Velasco regimes as they imposed new structures for water distribution based on both the Inca and colonial systems. *Campesinos* in the highlands resent the abrupt action of coastal authorities, who violate the cosmology of their ancestors. Demographic shifts increase pressure for a new irrigation system. Consequently, water distribution systems become rationalized bureaucratically rather than according to the needs of specific clans and moities.[88] Misunderstandings occur between elites urging national development and local farmers who are forced to change their terracing and other methods. In the case of Peru, authorities in the Andes and in Lima are occasionally able to negotiate their differences.

Still other areas are testimony to the conflict between agricultural sustainability and bureaucratic management. As noted in chapter 2, the Amazon flood plain is only recently undergoing development of its potential— forest diversity, fish freezing, raising fields above the floodplain, tapping the gene pools of wild and domestic plants, to mention several possible frontiers. A multifaceted approach must integrate private and public interests, cattle ranchers and farmers, and above all, community involvement.[89] Another instance of negligent planning is the El Cajón valley in Honduras. The construction of a dam in the 1960s flooded an area, which only in the 1990s came under development. The government is proceeding with a minimum of expertise. Noticeably absent are anthropologists and other social scientists who might offer advise on the human consequences of technical decisions.[90]

The Present and the Future: The Case of Mexico

Bordering a powerful and technologically advanced nation, Mexico increasingly feels the need to accelerate its industrial apparatus and improve the standard of living for its hundred million inhabitants. In 1993 leaders implemented NAFTA, followed by the enactment of the National Development Plan (1995–2000), with strengthening of the National Counsel of Science and Technology (CONACYT). The Salinas and Zedillo administrations strove to double the number of scientists and technicians by the year 2000. The full-time equivalent of research and related personnel rose from 269,000 in 1993 to 333,000 in 1998. Moreover, CONACYT expended its role in planning, promoting funding, international participation, and public awareness. With the advent of the twenty-first century new goals are set—ranging from providing computers to some ten million homes to redistribution of *maguiladora* production in the hope of lessening congestion and environmental degradation.[91]

Nonetheless, several interrelated problems—common to most Latin American nations—remain:

1. *Economics*. Financial instability and shortage of funds haunt technology development. As one example, Mexican scientists cite the all-too-frequent devaluation of the peso as a deterrent to future planning. Indeed, fiscal constraints force preference for quick results rather than basic research and theory. In developed countries long-term planning is more often a reality than a dream.
2. *Politics*. President Vicente Fox had difficulty bringing his fractured legislature with him in establishing priorities and implementing them

with financial support. Although Fox found better rapport with entrepreneurs than did PRI leaders, the business sector fails to grasp the importance of their obligation to support research and development.

3. *Regional diversity.* The disparity between the relatively prosperous north, the industrialized center, and the impoverished south has implications for technology. For instance, Chuapas has approximately the same rate of illiteracy and poverty as Honduras. In his 2000 electoral campaign Fox spoke of his Plan Puebla-Panama with the goal of creating an international transit and industrial infrastructure, shifting the economy from north to south, and extending private ownership of the region's natural resources—all with questionable ecological consequences. However, because of funding problems and indigenous opposition, Fox's plan for changing the marginal areas of his country also met with frustration. Other priorities become salient, for one, the necessity of increasing technical and professional training of the young.

4. *The international network.* Mexican research, development, and industrialization cannot progress in a vacuum. In regard to the most immediate neighbors, Fox hoped for more aid from the United States, as inspired by the close relationship with George W. Bush in early 2001, but after September 11, the White House had more immediate interests. Further, integration—largely sharing technology—with the Central American Common Market as well as Caribbean nations, notably Cuba, was initiated in the mid 1990s. Other directions of cooperation are with the Organization of American States, especially the Regional Program of Scientific and Technological Development. Mexico is also active in technological transfer with the Andean nations. These negotiations move slowly because of fiscal uncertainties, leadership shifts, and differing needs.[92]

Cyberspace and Technological Control

Latin America was slow to accept the Internet, but is now adopting it at an accelerated tempo. The Latin American computer market increased 21 percent in 1995. During the following year computer use sextupled in many countries, in advanced ones, even more—in Brazil the market grew 2,333 percent between 1995 and 1996. Also, by 1996, forty-five thousand "servers" or host computers were available in Latin America.[93] Several cities have versions of Silicon Valley, Guadalajara alone employing nearly one hundred thousand cyber technicians. The implications of this digital revolution are, of course, immense. For one, the concept of space and communication is drastically expanded. Never before has Latin America felt itself a genuine community, with the expansion of "social space" it is part of a global society. Various aspects of the information society are taking place—voice recognition simulation, artificial intelligence, virtual reality, and supercomputers.[94] Many examples abound in this cybernetic revolution.[95] Major newspapers and databases are now online. Web surfers can obtain considerable data about Peru in English, Spanish, and Quechua through the Peruvian Scientific Network (RCP). A Temporary Autonomous Zone, devised by the University of Texas, presents testimonies of masked EZLN members from Chiapas. E-mail is now replacing phone calls and at lower cost. Ramifications of the Internet revolution affect almost all aspects of society from industrial management to political strategies, from education to medicine. As elsewhere in the world, cyberspace now drives a compulsive technological revolution for Latin America.

The region is aware of both the positive and the negative aspects of cyberspace. Commerce, industry, government, and the military are entering a new frontier in this information technology. Authoritarian regimes look to the Internet for possibilities of surveillance. Individuals and groups already display an array of unsavory or dangerous messages. On the other hand, the Internet can work in tandem with education in order to effect appropriate social change. Larger countries, such as Brazil, are already a vast social laboratory exploring the frontiers of cyber technology.[96]

What will be the eventual effects of the digital age? For one, television and the Internet are driving Latin America and other developing areas toward subordination of local authorities to global structures.[97] The era of fiber optics in telecommunications elicits questions of social control, notably in the expansion of large enterprises such as Telmex in Mexico.[98] In the frenzy for virtual reality not only elites but the general population could be caught in this "technology of self-construction" and a globalization of time and space. In other words, cyberization hints at postmodernism and documents the pessimism of the postmodernists (chapter 5).

Latin America confronts some conflict-laden dimensions of science and technology. As literature is not constrained by logic, philosophy, nor science, magic and technology occasionally merge and then move apart. In García Márquez' *Macundo*, technological ironies creep into the banana factory and its foreign owners as well as in the mysterious antics of government bureaucrats as they indulge in trickery and fantasy. Along with García Márquez, both Isabel Allende and Octavio Paz refer critically to the effects of media, transport, and military technology on social actors.[99]

In summary, in comparison with Anglo-Americans, Latin Americans may have a measured response than do the super-industrialized areas of the world, or may not. Technology can promote or inhibit humanistic values. In specific reference to the computer, dangers of surveillance are all too apparent in view of the region's political history. At the very least, the Internet endangers worker autonomy and physically isolates individuals from peers and community.[100] Technology demands change in behavior and norms; it requires an intervention across various sectors of a nation. Human history is replete with alternating themes of chaos, continuity, and reform. The digital revolution is unlikely to be an exception.

Chapter Eleven

Social Structure and Change:
Latin America and the World

It is appropriate to place Latin American societies in a conceptual framework that calls for an analysis of the nature of social change and how it applies to the twenty republics. All complex societies display intervals of stability and crisis, harmony and conflict. Latin American nations are no exception, but are characterized by a greater degree of conflict, at least in recent decades, than are most Western nations. In this analysis we return to the issues of previous chapters—class, ethnicity, and the institutional fabric in order to understand the parameters of social change. Obviously, economic and political developments are powerful driving forces in shaping a society in its orientation to modernization, globalization, and democracy. Finally, we examine the place of Latin America in the world at large, particularly its relationship to the United States, as it is impossible to understand the area without considering the pressure of its northern neighbor.

The Relevance of Social Theory

Undoubtedly we can understand societies better if we consider theories focusing on the nature of society. Philosophers have dealt with this question for over two thousand years. Nineteenth-century thinkers were especially eager to establish the "grand design." Through the twentieth century, speculative schemes gave way to more systematic and empirical approaches, already anticipated by the works of Karl Marx, Emile Durkheim, Max Weber, among others. As noted in previous chapters, although various theories such as symbolic interaction and exchange offer significant insights, the two principal theories are functionalism and conflict. Of these two models, conflict-oriented theories are more applicable to Latin America. Neither functionalism nor conflict and their variations exhaust the range of social theories. Macro approaches appear in systems theory, micro in ethnomethodology, Fusions of the two models are found in neofunctionalism (a revised treatment to correct the inadequacies of earlier functionalism) and its blend of structure–function and conflict, feminist theories, and poststructuralism, to mention several contemporary conceptual viewpoints.

Functionalism

Functionalism, or structural-functionalism, as it is often called, involves several principles: (1) society is a system composed of actors, operating as an equilibrium, (2) as society is a "self-maintaining system," certain needs must be met if homeostasis is to be preserved, and (3) subsystems and cultural patterns are relevant in this social system.[1] In its various interpretations— Durkheim, Weber, and Talcott Parsons—functionalism is directed to the scientific analysis of the social system. For example, in Latin American society, the hypothesis that migration to the city is followed by higher rates of social disorganization and anomie—a statement based on what Durkheim labeled as "social facts." Durkheim also saw society as an integrated system. Religion, the family (both extremely important in Latin America), and other institutions are social systems. Other sociologists contributed too, notably Max Weber and his stress on social values and Talcott Parsons, who among other concepts outlined "pattern values," for example his note on particularism–universalism and ascription–achievement as noted in chapter 5.

However, the difficulty with functionalism, especially as espoused by Parsons, is the implicit assumption of a more-or-less continuous ongoing functions or processes in society, without sufficient acknowledgment of possible breakdowns in the system. Partly because of the concentration on Western, primarily North American, society, crisis or conflict does not receive adequate attention. Third World social systems behave even more unpredictably than in more developed areas, as elaborated in inset 11a.

INSET 11A   THE HAITIAN SIEGE: LESSONS IN
CONFLICT RESOLUTION

After decades of dictatorial rule by the Duvalier dynasty, Jean-Bertran Aristide was elected president in 1990. Troubling days lay ahead as Haitians struggled for the survival of democracy against military and landed elites. Several months after his inauguration, the military led by General Raoul Cédras removed Aristide from office on September 30, 1991. A familiar scenario followed: the Organization of American States (OAS) condemns the coup. A high-ranking delegation representing eight nations warns the Haitian government of impending economic consequences if it does not restore democracy. After over a year of turmoil, negotiations, and an economic embargo by the United Nations, the military verbally accedes to Aristide's return to power, but impediments continue.

The most dramatic event was the October 1993 seizure of the *Harlan County*, a cargo ship carrying 193 U.S. and 25 Canadian troops, who were to train the police and the military to build up the nation's infrastructure—a provision of a previous accord. The UN reestablished the embargo. Violence continued with assassinations of Aristide's followers, both laity and clergy. A year later, despite opposition by conservatives, President Clinton acceded to pressures for a solution of the crisis by dispatching a military intervention, thereby forcing the resignation of Cédras and his coterie. The United Nations lifted its embargo.

A frightful aspect was the violation of human rights, which has plagued Haiti for generations. Once the coup removed Aristide, beatings, torture, disappearances, and executions flared up despite the presence of over one hundred human rights observers. This tragedy was compounded by the embargo imposed by the UN and other international organizations. The effects of the embargo fell disproportionately on the poor, 90 percent of Haitians. As a result, thousands sought refuge in the United States. However, most of the refugees were intercepted by the U.S. Coast Guard and sent back to Haiti. Because of popular and congressional pressure for and against the exodus, both presidents, Bush and Clinton, wavered in their response to this crisis.

What lessons can the international community learn from this lengthy impasse in social justice? Several issues are relevant:

1. In dealing with complex situations, skilled negotiators, reporting to the OAS, UN, or other organizations are requisite.
2. Embargoes and sanctions are precarious, often resulting in hunger and unemployment for masses of the population. Each tightening of the embargo results in increased waves of refugees seeking haven in the United States, where because of socioeconomic grounds they are not acceptable. During the Cold War political refugees were occasionally welcome if they were fleeing a communist regime. Another danger of embargo is inconsistency, for instance, Air France continued its flights into Port-au-Prince, whereas other carriers suspended theirs, and the Unites States, despite prior agreements, chose to make exceptions in industrial trade.
3. The wealthy nations of the world must come to terms with the distinction between political and economic refugees and reevaluate the factor of race.
4. The use of coercive diplomacy, notwithstanding its risks, is in some instances the only means of effective change. International or multilateral action is preferable to unilateral.[2]

In 2004, Haiti was again beset with corruption and violence. President Aristide proved to be incapable of resisting the corruption and violence in his country. Still, conflict resolution is a principal means of approaching the schisms of a given society. Unfortunately in the Bush era the United States is reluctant to pursue this route.

Conflict Theories

Any equilibrium can be unstable. Competition and conflict arise in most segments of society, as abounds in Latin America—social stratification and mobility, ethnicity, gender, ideology including religion.

Several nineteenth-century thinkers built their social philosophy around conflict. Marx, the most systematized, had the greatest influence. Borrowing Georg Hegel's dialectic (thesis, antithesis, synthesis), conflict was intrinsic to change as successive economic forms evolved. It is not altogether farfetched to think of Latin America's transition from colonialism and dependency to find an antithesis in import substitution with a synthesis in neoliberalism and globalization. According to Marx, any synthesis would be revised by a successive dialectical process. Of course, any synthesis involves intricate behaviors with the actors moving from patterns of competition to those of cooperation or the reverse with a diversity of ideological, political, economic, and diplomatic models.[3] Marx's epistemology has special significance for Latin America. He held that the individual must question the basis of knowledge—what one assumes to be an unquestionable and absolute version of reality emerges from those who rule the society and its institutions. Whatever the shortcomings of Marx's economic theory, neo-Marxists remind us of his contribution to the sociology of knowledge. Latin America is no stranger to the problem of verbal façades and the difficulty of finding "true" reality (chapter 5).

As in probably every culture, individuals, groups, organizations, and the basic institutions (the economy, family, church, among others) in Latin America compete for authority, power, or even for survival. In this connection, a cluster of roles leads to redistributing resources, whereas another actor or group of actors desires to maintain the status quo, the conflict leading to a new form of social organization. Patterns of conflict vary by the amount of consensus expressed in coalition formation, as indicated in the disequilibrium between business leaders and the Institutional Revolutionary Party (PRI) in Mexico. Whereas in most of Latin America a cozy relationship flourishes between government and business, the PRI seldom appointed the economic elite to policy making, and generally government policy was less than friendly to business. Consequently, entrepreneurs and corporations organized, thereby achieving bargaining power with the state.[4] However, consensus building frequently fails to occur, conflict giving way to intensity and violence, as related to the degree and rate of social change. The path from severe frustration to aggression ending in verbal and armed conflict is seen from Chiapas to Ayacucho, from the highlands of Guatemala or the streets of Buenos Aires.

Lewis A. Coser offers a variation of conflict in turning to Georg Simmel's *organicism* or functional totality. For Simmel, the social system is disrupted by processes of deviations, dissent, and violence. The range of potential and actual conflict in Latin America varies from the struggle between established and marginal individuals, groups, and organizations, ideological schisms as between deductive versus empirical approaches, capitalism versus social welfare. A conspicuous point of tension is between the civilian and the military—even Venezuela, a country with a democratic structure for over a generation, became prey to military incursions in 1992 and beyond. In the first years of the twenty-first century waves of disruption and violence point to a deepening of conflict, most conspicuously in the Andean countries.

Conflict is occasionally characterized not only by chaos but by flexibility and adaptability, as reflected in the economic and political gyrations in Latin America, noted in chapter 4. Consequently, conflict is a variable and can move in a variety of directions. Moreover, conflict is not totally dysfunctional.[5] For example, by the 1990s even the ossified PRI had to accommodate to economic realities and new political forces. Again, conflict can move in a variety of directions. As an intriguing instance, in Lima, Peru, roaming comics (*cómicos ambulantes*) appear in streets, plazas, and parks with humorous critiques of society—its economic perils, modernization, occasionally assuming a utopian note. The most appealing ones appear on TV "talk shows." It is not clear whether they arouse or reduce both latent and manifest conflict.[6]

A ritualistic configuration of conflict can be seen in the political and military strategies in Latin American border and other disputes. One instance is the conflict between Argentina and Chile over the sovereignty of three islands in the Beagle Channel at the southern tip of South America. As the two nations desired frontage on both oceans, they squabbled over the specific border for decades. When Argentina and Chile became authoritarian military states in the 1970s, latent conflict became overt. By the late 1970s negotiation by the Vatican and other diplomatic agencies in the hope of a peaceful solution broke down and both nations increased their saber rattling. Inevitably the two adversaries had to reckon their strategic advantages as well as the expected military and economic costs. Argentina explored the possibility of enlisting support from Peru and Bolivia (enemies of Chile in the War of the Pacific, 1879–83), while Chile continually assessed its bargaining potential. With the return of democracy to Argentina in 1983, pressure for a peaceful settlement was finally resolved. As with other Latin American border disputes, between Ecuador and Peru, among others, the degree of accountability usually increases with democratic rule.[7]

Randall Collins, even more than most conflict theorists, envisions conflict as the basic direction of society. As societal arbiter, the state contains a near monopoly of significant resources—the military, police, power of taxation—forces and policies not too often in harmony in Latin America. What is meant by the state is, after all, the agency by which violence is organized.[8] Economic and other means of social competition and stratification guarantee a prolonged conflict, as portrayed in inset 11b.

INSET 11B   COLOMBIA: AN ENDURING SAGA OF CONFLICT

One can characterize every nation of the world as exhibiting degrees of both conflict and concord. This book repeatedly refers to disagreements and tensions throughout Latin America. At present, Colombia represents a tragic profile of entrenched conflict. In a world of both intermittent and prolonged episodes of violence—whether the Middle East, the Balkans, or Northern Ireland—Colombia exhibits a disturbing pattern in the dynamics of conflict—notwithstanding the reality that most individual behavior in Colombia is non-conflictual. Interestingly, since 1958 Colombia remains, if somewhat nervously, in the democratic camp, even if the political process is not fully institutionalized—passive participation in elections, questionable articulation of the three branches of government, and intermittent reports of presidential Machiavellian and corrupt practices, as in the case of President Ernesto Samper in his alleged contributions from drug lords to his 1994 election campaign.

In addition to one hundred years of intermittent conflict from the mid-nineteenth to the mid-twentieth century, Colombia experienced at least three periods of violence (with a certain degree of overlapping): (1) "partisanship" violence (1945–53), largely rural, between the Liberals and Conservatives, aggravated by the 1948 assassination of Jorge Gaitán, a charismatic political leader on the left, (2) "mafia" violence (1954–64) when the drug lords came into ascendancy, and (3) guerrilla revolutionaries (1961–89), variations of Marxist and Maoist ideological warfare.[9] By the 1990s phases two and three merged. Moreover, the conflict involves a three-pronged warfare. One constituent is the guerrilla operation numbering some twenty thousand members of the National Liberation Army (ELN) and the FARC Revolutionary Armed Forces of Colombia (FARC). Second are the military and paramilitary, such as Special Services of Private Vigilantism and Security (CONVIVIR). Third are the narcotraffickers. Rearrangements of coalitions emerge, particularly between guerrillas and the drug culture. Caught in the middle are peasants who find their livelihood threatened by the warfare surrounding their coca cultivation. The U.S. intervention, as highlighted in the 1.6 billion dollar package introduced in 2000, with further implementation in subsequent years, only adds to the confusion. Ironically the raison d'être of this tragic warfare is the satisfaction of North American quest for cocaine—which understandably Colombians seldom fail to mention. As the most recent twist to the Colombian tragedy is the near-landslide election in May 2002 of Alvaro Uribe, who promised increased military action against the guerrillas with, he hopes, increased aid from the United States and the United Nations.[10] Whether this aggressive action will bring stability remains questionable.

Social Change Theories and Latin America

Social science as "science" is committed to being value-free. This principle also applies to the elusive question of how societies change. Behavioral scientists have the right to recommend how a society should change but this aim places them squarely in a value commitment. The fundamental issue is the search for an adequate explanation of social change in the world community, including Latin America. It is my belief that conflict theory offers fewer difficulties than do functionalism and its variants, as this chapter indicates. Yet, in the trend toward international affiliations and the hope for more effective societies and their institutions, cooperation may gradually replace conflict. History points to cases of alternating periods of accommodation and conflict approaching violence. As one instance, in Argentina democracy prevailed through the first three decades of the twentieth century, with high levels of conflict resulting in authoritarian regimes (1930–55, 1966–73, 1976–83) interspersed by periods of uneasy democracy, at least phases in which concessions could be made between contending parties. Argentine democracy has seldom appeared stronger than it does under the presidency of Néaor Kirchner (2003–).

The approaches to analyzing social change hinge on foci or continua such as the search for trends or stages based on objective or subjective standards, or on factors and causes.[11] In a related context the theory of social change can be thought of dualistically as opposing approaches:[12]

1. Endogamous or exogamous—whether changes arise from without or within the society. To what degree are Latin America's problems internal or external? If to less extent than a half-century ago, debates surround the intervention of Europe and especially North America. Neoliberalism and particularly globalization stem largely from abroad.
2. Inevitable or contingent—Marx versus Durkheim and Weber. As social change can be contextualized as utopian and dogmatic or as statistical and empirical, both approaches are relevant to Latin America.
3. Methodological individualism or sociological realism—introspection and deduction versus empiricism. For Latin America, as for the rest of the world, idealized interpretations and goals have to be placed in the context of specific experiential events. Latin American intellectuals traditionally lean toward the contemplative and philosophical rather than the empirical. By the late twentieth century the intellectual approach turned more empirical.
4. Objectivity or commitment—Weber's distinction between value-free and value-relevant. Can one study Latin America or any complex human society without a personal investment? Or should one? A scientist must decide between subjectivity and objectivity.

Underlying these propositions is the principle, at least from the social scientist's viewpoint that the study of social change revolves about a systematic testing of hypotheses. Or to put the matter in more modest terms, Robert Merton's goal of assembling "theories of the middle range" is preferable to the all-embracing system building of Parsons, not to mention the classic social philosophers. This testing of a hypothesis or a cluster of hypotheses means examining overt responses as well as values and beliefs. An intriguing example is Everret Hagen's inquiry into Antioquia's economic ascent in Colombia. Antioqueños are considered an anomaly in Colombian cultural history. Many were originally, it appears, Basques and Jewish converts to Catholicism in contrast to the minor Spanish aristocracy who settled in Bogotá, Cartagena, and other centers. On arriving during the seventeenth century in what is today Antioquia in search of a fortune in mining, immigrants were soon faced with the dilemma that the mines were no longer profitable. Ever resourceful, Antioqueños turned to farming in coffee and tobacco. By the twentieth century they became heavily involved in textiles and other industries. Their entrepreneurial and hyperactive approach to economic affairs produced a stereotype and consequent joking about their business acumen among Colombians in regions marked by less assertive inhabitants. Although Hagen's thesis has been questioned, a research study using projective tests reveals Antioqueños' personality syndrome as having a "hands on" approach as compared to samples from other areas, notably Popoyán, whose subjects display a more passive or contemplative attitude.[13]

Theorists stress the role of conflict in social change. Still, conflict does not necessarily produce social change. At the same time, conflict is one means of initiating structural shifts with far-reaching implications for a given society. In this context, one recalls how conflict is interlaced in Latin American culture through all its history—conquerors and vanquished, ethnic and class stratification, clashes of authority in the colonial and national periods, and in the inter-institutional struggle for allegiance to the church, state, military, among other institutions. Since the 1960s new and more subtle conflicts surface. For instance, new technologies supersede vested ways of communication and industrialism, achievement-orientation challenges ascriptive selection, neoliberalism replaces earlier economic trends, and the political arena is redefining clientelism and populism. Again, we remind ourselves that these shifts are often functional even though laden with potential conflict or overt violence.

One aspect of conflict, both manifest and latent, is the drastic transformation of societal institutions. As one looks at Western history, one sees a striking decline of the church, which began to lose its authority and power with the schism into Roman Catholicism and Protestantism in the sixteenth century and increasing influence of science in the seventeenth century and beyond. The Latin American church experienced setbacks in the late colonial period, a gradual descent in the nineteenth century, and accelerated secular pressure in the twentieth century, especially the second half. Another institution suffering relative loss is the family, if less of a decline in Latin America than in most Western cultures. Economic, educational, political, and recreational institutions robbed the home of its central functions, particularly socialization, which was once the exclusive province of the family and the church. Perhaps it is more accurate to say that the family has changed its functions rather than losing them. At least, the family turns to new modes of socialization as opposed to the traditional authoritarian structure of the past. On the other hand, the economic institution— the flywheel of society—expanded its functions at a logarithmic scale in Latin America since the 1960s. Growth of political and educational institutions in scope and complexity affects the Latin American scene. Further, the totality of actors judge institutions as to whether they offer "social efficiency and collective welfare," which must be judged in the context of costs and benefits.[14] In other words, in the unfolding of Latin America the economy and the political order are at the core of social change for the present and the future. In regard to political power, the church and more recently the military have lost significantly. Generally, these changes are evolutionary, in contrast to more abrupt or violent change, as seen in inset 11c.

INSET 11C   REVOLUTION AS SOCIAL CHANGE

Social change can occur by evolution or revolution, or in varying degrees of both. The term "revolution" conveys different meanings, but however one defines the term, Latin

America had at least four during the last century—Mexico, Bolivia (however incomplete), Cuba, and Nicaragua. Only the Mexican Revolution conformed to the scope of violence of the French and Russian Revolutions, different though the three revolutions were. Still other countries veer toward radical change, often in the form of social movements directed to (1) macro or global changes such as socioeconomic justice, (2) quality of life as in the ecology, or (3) specific groups or subcultures, whether gender or ethnic. Latin America reflects all three.[15] Movements can be thwarted by the military, an authoritarian regime, or frozen by some other structural impasse. It behooves us to examine the anatomy or phases of revolutions and quasi-revolutions.

There is probably no final definition of revolution, but some aspects are clear. It is a process in which the structure of a society drastically changes. Revolution involves macrolevel societal entities such as the state or social class. The process is seldom complete unless both institutionalization and consolidation take place.[16] Jan Knippers Black looks at revolution as an ever-shifting relationship and ongoing process in Latin America with its three social classes, which can be viewed as a pyramid with the upper class at the top and the lower at the base. The strength and proportion of the upper, middle, and lower become stronger or weaker in the sequence of the three phases: evolution, revolution, and counterrevolution.[17]

In examining social change, the first phase, *evolution*, was for four centuries the elite's domination over the bourgeoisie and more harshly over the vast peasantry. The structure varied, and still varies, between the tightly hierarchical pattern of Mesoamerica and Andean countries as compared to a less rigid structure with a larger and more urbane middle class in the cone countries. Through an evolutionary process the middle class incorporates organized labor into their ranks as in, for instance, Costa Rica, Venezuela, Argentina, and Uruguay, if with periodic strains between the class sectors.

Revolution or "collective defiance" may begin in spontaneous rebellions that occur when conditions become intolerable.[18] Revolutions occur under a constellation of factors beyond the economy and social classes—strength or weakness of the state and its multiplex apparatus (particularly the military), the media, and other institutions. The relationship between city and countryside is important. Unlike most European prototypes Latin American revolutions tend to emerge in the rural. Further, a revolution, if successful, involves leadership transfer. For instance, in Mexico charismatic leadership appeared briefly in Francisco Madera, but as the revolution continued Emilio Zapata carried the banner of *tierra y libertad* (land and freedom) in his seven-year struggle (1912–19) against the *carranzistas* (followers of the corrupt president, Venustiano Carranza, who finally demanded the assassination of Zapata),[19] a succession of leaders and political gyrations, followed by a degree of stabilization in the late 1920s. Moreover, shifts of power, symbolic and monetary goals, and redistribution of resources are illustrated in Latin America. The Mexican Revolution meant defeat of classical colonialism, including dismembering of the church, transformation of national elites, and land reform (notably in the 1930s), all at the cost of approximately a million lives, far beyond the toll of other Latin American revolutions.

Revolutions have their special targets: intractable ownership of land and mines in Bolivia, the regressive Batista regime, and U.S. domination in Cuba, the despotic Somoza hegemony for Nicaragua, the Chiapas struggle or quasi-revolution against unscrupulous landlords and government officials. Yet, all have varied goals, ethnic relationships, agrarian reform, social welfare, and most critical, a restructuring of the class system. Institutionalization of political, economic, educational, and other organizational infrastructures became problematic in Mexico. Clientelism and party organization were redefined in the 1920s and again in the 1940s. During these decades the PRI was institutionalized but the revolution in a diminished form was consolidated, that is, it acquired popular support, meaning, and legitimacy, at least until the 1990s.

As a third phase of social change *counterrevolution* is the response to an incomplete revolution when economic elites and upper–middle classes, supported or led by a military, push to restore the former power structure. Hyperinflation and other aspects of a dilapidated economy can also precipitate counterrevolution. The degree of violence varies, but death, imprisonment, or exile is often the fate for political reformers, labor organizers, student agitators, and journalists. Examples are Guatemala (led by U.S. intervention) in 1954, Brazil in 1964, Chile and Uruguay in 1973. Argentina had recurrent episodes (the most dramatic were in 1966 and 1976) and with Chile represents the most violent counterrevolution. As the military and paramilitary are usually in control, the church is the one institution that may offer, conspicuously in the Pinochet case, a refuge. Eventually, *decompression* occurs as suppression eases, and the middle and occasionally the upper classes call for a return to normalcy.[20]

Revolution or Civil War?

Violent social movements appear in Latin America but more as truncated revolutions. Guatemala, El Salvador, Colombia, and Peru are diverse examples of peasant wars against the status quo. We may look at two of these:

As a kind of coffee kingdom dominated by a score of landowner families, El Salvador has a tragic history of violence. During the Great Depression coffee prices fell and pickers moved to the political left. Consequently, in 1932 General Maximiliano Hernández Martínez orchestrated the execution of no less than thirty thousand peasants. After this cataclysmic episode, political deviance remained buried until the 1960s. Finally, the left, frustrated by the inability of its marginalized political parties to effect significant changes, organized (with the help of disaffected military officers) an abortive reformist coup in 1979. As military death squads intensified their action in the 1980s, the United States sponsored a military–Christian-Democrat coalition in order to provide a degree of legitimacy. With the mobilization of the FMLN (Farbundo Martí National Liberation Front) a civil war was launched, which was to last over a decade. The FMLN fought mainly a war of resistance as the opposition offered a democratic façade with the landowning elite in control.[21] Finally in 1989 both sides saw the futility of the struggle and in 1992 reached a formal peace agreement under the auspices of the UN and the OAS. By 1997 the FMLN won parliamentary and local elections. As a concluding note to the Salvadorian tragedy, two of the generals apparently responsible for much of the violence during the 1980s, including the murder of four North American nuns, were offered residence in Florida![22]

The civil war in Guatemala could scarcely qualify as a revolution. It was essentially the massacre of nearly two hundred thousand peasants, predominantly Mayans, intermittently between 1960 and 1996, most of the violence occurring in the 1980s. In addition, an even greater number fled to Mexico and the United States. The genesis of the conflict was the oligarchy's employment of the military and paramilitary to repress demonstrations by exploited peasants, many of who joined guerrilla action under the umbrella National Revolutionary Unity (URN). The precipitating factor of increased violence began with the devastation of the 1976 earthquake (which killed twenty hundred people). In the wake of this catastrophe came foreign aid missions, propelled by the humanitarianism of the Medellín bishops' conference (chapter 8). As recalcitrance increased, these missions encouraged resistance. Members of indigenous grassroots organizations along with labor leaders, Christian Democrats, Social Democrats, and United Front of the Revolution party members, journalists, professors, and students were murdered in the early 1980s. Guerrilla activity continued to increase in the northern highlands and to a lesser extent along the Pacific coast. The war was based on both class and ethnicity. Many Mayan communities felt that the state used the war as a means of destroying the indigenous population.[23] The most intense campaign of violence was in 1984 with mass rape and murder of several thousand *campesinos* and villagers. Three presidents—all generals, Romero Lucas García, the "born-again" evangelist Efraín Ríos Montt, and Humberto Mejía Victores—oversaw the carnage. In the mid-1980s repression was institutionalized in "civil patrols" (Patrullas de Auto-defensa Civil), which aided the military in controlling the rural population. After 1986, military violence became less acute and in 1996 the Guatemalan government signed a peace accord with the guerrilla forces. However, violations of Mayan civil rights continue. General Montt even became a leading candidate for the 2003 presidential election.

The reign of violence in these two countries as well as other quite different settings—for example, guerrilla wars in Colombia and revolutionary social movements (notably the Sendero Luminoso) in Peru—illustrate the extent and depth of conflict in Latin America. Through revolutionary activity *salvadoreños* gained at least a few of their goals, for example, agrarian reform programs. Latent conflict remains, in rural areas of most of Central America and the Andes. For instance, Peru's President Alejandro Toledo is not too successfully struggling toward betterment of its poor. To repeat, revolutionary goals, mass defiance, civil disobedience, all are interlaced with gender, ethnic, class, religious, economic, and political variables.

The Economy and the Future

Latin America finds itself in a paradox, if not a convulsion, of competing economies struggling for socioeconomic improvement or even survival. In Peru, for instance, the informal sector accounts for 60 percent of labor

(95 percent of transport personnel), 40 percent of the GNP. A formidable part of the informal sector is the drug empire—twenty thousand farmers in cocaine production (60 percent of the world's crop), administered by some five hundred traffickers, often labeled as "criminals."[24] Could one not define the drug flow as globalization at its most rugged and intense expression?

Neoliberalism and Globalization in the Twenty-First Century

The saga of Western society, particularly Latin America, over the last five centuries is staggering: from the stage of discovery, exploration, and exploitation to one of industrialism and emergence of the nation state, and finally to a market-driven economy based on a global model. As noted in chapters 2 and 4, globalization itself is a bewildering process. Because of a shifting hierarchy of labor and related costs, workers remain in an ambiguous and anxiety-ridden suspension as to their employability. As one instance, shoe and textile manufacturing moved from north to south in the United States, then to Mexico and Brazil, only to find the labor market more economically attractive in Southeast Asia. In a broad sense, the world economy historically shifted from mercantilism and colonization to import substitution and finally a "free," world market economy—Latin America is still exporting cheap goods or raw materials with the burden of importing more expensive items from the major industrial powers. Today, globalization through neoliberalism exhibits several trends between 1975 and 2000: (1) wide dispersal of production systems and centralization of corporate power structures, (2) more than doubling of the world's wage-labor population between 1975 and 2000 with most of this growth in the Third World, (3) far-reaching demographic shifts as thousands from the Third World migrate to the First World, for example, the ratio of Hispanics rose from a twelfth to an eighth of U.S. population in the last fifth of the twentieth century, (4) urbanization rising geometrically in many areas as factories and *maquiladoras* absorb a seemingly endless supply of cheap labor, and (5) political consequences of popular opposition to both the ecological and exploitive aspects of globalization.[25]

The societal, specifically the political, aspects of neoliberalism and globalization are also impressive. More than ever, the market economy becomes the driving force of a society. Governmental decision making stressing capital formation and market consciousness along with the dismissal of protest portend a social crisis. Demonstrations beginning in Seattle in 1999 document the anxiety and hostility surrounding the IMF, NAFTA, World Trade Organization (WTO) for their potential assaults on local markets and labor as well as on the environment. Latin Americans are aware of the enormous demographic effects of globalization in the movement of workers across boundaries.[26] Many Latin Americans feel that a more balanced economic, and technological orientation would result if the European

Union played a larger role in order to counteract U.S. political and economic dominance.[27]

It is appropriate to view neoliberalism in a broad context and specifically in Latin America. As noted in chapter 4, collapse of the oil boom left Mexico and Venezuela in disaster as they and other nations scrambled to meet their debt obligations with the World Bank and IMF, autocratically driving Latin America into a zero-sum situation. At the same time, it was Chile, also in the 1980s, that led the way to a "free market" economy. Inspired by Chile's success, which continues into the twenty-first century, nearly all nations moved toward neoliberalism, despite the duress forced on the working class—largely because of First World demands. Even *sandinista* Daniel Ortega in his 2001 presidential campaign embraced a free market economy. Still, Nicaraguans chose businessman Enrique Bolaños (who significantly had political support from Washington) as he might more likely lift them out of their misery, possibly the second poorest country in Latin America (in 2000 the GNP per capita was $410 for Nicaragua and $380 for Haiti, with Honduras and Bolivia not far behind).

Whatever its perils, neoliberalism became the modus operandi. Clientelism assumes new meanings as Latin American nations attempt to pay off their foreign debt. In the harsh reality of meeting the balance of payments, several nations make progress, others fall further behind. Argentina reached a total indebtedness of 132 billion dollars in late 2001. After widespread riots came a change in government followed by an announcement of defaulting on the debt with inevitable inflation and a possible reversal of neoliberalism. This virtual meltdown of the economy occurred in a country with a standard of living in the 1920s comparable to Canada and Australia. Default on national debt usually has ripple effects through the hemisphere. Ironically it was the IMF recommendation to Argentina to privatize its social security fund that partly led to the fiscal breakdown. The stringent belt-tightening only made the patient worse. However, shrewd action on the part of President Néstor Kirchner turned the economy around resulting in 8 percent growth for Argentina in 2003 and 2004.

Governments waver and struggle between ruthless decrees of the international financial community and popular dissatisfaction at home. Several administrations resist international pressures, as did Siles Zuazo in Bolivia (1982–85), García in Peru (1985–90), and Ortega in Nicaragua (1979–90). For a few nations such as Uruguay and Costa Rica the crisis was less painful, keeping the IMF at bay by passing measures that met with popular protest, yet the democratic process survived. Others, caught in the web of hyperinflation and unable to meet payments turn to deflationary plans, for example, in the 1980s the Austral plan of Argentina's Alfonsín and the *Cruzado* of Brazil's Sarney—initial successes followed by disaster. As the crisis continues, several countries resort to strong-arm tactics with "draconian neoliberal shock plans."[28] Among these were Paz in Bolivia, Salinas in Mexico, Pérez in Venezuela, Menem in Argentina, and Fujimori in Peru. Controlling prices and inflation along with cutting wages, tariffs, and subsidies were the ingredients with appalling effects for the working class in the 1980s.

The change in the economic landscape in the 1990s is seen in the amount of Latin America's 1992 trade surplus—$243 billion ($534 for every individual in the region) of which nearly all the surplus was to service debt payments. Inspite of the short- and long-term impacts of globalization, economic data were less harsh in the 1990s than in the 1980s. Still, macroeconomic trends are costly. Privatization, foreign investment, and debt place a severe burden on the society. If the urban population has mixed gains the rural suffers deeply as dramatized in Mexico, where the transition from peon (hired labor) to an autonomous *campesino* remained intermittently frozen from the revolution to the end of the century.[29] In a broader sense, free trade and globalization have inimical effects for the rural population.

Difficult though it may be to assess the net advantage or disadvantage of neoliberalism, the upper segments of society gain the most. The number of Latin American billionaires grew sevenfold between 1987 and 1994 (from six to forty-two), dramatizing the extreme stratification of the region. It is estimated that a 2 percent tax on the income of the wealthiest fifth of the population would raise over half the indigents to above the poverty line.[30] The Comisión Ecnómica para América Latina (CEPAL) reminds us that wage rates tend to fall after any economic shock wave, the major countries (Argentina, Brazil, Mexico) suffering as much as the less advanced ones. In order to foster solvency and control inflation Mexico's Ernesto Zedillo (1994–2000) discouraged capital flight, decreased domestic demand, yet attracted external financing, restoring international reserves, and stabilizing exchange rates[31]

Socioeconomic Models and Transformation of the 1990s

The accelerated rate of socioeconomic change of the 1980s and 1990s not only affected capitalist nations but also socialist ones. The classic case is Cuba's metamorphosis. In its first three decades socialist Cuba enjoyed a relatively bountiful relationship as the Soviet Union replaced the United States as a stable patron of the sugar crop. After the Soviet breakup the Cuban economy declined by a third between 1989 and 1993 as underlined by the desperate exodus of refugees to the mainland. Gradually, Cuba followed China into what is labeled as "market socialism."[32] By 1994, as noted in chapter 4, Castro permitted a number of bold changes: opening of foreign investment, partial acceptance of a dollar economy, decentralization of the state economy, including transfer of state farms to cooperatives, and emergence of private business operations. These changes, especially the dollar economy, has the effect of restoring a class system based on access to dollars with their purchasing power.[33] Probably the most dramatic change is the growth of the tourist industry. The number of hotel rooms and airline traffic increased fifteenfold between 1989 and 1999. The attempt of the United States to isolate Cuba seems ever more ludicrous as nearly all European and Latin American nations maintain normal trade relations with Cuba. Even Israel, which along with Uzbekistan voted with the United States for the embargo at the UN, sells arms to Cuba.

A rather different development is the Nicaraguan experiment. In its pluralistic approach to socialism Nicaragua was never as rigid as Cuba. First of all, Sandinista leadership was not quite as abrasive as the Castro approach. Second, Sandinistas were sensitive, at least periodically, to the heterogeneous population, including its Afro-Caribbean and indigenous (notably Meskito) minorities. Third, and most important, the government, conscious of its huge debt and unemployment, feared that excessive socialism might not be the solution and would alienate the country further from the Western orbit.[34] In other words, the Contra and U.S. intervention induced a reticence about ending the private economy. For instance, peasant dissent led to a reduced focus on state farms. Briefly, the public sector of the economy increased from 15 percent under Somoza to 45 percent through the last half of the Sandinista regime (1979–89)—scarcely a model of classic communism.[35] In 2005 Nicaragua with most of Latin America leans toward neoliberalism, but is encountering discontent among its workers because of the benefit primarily going to the economic elites.

In countries with more traditional economic structures, neoliberalism and globalization proceed according to financial and political realities. As Latin America's largest power, Brazil reflects these variables on a grand scale. Stabilization became a major focus in the mid-1990s. As a result of Cardoso's revised *Real* plan, inflation declined from 2,500 percent in 1993 to 4 percent in 1997. The issuing of the *Real* as a new currency brought the advantages of debt reduction and parity with world currencies, permitting an import capacity and strengthening its bonds with First World markets, as well as encouraging closer ties to MERCOSUR (the free trade association of the cone countries). The success of this policy paved the way for Cardoso's reelection and his "liberal conservatism."[36] Unfortunately, economic elites, regionalism, and a fractious, regressive legislature did little to advance his economic and social reforms. The principal benefactors of the Brazilian deflation were members of the lower class, however impoverished they still remain. Indeed, a quasi-boom fell upon several countries in the 1990s. Among other effects was an outflow of capital, foreign investment also poured in, rising sixfold from 1990 to 1997, a net inflow reaching an estimated twenty-six billion dollars by 1997. Stock markets showed a renaissance with this inflow. Corporate and government bonds add to the seeming prosperity. Privatization continues, seemingly unabated.

In the October 2002 presidential election Lucio Ignácio Lula da Silva of the Brazil's Workers Party won a near-landslide victory. Despite his proletarian background, he supports neoliberalism and free trade agreements as well as curbing welfare costs, not to the approval of his left-wing adherents. Nor are the conservatives happy either. Still, in view of a five percent growth of the economy, his pushing a free trade policy for the cone countries, and a nearly 50 percent approval rating, "Lula" might win reelection in 2006.[37] However, corruption charges in 2005 have dampened the future chances of both "Lula" and the Workers Party.

A Changing Political Order

During the neoliberal onslaught the political process is no less complex than the economic. Democracy is almost always tenuous but it became more secure in Latin America during the 1980s, notwithstanding the economic debacle. Major nations moved from authoritarian to more representative patterns, and Mexico's venerable PRI came under ever deeper scrutiny. It is an open question as to the relationship between economic advance and democratic trends—or whether democracy is more stable in developed than in developing nations.[38] People, it appears, gradually become more sensitive to intricacies, or at least the crises, in the political process. Economic stagnation and pauperization in the first decade of the twenty-first century are producing a torrent of distress and rebelliousness in at least a dozen countries in the hemisphere.

Redefinition of Executive Power

With the decline of the *caudillo* system over the last half-century, new political patterns are evolving. As legislative and judicial functions are even less institutionalized than the executive, the legacy of a strong president remains. Yet, in the 1990s the mass media, public opinion, and a restless congress enjoy a growing influence. Two presidents, Collar and Fujimori, were impeached, Menem was threatened with impeachment, and three presidents of Ecuador were removed from office, followed in 2005 by Mesa's resignation in Bolivia. Corruption (largely kickbacks on public projects), clientelism, authoritarianism, defiance, and an overly close affiliation with U.S. economic interests are among the pitfalls. The Weberian idea of rational bureaucracies holds only recently—and marginally—for Latin America; executives can now be held accountable. Military governments represent a "bureaucratic authoritarianism," but they are gradually passing from the scene.[39] Moreover, economic decision making, including neoliberalism, demands more information processing at the top. Government leaders must avail themselves of economic analysts and planners. Public opinion surveys reveal that most Latin Americans, especially in the lower ranks, want a strong leader, even a dictator, who will deliver, that is, meet their economic needs.

Other variables affect the governmental apparatus, including the presidency. For one, as labor organizations decline in strength the working class is less potent, the middle class asserts more power, as witnessed in the Mexican election of 2000 with its challenge to the PRI. That is, a vibrant populism is based on an intricate clientelism, increasingly linked to the urban middle class. A higher educational level and a growing pluralism are additional factors.[40]

However, no set pattern operates among Latin American executives; the case of Venezuela is instructive—a nation with a strong party organization, albeit with intraparty conflicts. Nearly 90 percent turned out at the polls in the 1960s and early 1970s, followed by startling declines thereafter. Reforms improved the party process at local and regional levels, but the national fabric became chaotic.[41] External debt, declining oil revenues, and breakdown of its two-party structures *Acción Democrática* and *Copei* left an open field for new sociopolitical movements. Into this vacuum moved army officer Hugo Chávez, who had led a military revolt against President Pérez in 1992.

Political Parties and the Legislative Process

A major problem of the governmental process, notably political parties, is the lack of institutionalization. Parties are caught in a continuing struggle for their roots and loyalties. Conflict is further confounded by economic crises, class cleavages, and personality clashes. In addition are the perennial problems of finance, organization, and leadership. Novel social movements with potent slogans become a powerful magnet for expressing public distress and may develop into effective political organizations, as noted in chapter 4. New issues—a market economy and human rights—are alternating facets of the political landscape.[42]

Idiosyncrasies of party ideologies and alliances exist in every nation. For one, Uruguay until the 1960s offered possibly the most advanced political structures in the hemisphere with its political balance between the mildly conservative Blancos and the liberal Colorados. A desperate economic breakdown precipitated the Tupamaro revolt followed by military rule in 1973. In this frozen political landscape a kind of anomie set in; consequently, the working relationship between the two parties wore thin. New parties emerged, most important was the Frente Amplio (Broad Front). The 1984 election displayed considerable polarization with some duplicity in party behaviors, including reversals—Blancos became more leftist than Colorados. Institutional reforms (for one, a legislative majority for the winning party) and new leadership made for more stability in the election of 1989, yet, polarization remains.[43] However, the Broad Front reached its full stride with the charismatic Tubaré Vázquez. Symbolic of the political transformation of Uruguay is the entry of former revolutionaries, the Tupamaros, into politics. They like the populist Vásquez represent the pragmentism of the leaders of Uruguay's neighbors, Brazil and Argentina.[44]

In assessing the political process we may return to the functionalist viewpoint: how effective are governments in meeting the needs of the people? Do "democratic" governments have the symbolic and material equipment to deliver resources to the two-thirds of the population who live in—or on the edge of—poverty? Social change emerges through both evolution and revolution at an accelerated rate. Before readdressing these questions it is fitting to view Latin America in its relationship to the world at large, particularly the presence of the superpower to the north.

Latin America and the United States

The region should be viewed in a widening context. In fact, Latin America has been intermittently the victim of dominance by Europe, predominantly Spain, in the colonial period, and by Britain and even more by the United States in the national period. Through the greater part of the twentieth century, major Asian and European countries invested heavily in Latin America, which in turn enjoyed a degree of trade reciprocity with both Western and Eastern Europe. However, trade relations with the European Union can be very complex as some twenty different countries often fail to reach a consensus.[45] Latin America is now a major actor in the UN. Consequently, many observers feel that one of its nations, presumably Brazil, should have more of an official presence, possibly a permanent seat on the Security Council. Certainly, the second half of the twentieth century found Latin America extending its limited hegemony to wider horizons. The OAS strengthened its functioning beyond the feeble U.S.-dominated Pan American Union of pre–World War II. Over the last fifty years the OAS has grown in strength, functioning, and complexity, as seen in figure 11.1. Also, a number of trade organizations arose in recent decades, not only within the hemisphere but in an expanding global setting. Moreover, in trade relations Latin America is involved in a host of organizations such as GATT (General Agreement on Trade and Tariffs). Membership in these associations underlines the growing trade Latin America has with the world. For instance, trade with Japan nearly doubled in the 1990s, with exports exceeding imports in several countries.

Still, Latin America's history and likely its future show the overpowering presence of the United States. This uneasy linkage has a lengthy heritage. In 1823 President James Monroe, mainly through the influence of his secretary of state (John Quincy Adams), proclaimed that his country should act as protector of the newly independent lands to the south. According to the original intention, both Latin America and the United States were to be isolated from European powers. However, as the United States grew in strength, intervention replaced isolation. It is pertinent that Monroe had a knowledge of Spanish, whereas few of his successors did. The relationship between the United States and its southern neighbors usually varied from indifference to hostility and outright occupation. The expansionist fervor of the Polk era found its way to the Mexican War and expansionism into the entire West. Another example of "manifest destiny" was the State

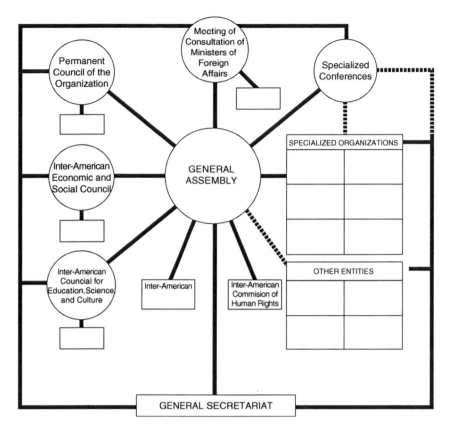

Figure 11.1 Structure of the organization of American states.
Source: OAS Digest.

Department's attempt to annex the Dominican Republic—a movement fortunately forestalled by Charles Sumner, chairman of the Senate Foreign Relations Committee. In the first third of the twentieth century the United States intermittently occupied Cuba, the Dominican Republic, Haiti, and Nicaragua. Even in immediate postwar years the U.S. State Department, the Voice of America, and the CIA pressed their agenda, both subtly and boldly, on Latin American nations, including major ones such as Brazil.[46] From the 1950s through the 1980s the Cold War dictated a policy of confrontation through a series of interventions.

The Cold War and U.S. Intervention

With the 1949 fall of China to a communist regime, the Koran War, both the administration and Congress lost no time in establishing a sustained campaign against what they perceived as Soviet expansion. In 1954 during the Eisenhower administration the United Fruit Company, aroused by President

Juan Arbenz's agrarian reform program in Guatemala, pressured the State Department and the Central Intelligence Agency to act. Guatemalan troops in concert with the oligarchy, orchestrated by the CIA, supplanted Arbenz with a reactionary regime. Although this intervention deeply offended Latin Americans, "Operation Success," as it was called, became the model for future military moves to "cleanse the hemisphere of Communism."

However much a perceived threat socialist Juan Arbenz may have been, Fidel Castro was regarded as the most dangerous force in the Americas. After his defeat of the corrupt Batista regime in January 1959, Washington was uncertain about Castro's ideology and goal, but by 1960 his Marxism became a reality. The Eisenhower administration bequeathed to John Kennedy a plan for the invasion of Cuba. In 1961 the CIA equipped some twelve hundred Cuban exile mercenaries to mount what was to be the ill-fated "Bay of Pigs" episode. The invaders encountered sixty thousand Cuban troops, and within seventy-two hours were captured and imprisoned, later released when the United States provided Cuba with fifty million dollars worth of medical supplies. The results might have been different had Kennedy (who had been misled by the CIA's underestimation of the strength of Cuban forces) had sent in U.S. armed forces, but he stood by his word not to employ them. The Kennedy administration turned to the CIA for alternatives for the "Cuban problem," such as an attempt at assassination, only to provoke greater Cuba–Soviet Union bonding. The conflict between the United States and Cuba reached a climax with the revelation of the presence of Soviet missiles in Cuba in October 1962. After a nervous week of threat and negotiation the Soviet Union agreed to remove the missiles in Cuba after the United States consented to remove its missile deployment in Turkey, a promise not to invade Cuba, and to make no further attempts on its leader's life. Liberals in Latin America were not happy with Kennedy's actions, but his establishment of the Alliance for Progress and the Peace Corps greatly reduced ambivalence throughout the region from 1961 to Kennedy's assassination in 1963. Unfortunately, the Alliance proposed socioeconomic reforms that displeased the oligarchy and aroused little enthusiasm in the U.S. Congress.

A paranoid fear of communism continued to dominate the U.S. approach to foreign relations. President Lyndon Johnson was no exception, as shown in the catastrophic excursion in Vietnam. In the Dominican Republic's transition to democracy following the assassination of Rafael Trujillo and the end of his brutal dictatorship (1930–61), the country was trapped between the right and the left. In 1965 Johnson became concerned about the possible rise of a second Cuba. In a questionable election a conservative junta replaced liberal president Juan Bosch with one of their own. As a counter coup attempted to restore Bosch, disorder bordering on civil war followed. Johnson responded by dispatching a total of twenty thousand troops. The Johnson Doctrine triumphed and conservatives, principally Juaquín Baguer, ruled until 1996.

Through the Cold War the CIA was active in its campaign to rid the hemisphere of any sign of communism, including the pursuit and murder of charismatic Argentine agitator Che Guevaro in 1967. Possibly the most dramatic interference in a constitutional government was the U.S. government's activity in Chile. In 1964 the CIA through an intermediary gave the Christian Democrats one million dollars for their electoral campaign in order to prevent a Marxist victory, a scandal that discouraged further incursions by the United States during the Johnson presidency. However, President Richard Nixon and his national security advisor Henry Kissinger became increasingly impatient during the Allende regime (1970–73), particularly the nationalization of U.S. holdings. As a result Nixon canceled most aid programs (but not the military) and urged the World Bank and IMF to tighten their aid. On September 11, 1973 Allende was ambushed into a suicide (or murder) by the Chilean armed forces, an event in which the CIA and multinational corporations played at least an indirect role.

Because of Central America's proximity to the U.S. mainland and especially the Panama Canal, a prolonged and indefensible intervention bore down on the area throughout the 1980s. President Reagan was driven by his fear of the spread of "communism," with no recognition that the drift toward Marxism arose more from the insufferable conditions of the rural and urban poor than from Cuban and Soviet infiltration. Whereas President Jimmy Carter leaned toward a human rights perspective (not always consistently), Reagan insisted that stamping out any leftist government or revolutionary movement was in the "U.S. interest"—however questionable its meaning. As the 1980 electoral campaign voiced this doctrine, the Latin American oligarchy were ecstatic at Reagan's victory at the polls. Let us examine the U.S. involvement in Central America:

El Salvador's tragedy dates from over a century ago when an "aggressively expansionist agro-export" oligarchy displaced the peasant economy. Increasingly, the state and military were pledged to accommodate the landed elite.[47] U.S. support of El Salvador's military was emphatically shown in the role of Roberto d'Aubuisson, who like many Latin American officers was trained in the U.S. Army School of the Americas at Fort Benning, Georgia. D'Aubisson commanded the "death squads," which massacred thousands of peasants and assassinated Bishop Oscar Romero in 1980. The Washington administration simply closed its eyes to the carnage—or rationalized it as a means of combating communism.[48] Congress, recalling the Vietnam debacle, was suspicious of White House's intentions and insisted on attention to human rights. Reagan spokesmen consequently moved to provide a façade of civility, and with this leverage persuaded the Salvadorian right to offer centrist candidate Christian Democrat José Napoleón Duarte for the presidency.[49] Only at the end of the Cold War was there an acknowledgment of the need to consider human rights.

The same mentality prevailed during the Guatemalan reign of terror. As early as the 1960s President Johnson sent the Green Berets from Fort Bragg, North Carolina, to serve in counterinsurgency action. Under the MDA (Military Defense Assistance Act) the U.S. military provided equipment and training to Guatemalan forces. Years later President Carter soon became aware of the severe human rights violations and ended military aid. In the 1980s the situation was much graver. One relevant development was President Lucas García's reported contribution of five hundred thousand dollars to Reagan's election campaign. The gift was not forgotten. Reagan's staff—Vice President George Bush and UN ambassador Jeanne Kirkpatrick, among others—were seemingly no more offended by human rights violations in Guatemala than elsewhere. Although Congress voted for curtailment of military aid, the amount of economic and military aid rose yearly.[50]

When a socialist regime came to power in 1979, Nicaragua was destined to be *the* major problem of the Americas. This grim history must be viewed in the context of the Somoza dictatorship (1936–79). A cozy relationship continued between Managua and Washington until the advent of the Carter presidency. Bolstered by Somoza's lobby (a public relations firm employing a U.S. ex-congressman and former government officials), conservative members of Congress saw Somoza as a loyal friend and a committed anticommunist. As reform movements, notably the FSLN (Sandinista Front for National Liberation), grew in strength through the 1970s the suppression of human rights was intensified. Moreover, the United States became less enthusiastic about the Somozas when it was revealed that a large share of relief funds sent to Managua after the 1976 earthquake ended in the pockets of the Somoza family. The murder of editor Pedro Chamorro in 1978 further disturbed President Carter, who gave his blessings to Somoza's downfall in 1979. However, with the more than occasional Marxist tone of the *Sandinistas*, Carter and especially Congress turned sour. As aid from Washington declined, economic problems mounted. Not surprisingly, Nicaragua looked to Cuba and the Soviet Union for help. Predictably, the U.S. administration stressed the Sandinistas' property rights violations but ignored their humanitarian advances in health, education, and religious tolerance.[51]

With the 1980 election of Ronald Reagan, a military confrontation was predictable. Through covert means the White House funded the Contras (mainly composed of former henchmen of Somoza, including his National Guard), who operated largely from Honduras, carrying out attacks against the Sandinistas. At one point the United States mined Nicaragua's principal harbor, resulting in a condemnation by the World Court. The White House staff promptly responded that the United States was simply acting out of security needs (as though Nicaragua were on its border). Meanwhile, the Contadora, a committee composed of leaders of several Latin American countries (Colombia, Mexico, Panama, and Venezuela), headed by Costa Rica's president Oscar Arias was searching to find a peaceful solution for the Nicaraguan crisis. Fearful of the outcome, the Reagan administration made various efforts to sabotage the Contadora process. Nonetheless, a compromise was found and a free, well-monitored election took place in 1990. Reagan's policy, especially his support of the Contras, aroused considerable dissonance among Latin Americans.[52]

After the end of the Cold War in 1989, the U.S. attitude shifted to benign indifference except when business interests are concerned. Still, intervention was not ruled out. The most violent episode was the invasion of Panama and the capture of General Manuel Noriega in 1989. Moriega, a kind of rogue elephant gone astray, had worked as a CIA agent while apparently offering his services to Cuba as well as being active in drug trafficking. Once in the presidency (following the death of General Omar Torijllos in a somewhat mysterious plane crash in 1981) Noriega provided resources, notably a landing area for planes serving the Contras. His role of double agent finally became known to Reagan and Vice President George Bush, but his removal would have to wait for the cessation of hostilities in Nicaragua. In 1988 when a U.S. court indicted Noriega on drug charges he became ever more defiant. President Bush was beseeched by the U.S. press to arrest and bring him to justice. Finally, in order to protect his image as resolute, Bush ordered an invasion ("Operation Just Cause") consulting neither the OAS nor any Latin American leader. At the cost of dozens of U.S. casualties and over five hundred Panamanian lives, Noriega was captured, tried for drug trafficking, and sentenced to forty years in a Miami prison. Bush's unilateral action was widely criticized by Latin American and other nations.

During the Clinton presidency the United States remained relatively indifferent toward Latin America as compared to Europe and Asia, where diplomatic and economic ties are traditionally vibrant. One exception is, because of the drug problem, Colombia (which after Israel and Egypt is the largest recipient of U.S. security aid). Also, for a variety of reasons—proximity, immigration, trade, security—the Mexican relationship was of high priority. The 1993 enactment of the NAFTA agreement, which links Canada, Mexico, and the United States, may eventually be extended to nearly all the hemisphere as a free trade zone. NAFTA remains, with its tenuous consequences, a relatively new chapter in the U.S. attempt to bond with Latin America.

In January 2001 the presidency of George W. Bush began almost as a return to isolationism in contrast to Clinton's commitment to international linkages and nation building. Bush's appointments to key positions relating to Latin America, for example, Robert Noriega, Roger Pardo-Mauer, and Elliott Abrams, hark back to the Reagan era. Bush, possibly aware of Mexico's troubled past (see inset 11d), displayed a verbal interest in Latin America, his first state dinner honoring Vicente Fox, with whom he felt a congeniality as a fellow business entrepreneur and because of their association when Bush was governor of Texas. After the September 11, 2001 catastrophe, the administration concentrated on national security, which in turn focused attention to other world areas. Although Bush has visited briefly in several Latin American nations, his vision of inter-American affairs, as for other areas of the world, remains unclear except for commercial issues and his determination to wage war against terrorism.

INSET 11D   MEXICO: AN UNEASY PRESENCE

Mexicans quip that they are "so far from God, so close to the U.S." As the geographically closest Latin American neighbor to the United States, Mexico commands special attention. The relationship has only in recent decades become amicable, and even today Mexico, like Canada, is diligent about protecting its autonomy. Historically, the bitter heritage of the Mexican War of the 1840s was somewhat mitigated by the U.S. concern with the French occupation of Mexico (1862–67). The U.S. Civil War prevented any military action, but once over Washington made clear its intentions, indirectly helping the guerrilla assaults during the Benito Juárez presidency. All this was a prelude to the U.S. investments in Mexico during the Díaz regime (1876–1911). Trade between the two countries rose from $9 million in 1870 to $117 million in 1910.[53]

Relations during the Mexican Revolution drifted from modest idealism in the United States, including support for Francisco Madera, to hostility in the era of Carranza and Zapata. Relations during the Woodrow Wilson presidency (1913–21) deteriorated into a self-defeating policy. Part of the problem was the limited ability of U.S. representatives, few of whom spoke Spanish. Still worse, they appeared to place business interests over diplomacy and justice, directly or indirectly aiding the counterrevolution. In three episodes U.S. troops intervened during the Mexican Revolution.

In the postrevolutionary period, relations between the two countries remained ambivalent. In 1929 President Herbert Hoover announced a "Good Neighbor" policy, which as implemented by Franklin Roosevelt, made for a less confrontational climate (dampened significantly by Mexico's 1938 expropriation of foreign-owned oil companies). World War II ushered in a closer bonding, as Mexico provided necessary material resources and declared war against the Axis in early 1942 (as did Brazil later that year, followed by other republics, but not Argentina, which had Axis connections).

The postwar period, marked by sizeable industrial growth on both sides of the border, differed markedly in economic development. The self-image of the United States is one of superpower while Mexico and its neighbors remain conscious of their state of dependency.[54] Also, few observers could escape the perception that Europe received massive aid from the Marshall Plan in the late 1940s and to the Middle East in later decades, whereas most of the developing world was left to its own resources.

The Cold War and other political and economic entanglements kept both Mexico and the United States in a delicate equilibrium. The Cuban Revolution moved the two nations into a precarious position. Mexico recognized the Castro regime, as did several other Latin American nations, but continued its traditional balancing act with the United States by refraining from too close a relationship with either side. A complicating factor was the slackening economic growth in Mexico, including a severe decline of the peso in 1976— a precursor to more devastating declines in the 1980s. Trade barriers were another ongoing point of friction. On the diplomatic front Mexico, especially during the Echeverría presidency (1971–77), heightened its sensitivity to Third World problems, from the fate of the Palestinians in the Middle East to the U.S. challenges to Allende's Chile.

Although the U.S.–Mexican relationship gradually became more friendly over the last several decades, problems remain. A continuously troubling issue is the *bracero* (field hand) question. U.S. farmers welcome the cheap labor (with no social benefits) of Mexican migrants. As early as 1951 Mexico made clear its preference for its own agencies to contract work in the United States. The problem of immigration continued throughout the remainder of the century into the Fox presidency. With security concerns following the September 11 tragedy, immigration and the status of illegal aliens became an even more delicate issue. Further, at the beginning of the twenty-first century a major problem is the control of narcotrafficking and its inherent corruption. But fundamental to the U.S.–Mexico bond is President Fox's verbal insistence on freedom in trade (as represented by NAFTA), a rational immigration policy, and a more equitable political process, notably free elections, that is, reduced status of the PRI—goals that appear far from settled.

A Summing Up

In reflecting on the dimensions of change in Latin America we return to the theme of how societies function, that is, how Latin America compares with other world areas. Again we remind ourselves of the many Latin Americas. All generalizations must be refined by time and place. Even the smallest country represents a cross-section of societal variables. That is, European, indigenous, and mestizo cultures are moving in varying degrees to modernity, which itself can be defined in diverse categories, as chapters 4 and 5 noted. Moreover, in view of advances in transportation, communication, science, and industrialization, much of the world is gradually arriving at a "superculture." In this context we are concerned with the quality of life in both developed and developing societies. Further, Latin America is making significant progress in health, life expectancy, educational attainment, including literacy. Particularly at point is the stability of democracy in Latin America.

Conflict and Violence

This chapter also explored the degree that conflict is or is not a catalyst for change and development. Implicit in this discussion is the role of conflict in reaching collective goals, as they involve availability of resources along with the individual and social costs. Motivations and the degree of rationality are also relevant. A myriad of other factors enter the equation: ideology, flexibility, leadership, to list a few. These factors are relevant to a seemingly stable society as well as one in conflict or undergoing orderly transformation.

The most distressing periods of conflict and violence appear to be of the past—the horrendous decades from the 1960s to the 1980s. Yet, in 2004 and 2005 Colombia, Bolivia, Peru, and Venezuela, for example, are evidence that violence remains an option, if not a compulsion. In still other countries violence is a potential as Central American and Andean nations testify. Even starvation is a form of violence. Violence emerges at the inability to meet rising expectations. The frustration–aggression hypothesis is still applicable. According to a study of twenty Latin American countries, the rate of violence is negatively correlated with several variables: social cohesion, institutionalized political parties, open communication channels, and several specific indices such as the amount of educational budgets. Costa Rica has the lowest record of violence. It is also the only Latin American country without a military contingent (depending on a national police force to maintain order, thereby permitting funds to underwrite education and other social needs). Violence does flair up in Latin America's democracies, but less than in authoritarian political systems. On the whole, repression, violence, and revolutionary activity declined after the end of dictatorships in the mid-1980s.

Economic Progress

As repeatedly implied throughout the text, economic betterment is essential to Latin America's development. Both international and national programs, including NGOs, are oriented to this goal. An interesting case study is the Strategy for the Reduction of Poverty (ERP) in Honduras, where Hurricane Mitch in 1999 heightened attention to its tragic socioeconomic situation. The IMF, UN, Organization for Economic Co-operation and Development (OCED) in 2000 approved a fifteen-year project at an estimated cost of $2,666 million with specific targets—reducing poverty by a fourth, increasing education in all three cycles, raising the ratio of secondary school graduates entering the labor force, health measures, long-term considerations such as the environment, but most important, stimulation of economic growth. Translating macroeconomic goals into microeconomic performance depends on a complex of variables, particularly the political culture. Rural economic growth involves land tenure reform. These strategies are, of course, contingent on factors well beyond Honduras's control—the world economy, the commitment of the lending agencies, and the capacity for institutional change.[56]

The Struggle for a Stable Democracy

Notwithstanding the trouble spots in Latin America, the prospects for representative governments seem brighter than ever before in Latin America's history, at least in a long-term perspective. This situation must be assessed in the sequence of not altogether consistent events over the last quarter of the twentieth century. One, the calamitous economic debacle of the 1980s saw a return to democratic growth. Two, the transfer of state economies to private enterprise and open markets becomes an invitation to restructuring political processes, if to uncertain ends. Particularly vulnerable are the parties oriented to the left. Nonetheless, parties to the left still maintain their strength. Three, the political apparatus is redefined in the aura of electronic communication, as television assumes the fulcrum in political leadership. Four, educational expansion and a growing middle class induce increased literacy and civic participation. Five, international processes, political and economic, impinge on the Latin American scene.

Throughout Latin America there are questions about the viability of democracy. A Chilean historian speaks not only for his own country when he claims that demographic countries cater to opportunism rather than following a course of integrity combined with pragmatism.[57] When respondents are asked to evaluate democracy, many feel ambivalent, saying "what has democracy done for me," or "sometimes an authoritarian order is justified." The range of support for democracy varies from roughly 55 percent in Brazil to 85 percent in Costa Rica.

International Relations, the United States, and the Future

Latin America is enormously widening its horizons, with growing ties to Asia and Europe. Despite remnants of colonialism, partnership is gradually replacing subservience. Internally the area has established a number of trade organizations reducing cross-national tensions of the past. Over the decades Latin America is assuming more autonomy vis-à-vis the United States. Still, the region is confused by the unfocused and inconsistent policies of the George W. Bush administration, which remains relatively detached from a number of crises, for instance Haiti's socioeconomic disaster.

Still, the U.S. presence continues to loom large, especially in the economic sphere. One phenomenon, "delocalization" or the displacement of local banks, firms, and industries by larger entities from abroad, spells disaster for local enterprise.[58] Unfortunately, Latin American financial and business interests do not have the same privilege of entry into the United States or several other Western countries as can, say, East Asian operations. Another complicating problem is the contradictory policies of the Bush agenda— pledging free trade but later placing tariffs and limits on a number of items, from steel to wood pulp. Significantly, the gross production of Latin America with its half billion people is approximately only a third of the U.S. GNP with 293 million population or the European Union of over 400 million.

The problem of this economic and political disequilibrium lies disproportionately with the United States. Ever since mid-twentieth century the United States dominates the world as the one superpower, for instance, number "1" on the international telephone circuit, whereas all other nations have two or three digits; also, on the Web, with no suffix, but other nations must be identified ("uk" for Britain, "br" for Brazil, and so forth). Beyond these symbols is its economic superiority, the U.S. GNP is over two times that of its closest rival, Japan. This dominance appears to lead to self-righteousness and even arrogance rather than restraint, to a unilateral rather than a multilateral approach.

This self-importance and exuberance came to a shattering realization of vulnerability on September 11, 2001. The European Union acknowledged the attack on one of its members is an attack on all and the OAS, at the suggestion of Brazil's Cardoso, declared the same principle. This feeling of support around the world gradually vanished as the Bush administration and Congress began their unilateral campaign against terrorism, particularly the war against Iraq. An unfortunate aspect of the U.S. antiterror program is its effect on civil rights in other countries. As one instance, the Guatemalan government, seemingly with the approval of the U.S. administration, renewed attacks on Mayan Indians on the pretext of national security measures. In addition, the U.S. administration is loosely using the label "terrorism" to various operations, such as illegal migration, money laundering, and drug trafficking ("narcoterrorism").[59]

As the struggle against terrorism proceeds, it is hoped that the United States may reevaluate its goals and priorities in the wider world. The federal government must acknowledge its penchant for acting "in the nation's best interest" or expediency, as, for instance, its approach to Latin America during the Cold War. The United States needs to examine more judiciously its strengths and weaknesses. On one side are its tremendous assets—a favorable standard of living for 80 percent of its people, high attainment in science, technology, health, and the arts, among other pluses. No less critical, the United States must face the negative aspects, several of which place the nation closer to the Third World than the First World—extremely steep economic stratification with ethnic overtones, high level of violence (including the death penalty), an uneven democracy (low voter participation, questionable campaign financing, telemedia dominance, lobbyist power), a large penal population of over two million (not least because of a flawed drug policy), to cite several troublesome indices.

Positive and negative characteristics can be identified with other national cultures, but the burden of self-examination rests with the United States, the only superpower, at this crucial juncture of history. One wonders whether in future decades the United States will restore its former image of international understanding as implemented in its aid program, which was offered during the Cold War as a means of counteracting the threat of Marxism. At present the United States is seventeenth among Western nations in per capita humanitarian aid. In other words, according to the Cato Institute, less than a fifth of 1 percent of the GNP is committed to international nonmilitary aid. Increased aid to Latin America, Africa, Middle East, and Asia could restore a benevolent image of the United States—and thereby reduce indirectly the threat of terrorism. In other words, the U.S. international policy might in the long run be more successful if based on what is right rather than what is expedient. Such a change will be welcome in Latin America, which has been gradually moving toward wider humanitarian goals.

In the meantime Latin American republics struggle with their economic and sociopolitical dilemmas. The U.S. economic slowdown and its militarism during the early years of the twenty-first century have repercussions for its southern neighbors, notably in Mexico (90 percent of its exports go to the United States). Further, President Vicente Fox seems unable to find a consistent configuration of policies and decision making, hardly facilitated by political rigidities in the United States. Violence continues in Colombia with the peace process stagnant. An Argentine recovery is promising but seems uncertain in view of nearly a 20 percent rate of unemployment and over 20 percent of its population living below the poverty line. At present, leadership from the United States is uninspired and ambivalent. One may hope that a more dynamic outlook, akin to the Kennedy and Carter regimes, may return.

Notes

Chapter One   Latin America:
Change and Diversity

1. Aline Heig, "Race in Argentina and Cuba, 1880–1930: Theory, Policies, and Popular Reaction," in Richard Graham (ed.), *The Idea of Race in Latin America, 1870–1914* (Austin: University of Texas Press, 1990), pp. 17–69.
2. Anthony McFarlane, "African Slave Migration," in Simon Collier, Harold Blakemore, and Thomas E. Skidmore (eds), *The Cambridge Encyclopedia of Latin America and the Caribbean* (New York: Cambridge University Press, 1992), pp. 138–142.
3. Martin W. Lewis and Kären E. Wigen, *The Myth of Continents: A Critique of Metageography* (Berkeley: University of California Press, 1997).
4. Jon S. Vincent, *Culture and Customs of Brazil* (Westport, CT: Greenwood Press, 2003), pp. 20–21.
5. Peter Wade, "Race and Nation in Latin America: An Anthropological View," in Nancy P. Appelbaum et al. (eds), *Race and Nation in Latin America* (Chapel Hill: University of North Carolina Press, 2003), pp. 263–282.
6. Néstor García Canclini, *Culturas Híbridas: Estrategías para Entrar y Salir de la Modernidad* (Mexico, DF: Editorial Gribalbo, 1990), p. 69.
7. Keith Jenkins, *Re-thinking History* (New York: Routledge, 1991), p. 17.
8. Julian H. Steward and Louis C. Faron. *Native Peoples of South America* (New York: McGraw-Hill, 1959), pp. 16–17.
9. Beatriz Manz, *Paradise in Ashes: A Guatemalan Journey of Courage, Terror, and Hope* (Berkeley: University of California Press, 2004).
10. Caroline Moser and Cathy McIlwaine, *Violence in a Post-Conflict Context: Urban Poor Perception in Guatemala* (Washington, DC: The World Bank, 2000).

Chapter Two   Geography:
Barriers and Challenges

1. Preston E. James, *Latin America*, 4th ed. (New York: Odyssey Press, 1969), p. 20.
2. Tom L. Martinson, "Physical Environments of Latin America," in Bryan W. Blouet and Olwyn M. Blouet (eds), *Latin America and the Caribbean: A Systematic and Regional Survey* (New York: Wiley, 1993), pp. 1–33.
3. Clifford T. Smith, "Geography," in Simon Collier, Harold Blakemore, and Thomas E. Skidmore (eds), *The Cambridge Encyclopedia of Latin America and the Caribbean* (New York: Cambridge University Press, 1992), pp. 33–44.
4. Roberto Lobato Corrêa, *Trajetórias Geográficas* (Rio de Janeiro: Bertroud, 1996), p. 299.

5. Alberto C. G. Costa, Conrad P. Kottak, and Rosane M. Prado, "The Sociopolitical Context of Participatory Development in Northeastern Brazil," *Human Organization*, 1997, 36(2), 138–151.
6. Judith Lisansky, *Migrants to Amazonia: Spontaneous Colonization in the Brasilian Frontier* (Boulder, CO: Westview Press, 1990).
7. Peter Crabb, "Water: Confronting the Critical Dilemma," in Ian Douglas et al. (eds), *Companion Encyclopedia of Geography* (London: Routledge, 1996), pp. 526–552.
8. Brian W. Ibery and Ian Bowler (eds), "Industrialization and World Agriculture," in Douglas et al., *Companion Encyclopedia*, pp. 228–248.
9. Smith, *Geography*, p. 11.
10. Peter W. Rees, "Transportation," in Bryan W. Blouet and Olwyn M. Blouet (eds), *Latin America and the Caribbean: A Systematic and Regional Survey* (New York: Wiley, 1993), pp. 81–120.
11. Elinor G. K. Melville, *A Plague of Sheep: Environmental Consequences of the Conquest of Mexico* (New York: Cambridge University Press, 1994), p. 88.
12. Dan Kloster, "*Campesinos* and Mexican Forest Policy during the Twentieth Century," *Latin American Research Review*, 2003, 38, 94–126.
13. Larry Rohter, "Mapuche Indians in Chile Struggle to Take Back their Forests," *New York Times*, August 11, 2004, A3.
14. María Guadalupe Rodrigues, "Environmental Protection Issue Networks in Amazonia," *Latin American Research Review*, 2000, 35(3), 125–153.
15. R. O. Bierregaard et al., *Lessons from Amazonia: The Ecology and Conservation of a Fragmented Forest* (New Haven, CT: Yale University Press, 2001).
16. Nigel J. H. Smith, *The Amazon River Forest* (New York: Oxford University Press, 1999), pp. 155–164.
17. Charles D. Brockett and Robert R. Gottfried, "State Policies and the Preservation of Forest Cover," *Latin American Research Review*, 2002, 37(1), 7–40.
18. Luis A. Vivanco, "Spectacular Quetzals, Ecotourism, and Environmental Futures in Monte Verde, Costa Rica," *Ethnology*, 2001, 40(2), 79–92.
19. Lourdes Arizpe, Fernanda Paz, and Margarita Velásquez, *Culture and Global Change: Social Perspectives of the Lacandona Rain Forest in Mexico* (Ann Arbor: University of Michigan Press, 1996), pp. 19–35.
20. Larry Sawers, "Income Distribution and Environmental Degradation in the Argentine Interior," *Latin American Research Review*, 2000, 35(2), 3–33.
21. Brian Long, "Conflicting Land-Use Schemes in the Ecuadorian Amazon: The Case of Sumaco," *Geography*, 1992, 77, 336–347.
22. Anthony Bebbington, "Modernization from Below: An Alternative Indigenous Development," *Economic Geography*, 1993, 69, 274–285.
23. Roger A. Clapp, "Creating Competitive Advantage: Forest Policy in Chile," *Geography*, 1995, 71, 273–280.
24. Ricardo Lagos, *Después de la Transición* (Buenos Aires: Grupo Editorial Zeta, 1993), pp. 146–157.
25. Carlo J. Bonura, Jr., "The Occulted Geopolitics of Nation and Culture," in Gearóid O. Tuathail and Simon Dolby (eds), *Rethinking Geopolitics* (New York: Routledge, 1998), pp. 86–105.
26. Jonathan R. Barton, *A Political Geography of Latin America* (New York: Routledge, 1997), pp. 9–11.
27. James G. Blight and Philip Brenner, *Sad and Luminous Days: Cuba's Struggle with the Superpowers after the Missile Crisis* (Boulder, CO: Rowman & Littlefield, 2002), pp. 176–178.

28. I. Wallerstein, *The Modern World System* (New York: Academic Press, 1974).
29. Peter J. Taylor, *Political Geography: World-Economy, Nation-State and Locality* (New York: Longman, 1985), pp. 87–91.
30. Rob Wilson and Wimal Disanayake, *Global/Local: Cultural Production and the Transnational Imaginary* (Durham, NC: Duke University Press, 1996), pp. 81–82, 147–151.
31. Christopher Chase-Dunn et al., "Trade Globalization since 1795: Waves of Integration in the World-System," *American Sociological Review*, 2000, 65, 77–95.
32. Nestor García Canclini, *Consumidoes y Ciudadanos: Conflictos Multiculturales de la Globalización* (Mexico, DF: Editorial Grijalbo, 1995), pp. 151–154.
33. Suzanne Schech and Jane Haggis, *Culture and Development: A Critical Introduction* (Malden, MA: Blackwell, 2000), pp. 58–62.
34. Robert J. Holton, *Globalization and the Nation-State* (New York: St Martin's, 1998), pp. 91–95, 161–167.
35. Pablo González Casanova, *Los Militares y la Política en América Latina* (Mexico, DF: Océano, 1988), pp. 85–88.
36. Peter M. Slowe, *Geography and Political Power* (New York: Routledge, 1990), pp. 27–33.
37. Barton, *Political*, pp. 77–83.

Chapter Three An Unfolding Heritage

1. Pablo E. Pérez Maraína, *La Colonialización: La Huella de España in América* (Mexico, DF: Biblioteca Iberoamericana, 1990), p. 57.
2. Arli Oweneel, "From Tlahtocoatl to Gobernadoryol: A Critical Examination in 18th-century Central Mexico," *American Ethnologist*, 1995, 22(4), 330–345.
3. Américo Castro, *The Structure of Spanish History* (Princeton, NJ: Princeton University Press, 1954).
4. Ross W. Jamieson, *Domestic Architecture and Power: The Historical Architecture in Colonial Ecuador* (New York: Plenum Publishers, 1999), pp. 189–210.
5. M. A. Eugenio Martínez, *La Illustración (Siglo XVIII): Pelucas y Casacas en los Trópicos* (Mexico, DF: Biblioteca Iberoamerican, 1990), p. 90.
6. Stephanie Merriam, "The Counter-Discourse of Bartolomé de las Casas," in Jerry M. Williams and Robert E. Lewis (eds), *Early Images of the Americas: Transfer and Invention* (Tucson: University of Arizona Press, 1993), pp. 149–162.
7. Steve J. Stern, "Early Spanish Accommodation in the Andes," in John E. Kicza (ed.), *The Indian in Latin American History: Resistance, Resilience, and Acculturation* (Wilmington, DE: SR Books, 1993), pp. 21–34.
8. Mark A. Burkholder and Lyman L. Johnson, *Colonial Latin America* (New York: Oxford University Press, 1990), pp. 115–116.
9. Severo Martínez Pelaez, *La Patria del Criollo* (Guatemala City: Editorial Universiario, 1973), pp. 573–577.
10. Fiona Wilson, "Reconfiguring the Indian: Land–Labour Relations in the Postcolonial Andes," *Journal of Latin American Studies*, 2003, 35, 221–247.
11. E. Bradford Burns, *The Poverty of Progress: Latin America in the Nineteenth Century* (Berkeley: University of California Press, 1980), pp. 91–92.

12. Howard J. Wiarda and Harvey F. Kline, *Latin American Politics and Development*, 3d ed. (Boulder, CO: Westview Press, 1990), pp. 34–35.
13. Paul H. Lewis, *Political Parties in Paraguay's Liberal Era* (Chapel Hill: University of North Carolina Press, 1993), pp. 144–147.
14. Guy P. C. Thomson, "Pueblas de Indios and Pueblas de Ciudadanos: Constitutional Bilingualism in 19th Century Mexico," *Bulletin for Latin American Research*, 1999, 18(1), 89–100.
15. Anita Brenner, *The Wind That Swept Mexico* (Austin: University of Texas Press, 1971, 1996 edition), pp. 8–11.
16. Robert A. Hayes, *The Brazilian World* (St. Louis: Forum Press, 1982), pp. 4–5.
17. Viviane Brachet-Marquez, *The Dynamics of Domination: State, Class, and Social Reform in Mexico 1910–1990* (Pittsburgh: University of Pittsburgh Press, 1990).

Chapter Four   Society, Economy, and Government

1. Enrique González, *El Laberinto Cultural Venezolano* (Caracas: Editorial Tropykos, 1997).
2. Franciscco Romero Estrada, "Factores Que Provocarn las Migracioness de Chinmos, Japmes y Coreanos hacia México: Siglos XIX y XX," *Revista de Ciencias Sociales*, 2001, 43, 141–153.
3. Alexander S. Dawson, "From Models for the Nation to Model Citizens: *Indigismo* and the 'Revindication' of the Mexican Indian," *Journal of Latin American Studies*, 1998, 30, 279–308.
4. Natividad Gutiérrez, "What Indians Say about *Mestizos*: A Critical View of a Cultural Archetype of Mexican Nationalism," *Bulletin of Latin American Research*, 1998, 17, 285–301.
5. Neil Harvey, *The Chiapas Rebellion: The Struggle for Land and Democracy* (Durham, NC: Duke University Press, 1998), pp. 339–340.
6. Mzría C. and Obregón R., "La Rebelión Zapatista en Chiapas: Antecedents, Causas y Desarrollo de su Primera Fase," *Mexican Studies*, 1997, 13(1), 149–163.
7. Victoria Sanford, *Buried Secrets: Truth and Human Rights in Guatemala* (New York: Palgrave Macmillan, 2004).
8. Robert Albro, "Introduction: A New Time and Place for Bolivian Politics," *Ethnology*, 1998, 37(2), 99–118.
9. David Cleary, " 'Lost Altogether to the Civilized World': Race and *Cabanagem* in Northern Brazil, 1750 to 1850," *Comparative Studies in Society and History*, 1998, 10, 109–134.
10. John Burdick, "What is the Color of the Holy Sprit? Pentecostalism and Black Identity in Brazil," *Latin American Research Review*, 1999, 34(2), 109–131.
11. Ilka Boaventura Lerte, *Negros no Sul do Brasil: Invisibilidade e Terretorialidade* (Santa Catariina: Letras Contemorâneas, 1996).
12. Francisco Vidal Luna and Herbert S. Klein, "Slave Economy and Society in Minaus and São Paulo, Brazil 1830," *Journal of Latin American Studies*, 2004, 36, 1–18.
13. Michael J. Mitchell and Charles H. Wood, "Ironies of Citizenship: Skin Color, Police Brutality, and the Challenge of Police Brutality in Brazil," *Social Forces*, 1998, 77, 1001–1020.
14. France Windance Twine, *Racism in a Racial Democracy: The Maintenance of a White Supremacy in Brazil* (New Brunswick, NJ: Rutgers University Press, 1997), pp. 139–141.

15. Mary Weismantel and Stephen F. Eisenmanm, "Race in the Andes: Global Movements and Popular Ontologies," *Bulletin of Latin American Research*, 1998, 17, 121–142.
16. Winthrop R. Wright, *Café con Leche: Race, Class, and National Image in Venezuela* (Austin: University of Texas Press, 1990), p. 5.
17. Lourdes Martínez-Echazábal, "*Mestizaje* and the Discourse of National, Cultural Identity in Latin America, 1845–1959," *Latin American Perspectives*, 1998, 23(3), 21–42.
18. Scott H. Beck and Kenneth J. Mijeski, "*Idena* Self-Identity in Ecuador and the Rejection of *Mestizaje*," *Latin American Research Review*, 2000, 35(1), 119–137.
19. Gerardo Otero, "The 'Indian Question' in Latin America: Class, State, and Ethnic Identity Construction," *Latin American Research Review*, 2003, 38(1), 248–267.
20. Marshall C. Eakin, *Brazil: The Once and Future Country* (New York: St. Martin's Press, 1997), pp. 112–114.
21. Julio Ortega, *La Cultura Peruana: Experiencia y Conciencia* (Mexico, DF: Fondo de Cultura Económica, 1978), p. 37.
22. Luis C. Silva, "Crisis y Bajos Ingresos Obligan a Nacionales a Emigrar del País," *Unomásuno*, March 1, 2004, 19.
23. Les W. Field, "Constructing Local Identities in a Revolutionary Nation: The Cultural Politics of the Artisan Class in Nicaragua 1979–1990," *American Ethnologist*, 1995, 22, 786–806.
24. Alejandro Portes and Kelly Hoffman, "Latin American Class Structures: Their Composition and Change during the Neoliberal Era," *Latin American Research Review*, 2003, 38, 41–48.
25. John Humphrey, "Are the Unemployed Part of the Urban Poverty Problem in Latin America?" *Journal of Latin American Studies*, 1994, 26, 713–736.
26. Latin American Center, *Statistical Abstract of Latin America* (Los Angeles: University of California, 1999), p. 432.
27. Richard L. Williams and Kevin G. Guerrieri, *Culture and Customs of Colombia* (Westport, CT: Greenwood Press, 1999), pp. 5–6.
28. Charles Wagley, *Amazon Town: A Study of Man in the Tropics* (New York: Columbia University Press, 1953); Richard Pace, *The Struggle for Amazon Town: Gurupá Revisited* (Boulder, CO: Rienner Publishers, 1998), pp. 166–174.
29. Robert C. Williamson, "Social Class, Mobility, and Modernism: Chileans and Social Change," *Sociology and Social Research*, 1976, 56, 149–163.
30. Stephen D. Morris, "Reforming the Nation: Mexican Nationalism in Context," *Journal of Latin American Studies*, 1999, 31, 363–397.
31. Richard M. Morse and Jorge E. Hardoy, *Rethinking the Latin American City* (Baltimore: Johns Hopkins Press, 1992), pp. 31–35.
32. Mauricio Tenorio Trillo, "1910 Mexico City: Space and Nation in the City of the *Centenario*," *Journal of Latin American Studies*, 1997, 29, 123–139.
33. Néstor García Canclini, "Mexico: Cultural Globalization in a Disintegrating City," *American Ethnologist*, 1995, 22, 743–755.
34. Roger Cohen, "A Young Killer's Fear and Hope in 'a Very Dark Place,' " *New York Times*, April 29, 2000, A6.
35. Karen Spalding, "Class Structures in the Southern Peruvian Highlands, 1750–1920," in Benjamin S. Orlove and Glyn Custred (eds), *Land and Power in Latin America* (New York: Holmes & Meier, 1980), pp. 79–98.
36. Meriolee S. Grindle, "Agrarian Class Structures and State Policies: Past, Present, and Future," *Latin American Research Review*, 1993, 28, 174–187.

37. Tanya Korovkin, "Taming Capitalism: The Evolution of the Indigenous Peasant Economy in Northern Ecuador," *Latin American Research Review*, 1997, 32(3), 89–110.
38. Carlos I. Egregori, "After the Fall of Abimael Gúzmaan: The Limits of Sendero Luminoso," in Maxwell A. Cameron and Philip Mauceri (eds), *The Peruvian Labyrinth* (University Park: Pennsylvania State University Press, 1997), pp. 179–191.
39. Leon Zamosc, "Peasant Struggles of the 1970s in Colombia," in Susan Eckstein (ed.), *Power and Popular Protest: Latin American Social Movements* (Berkeley: University of California Press, 1989), pp. 102–131.
40. Peter Brown, "Institutions, Inequalities, and the Impact of Agrarian Reform on Rural Mexican Communities," *Human Organization*, 1997, 56, 102–110.
41. Steven M. Helfand, "The Political Economy of Agricultural Policy in Brazil: Decision Making and Influence from 1964 to 1992," *Latin American Research Review*, 1999, 34(4), 3–42.
42. *World Population Data Sheet* (Washington, DC: Population Reference Bureau, 1999).
43. Carlos Huneeus, "Technocrats and Politicians in an Authoritarian Regime: The 'ODEPLAN boys' and the 'Gremilists' in Pinochet's Chile," *Journal of Latin American Studies*, 2000, 32, 461–501.
44. Marcus J. Kurtz, "Chile's Neo-Liberal Revolution: Incremental Decisions and Structural Transformation, 1973–89," *Journal of Latin American Studies*, 1999, 31, 399–427.
45. Patrice Franko, *The Puzzle of Latin American Economic Development*, 2d ed. (Boulder, CO: Rowman & Littlefield, 2003).
46. P. Mzría and G. García, "Ajusste Económico, Democratización y Procesos de Privitización de los Espacios Públicos en Venezuela," *Revista Interamericana de Planificación*, 1998, 30(199), 77–91.
47. Lafislau Dowhor, "Decentralization and Governance," *Latin American Perspectives*, 1999, 25(1), 28–44.
48. Henry Veitmeyer and James Petras, *The Dynamics of Social Change in Latin America* (New York: St. Martin's Press, 1999), pp. 91–98.
49. Jorge G. Castañeda, "NAFTA at 10: A Plus or a Minus?" *Current History*, February 2004, 103, No. 870.
50. Elizabeth Elkin, "Science and Culture Clash in a Mexican Staple: Corn," *New York Times*, March 27, 2005, 10.
51. Amoury de Souza, "Redressing Inequalities: Brazil's Agenda at Century's End," in Susan Kaufman Purcell and Riordan Roett (eds), *Brazil under Cardoso* (Boulder, CO: Rienner Publishers, 1997), pp. 63–88.
52. S. P. Costa, *Não Será Nenhum País Como Esta* (Rio: Editora Record, 1992).
53. David Rock, *Authoritarian Argentina* (Berkeley: University of California, 1992), pp. 138–142.
54. Tita Rosenberg, *Children of Cain: Violence and the Violent in Latin America* (New York: Morrow, 1991), pp. 380–383.
55. Tito Drago, *Chile: Un Doble Secuestro* (Buenos Aires: Editorial Compatense, 1993).
56. *Actas Tupamaras* (Madrid: Editorial Revolución, 1982), pp. 249–251.
57. Graciela Medina Batgista, "La Democracia Liberal y Algunas de sus Vicisitudes," *Revista Paraguaya de Sociología*, 1999, 36(105), 159–178.
58. Jaime Salinas Sedó, *Desde el San Felipe: En Defensa de la Democracia* (Lima: Mosca Azul, 1997).

59. Tulio Halparín Donghí, *Historia Contempranea de América Latina* (Madrid: Alianza Editorial, 1981), pp. 314–316.
60. Saul Landau, *The Guerrilla Wars of Central America* (New York: St. Martins' Press, 1993), pp. 67–93.
61. Eric Selbin, *Modern Latin American Revolutions* (Boulder, CO: Westview Press, 1993), pp. 171–177.
62. Susan Eckstein, *Power and Popular Protest: Latin American Social Movements* (Berkeley: University of California Press, 1989), pp. 22–23.
63. Mitchell A. Seligson, "Trouble in Paradise: The Erosion of System Support in Costa Rica, 1878–1999," *Latin American Research Review*, 2002, 37(1), 160–185,
64. Catherine Hite, *When the Romance Ended: Leaders of the Chilean Left 1968–1991* (New York: Columbia University Press, 2000), pp. 195–199.
65. Manuel Moreno Fraginals, "Transition To What?" in Miguel A. Centeno and Mauricio Font (eds), *Toward A New Cuba? Legacies of a Revolution* (Boulder, CO: Rienner Publishers, 1997), pp. 21–25.
66. Debra Evenson, *Revolution in the Balance: Law and Society in Contemporary Cuba* (Boulder, CO: Westview Press, 1994), p. 22.
67. Louis A. Pérez, "Fear and Loathing of Fidel Castro: Sources of U.S. Policy toward Cuba," *Journal of Latin American Studies*, 2002, 34, 227–254.
68. Roberto Foliari, "Redemocratizar el Sistema Político," *Revista Paraguaya de Sociología*, 1998, 153(160), 149–161.
69. Janet Kelly, "Democracy Redux: How Real is Democracy in Latin America?" *Latin American Research Review*, 1998, 33(1), 212–225.
70. Manis Sodré, *O Brasil Simulado: Ensaios sobre Quotidiano Nacional* (Rio: Fundo Editorial, 1991), pp. 70–71.
71. Alan Knight, "Populism and Neo-Populism in Latin America, Especially Mexico," *Journal of Latin American Studies*, 1998, 30, 221–248.
72. Carlos de la Torre, *Populist Seduction in Latin America: The Ecuadorian Experience* (Athens: Ohio University Center for International Studies, 2000).
73. Laura Nuzzi O. Shaughnessy and Michael Dodson, "Political Bargaining and Democratic Transitions: A Comparison of Nicaragua and El Salvador," *Journal of Latin American Studies*, 1999, 31, 99–127.
74. John C. Dugas, "Drugs, Lies, and Videotape: The Samper Crisis in Colombia," *Latin American Research Review*, 2001, 36(2), 157–174.
75. Bruce M. Wilson, *Costa Rica: Politics, Economics, and Democracy* (Boulder, CO: Rienner Publishers, 1998), pp. 160–162.
76. John Peeler, *Building Democracy in Latin America* (Boulder, CO: Rienner Publishers, 1997), pp. 190–193.

Chapter Five Identity, Modernity, and the Arts

1. Danielle Pitta, "Dynamiques du Symbole dans la Médiaion Myhique," *Sociétes*, 2000, 4(70), 13–18.
2. Jaime Andres Peralta, *En Busca de América Latina* (Bogotá: Tercer Mundo Ediores, 1992), pp. 274–275.
3. John P. Gilliin. "Some Sigposts for Policy," in "Council on Foreign Policy" in *Soical Change in Latin America Today* (New York: Random House, 1960), 14–62.

4. Robert M. Levine, *Brazilian Legacies* (Armonk, NY: M. E. Sharpe, 1997), p. 81.
5. Michael Handelsman, *Culture and Customs of Ecuador* (Westport, CT: Greenwood Press, 2000), pp. 42–43.
6. Lawrence E. Harrison, *The Pan American Dream* (New York: Basic Books, 1997), pp. 142–143.
7. Rafael Falcón, *Salsa: A Taste of Hispanic Culture* (Westport, CT: Praeger, 1998), p. 68.
8. Ofelia Schutte, *Cultural Identity and Social Liberation in Latin American Thought* (Albany: University of New York Press, 1993), pp. 45–46.
9. Robert Levine and Ara Norenzayan, "The Pace of Life in 31 Countries," *Journal of Cross-Cultural Psychology*, 1999, 30(2), 178–205.
10. "Caza Zurros," *Noticias* (Buenos Aires), July 12, 2003, 1.
11. Anton Rosenthal, "Spectacle, Fear and Protest: A Guide to the History of Urban Public Space in Latin America," *Social Science History*, 2000, 24(1), 23–41.
12. Roberto DaMatta, *Carnivals, Rogues and Heroes: The Interpretation of the Brazilian Dilemma* (Notre Dame, IN: University of Notre Dame Press, 1991), pp. 32–35.
13. Andres L. Mteo, *Al Filo de la Dominicanidad* (Santo Domingo: Ediores Gtrinitaria, 1996), p. 300.
14. Constantin von Barloewe, *History and Modernity in Latin America* (Providence, RI: Berghahn Books, 1992), p. 116.
15. George Yudice, Jean Franco, and Juan Flores, *On Edge: The Crisis of Contemporary Latin American Culture* (Minneapolis: University of Minnesota Press, 1992), p. viii.
16. José Vasconcelos, *La Raza Cósmica* (Mexico, DF: Editora Calpe, 1994, reprinted edition).
17. Iscar Montero, *José Martí: An Introduction* (New York: Palgrave-Macmillan, 2004).
18. Mario Hernández Sánchez-Baarra, *Formación de las Nacioones Iberoamericanas (Siglo XIX)* (Mexico, DF: Biblioteca Ainteramericana, 1990), pp. 114–122.
19. Josefina Ludmer, *El Género Gauchesco: Un Tratado sobre la Patria* (Buenos Aires: Editorial Sudamericana, 1988).
20. Fernando O. Assençáo, *Historia del Gaucho* (Buenos Aires: Editorial Claridad, 1999).
21. Ruben G. Oliven, " 'The Largest Popular Culture Movement in the Western World': Intellectuals and Gaúcho Traditionalism in Brazil," *American Ethnologist*, 2000, 27(1), 128–146.
22. Enrique Florescanoi, *Memory, Myth, and Time in Mexico*, translated by Albert Bork (Austin: University of Texas Press, 1994), pp. 221–222.
23. Alejandro de la Fuente, "Myths of Racial Democracy: Cuba, 1900–1912," *Latin American Research Review*, 1999, 34(3), 39–74.
24. Alejandro de la Fuente, "Race, Ideology, and Culture in Cuba: Recent Scholarship," *Latin American Research Review*, 2000, 35(3), 199–210.
25. David W. Porter, *Gender and Society in Contemporary Brazilian Cinema* (Austin: University of Texas Press, 1999), pp. 103–107.
26. DaMatta, *Carnivals*, pp. 55–66.
27. Anonio Roséerio, *Carnaval Jiexa* (San Salvador: Corrupio, 1981).
28. J. Lowell Lewis, "Sex and Violence in Brazil: Carnival, *Capoeira*, and the Problem of Everyday Life," *American Ethnologist*, 2000, 26, 529–557.

29. John C. Chasteen, "The Prehistory of Samba: Carnival Dancing in Rio de Janeiro, 1840–1917," *Journal of Latin American Studies*, 1996, 28, 29–47.
30. Leopoldo Zea, "Convergenia y Especifidad de los Valores de América Latina y el Caribe," in Leopoldo Zea and Maario Megallón (eds), *Latinoamérica Encrucijada de Culturas* (Mexico, DF: Fondo de Cultura Económica, 1999), pp. 9–34.
31. Anita Herzfeld, "Language and Identity in Central America: History of Oppression, Struggle, and Achievement," in Julio López-Arias and Gladys M. Varona-Lacey (eds), *Latin America: An Interdisciplinary Approach* (New York: Peter Lang, 1999), pp. 43–55.
32. Eugenia Ibarra, "Identidades y Identidades Latinamericanas?" in J. R. Qauiesada and Mariía Zavala (eds), *500 Años: Holocausto ó Descumbriento?* (San José: Editorial de la Universidad Interamericana, 1991), pp. 263–273.
33. Guillermo Bonfil Batalla, *México Profundo: Una Civiliación Negada* (Mexico, DF: Editorial Grujalaba, 1987), pp. 194–197.
34. Herzfeld, "Language and Identity," pp. 43–68.
35. Robert C. Williamson, *Minority Languages and Bilingualism: Case Studies in Maintenance and Shift* (Westport, CT: Ablex Publishing, 1991), pp. 19–26.
36. José Augustín, *Tragicomedia Mexicana* (Mexico, DF: Editorial Planeta, 1990), pp. 356–357.
37. Ronald Ingelhart, *Modernization and Postmodernization: Cultural, Economic and Political Change in 43 Societies* (Princeton, NJ: Princeton University Press, 1997), p. 82.
38. Augustín Cuerva, *El Proceso de Dominación Política en el Ecuador* (Quito: Editorial Planeta, 1997).
39. Carlos Uribe Celis, *La Mentalidad del Colombiano: Cultura y Sociedad en el Siglo XX* (Bogotá: Editorial Nueva América, 1992), p. 200.
40. Alma Guillermoprieto, *The Heart That Bleeds* (New York: Random House, 1994), pp. 359–386.
41. Jeffrey M. Pilcher, Que Vivan los Ttamales: *Food and the Making of Mexican Identity* (Albuquerque: University of New Mexico Press, 1998).
42. Thomas E. Skidmore, "Raíces de Gilberto Freyre," *Journal of Latin American Studies*, 2002, 34, 1–20.
43. Isabel Rodríguez Vergara, *Colombia: Literatura y Cultura del Siglo XX* (Washington, DC: Organization of American States, 1995).
44. Floyd Merrill, *Sobre las Culturas y Civilaciones Latinoamericanas* (Lanham, MD: University Press of America, 2000), pp. 9–21.
45. Jorge Laraín Ibañez, *Modernidad, Razón y Identidad en América Latina* (Santiago: Editorial Andres Bello, 1996), pp. 179–201.
46. José de la Guarda, *Un Monumento Ecuatoriano* (Quito: Uiversidad Andina Simón Bolívar, 1996).
47. Sarah Radcliffe and Sallie Westwood, *Remaking the Nation: Place, Identity and Politics in Latin America* (London: Routledge, 1996), pp. 8–9, 54–67.
48. Rachel Carr, "Ritual, Knowledge, and the Politics of Identity in Andean Festivities," *Ethnology*, 2003, 42, 39–54.
49. Wanderley Messias da Costa, *O Estado e as Políticas Territoriais no Brasil* (São Paulo: Editora Contexto, 1988), pp. 69–71.
50. Scott Lash and Jonathan Friedman (eds), *Modernity and Identity* (Cambridge, MA: Blackwell, 1992), pp. 1–4.
51. José J. Brunner, *América Latina: Cultura y Modernidad* (Mexico, DF: Editorial Grijalbo, 1992), pp. 118–119.

52. Aruda Arminda do Nascimeno, "Metropole e Cultura: O Novo Modernismo Paulis em Meados do Seculo," *Tempo Social: Revista de Socioloía da USP*, 1997, 19, 39–52.

53. Enrique Medina López, "El Retrato Literato: Historia de Vida y Crónica en la Sociología de la Clase Media de Gabriel Careaga," *Sociología*, 1996, 11(22), 185–193.

54. Claudio Lomnitz, *Modernidad Indecisiva: Ensayos sobre Nación y Modernidad en México* (Mexico, DF: Planeta, 1998), p. 222.

55. Carolina Farías Campero, "La Posmodernidad y los Lenguajes del Arte: Propuestos del Fin de Siglo," in Zidane Zeraoui (ed.), *Modernidad y Posmodernidad* (Mexico, DF: Editorial Limusa, 2000), pp. 161–182.

56. David Lyon, *Postmodernity* (Minneapolis: University of Minnesota Press, 1994), pp. 6–10.

57. Norberto Ras, *Criollismo y Modernidad* (Buenos Aires: Academia Nacional de Ciencias de Buenos Aires, 1999), pp. 341–345.

58. Freddy Martinez Navarro, "Los Paradigmas de la Sociología y el Problema Contemporaneo," in Zeraoui, *Modernidad*, pp. 127–154.

59. Néstor García Canclini, *Consumidores y Ciudadanos* (Mexico, DF: Editorial Grijalbo, 1995), pp. 13–17.

60. Ricardo Lagos, *Después de la Transición* (Buenos Aires: Editorial Zeta, 1993) pp. 177–188.

61. Ronald Inglehart and Wayne E. Baker, "Modernization, Cultural Change, and the Persistence of Traditional Values," *American Sociological Review*, 2000, 65, 19–51.

62. Peter A. Szok, La Ultima Gaviota: *Liberalism and Nostalgia in Early-Twentieth Century Panama* (Westport, CT: Greenwood Press, 2000), pp. 119–121.

63. Inglehart and Baker, *Modernization*, 17–23.

64. June Nash, "Global Integration and the Commodification of Culture," *Ethnology*, 2000, 39(2), 2–141.

65. Laraín Ibañez, *Modernidad*, pp. 170–172, 208.

66. Marycela Córdova, "Modernidad Cultura y Devinir en el Mundo Actual," in Zeraoui, Modernidad, 135–160.

67. Eric Zolov, *Refried Elvis: The Rise of the Mexican Counter Culture* (Berkeley, CA: University of California Press, 1990), pp. 258–259.

68. Rafael Luccaarraíz, *Vueltas a la Patría* (Caracas: Editorial Eclepsidra, 1997), pp. 127–128.

69. Octavio Paz, *Alternating Current* (New York: Viking Press, 1967), pp. 201–202.

70. Keith Booker, *Vargas Llosa among the Postmodernists* (Gainesville: University Press of Florida, 1994), pp. 189–191.

71. Daniel Balderson, *Out of Context: Historical Reference and the Representation in Borges* (Durham, NC: Duke University Press, 1993), pp. 35–40, 116–119.

72. Gerald Martin, *Journeys through the Labyrinth: Latin American Fiction in the Twentieth Century* (New York: Verso, 1989), p. 143.

73. Walnice Nogueira Galvão, "O Epos da Modernização," *Luso-Brazilian Review*, 1994, 36(1), 1–14.

74. Fernando Alegría, *Nueva Histoira de la Novela Hispanoamericana* (Hanover, NH: Ediciones del Norte, 1996), pp. 209–213.

75. Wendy B. Faris, *Carlos Fuentes* (New York: Ungar Publishing Co., 1983), pp. 190–191.

76. Gordona Yovanovich, *Julio Cortázar's Character Mosaic* (Toronto: University of Toronto Press, 1991), pp. 222–227.
77. José Luis Méndez, *Cómo Leer García Márquez: Una Interpretación Sociológica* (Río Piedrez, PR: Editorial de la Universidad de Puerto Rico, 1989), pp. 217–220.
78. Harold Broom, *Gabriel García Márquez* (New York: Chelsea House, 1989).
79. David W. Foster, *Studies in the Contemporary Spanish Short Story* (Columbia: University of Missouri Press, 1979).
80. Donald L. Shaw, *The Post-Boom in Spanish American Fiction* (Albany: State University of New York Press, 1998), pp. 177–178.
81. Gordon Brothereton, "Indigenous Literature in the Twentieth Century," in Leslie Bethell (ed.), *A Cultural History of Latin America* (New York: Cambridge University Press, 1998), pp. 291–310.
82. Dan Versenyi, *Theatre in Latin America* (New York: Cambridge University Press, 1993), pp. 103–104.
83. Diana Taylor, *Theatre of Crisis: Drama and Politics in Latin America* (Lexington: University of Kentucky Press, 1991), pp. 152–156.
84. Jacqueline E. Bixler, "Remembering the Past: Memory-Theater and Tlatelcolco," *Latin American Research Review*, 2002, 37, 119–135.
85. Guillermoprieto, *Heart That Bleeds*; Sean Dillon, "Mexico's Troubadours Turn from *Amor* to Drugs," *New York Times*, February 19, 1999, A4.
86. Nancy Morris, "Cultural Interaction in Latin American and Caribbean Music," *Latin American Research Review*, 1999, 34(1), 187–200.
87. George List, *Music and Poetry in a Colombian Village: A Tricultural Feat* (Bloomington: Indiana University Press, 1983).
88. Simon Collier, Thomas E. Skidmore, and Harold Blakemore, *The Cambridge Encyclopedia of Latin America and the Caribbean* (New York: Cambridge University Press, 1992), p. 406.
89. David W. Foster, "Tango, Buenos Aires, Borges: Cultural Production and Urban Sexual Regulation," in Eva P. Bueno and Terry Cesar (eds), *Imagination Beyond Nation* (Pittsburgh: University of Pittsburgh Press, 1998), pp. 167–193.
90. Gilbert Chase, *Art in Latin America* (New York: Free Press, 1970), pp. 3–5.
91. Dorothy Chaplik, *Latin American Art: An Introduction to Works of the Twentieth Century* (Jefferson, NC: McFarland & Company, 1989), pp. 127–128.
92. David Craven, "Recent Literature on Diego Rivera and Mexican Muralism," *Latin American Research Review*, 2001, 36(3), 221–237.
93. Marta Traba, *Art of Latin America 1900–1980* (Baltimore: John Hopkins University Press, 1994), pp. 23–25.
94. Holly Barnet-Sanchez, "Frida Kahlo: Her Art and Life Revisited," *Latin American Research Review*, 1997, 32(3), 243–257.
95. Collier, *Cambridge*, pp. 434–435.
96. Joseph L. Scarpaci and Virginia Tech, "Architecture, Design and Planning: Recent Scholarship on Modernity and Public Spaces in Latin America," *Latin American Research Review*, 2003, 38, 234–251.
97. Claes af Geijerstan, *Popular Music in Mexico* (Albuquerque: University of New Mexico Press, 1976), p. 125.
98. Paul Almeida and Rubén Urbizagastéqui, "Cutemay Camones: Popular Music in El Salvador's National Liberation Movement," *Latin American Perspectives*, 1999, 26(2), 13–42.

99. María M. Jaramillo et al., *Literatura y Cultura: Narrativa Colombiana del Siglo XX* (Bogotá: Ministerio de Cultura, 2000).

100. Foster, "Tango, Buenos," p. 169.

101. Josaphat B. Kubayanda, *The Poet's Africa* (Westport, CT: Greenwood Press, 1990).

102. Marjorie Agostín, "*Passion, Memory, Identity: Twentieth-Century Latin American Jewish Writers* (Albuquerque: University of New Mexico Press, 1999), pp. xiv–xv.

103. David W. Foster (ed.), *Latin American Writers on Gay and Lesbian Themes: A Bio-Critical Sourcebook* (Westport, CT: Greenwood Press, 1994).

104. José Sánchez-Parga et al., *Signos del Futturo: La Cultura Ecuatoriana en los 80* (Madrid: Agencia Española de Cooperación Internacional, 1991).

105. Ruben Gallo, *New Tendencies in Mexican Art: The 1990s* (New York: Palgrave-Macmillan, 2004).

106. Aníbal Quijano, "Modernity, Identity, and Utopia in Latin America," in John Beverley et al. (eds), *The Postmodernism Debate in Latin America* (Durham, NC: Duke University Press, 1995), pp. 201–216.

107. Doris Sommer, *Proceed with Caution When Engaged by Minority Writing in the Americas* (Cambridge, MA: Harvard University Press, 1999), p. 247.

Chapter Six   Gender, Socialization, and Women's Roles

1. Jacques Lacan, *Feminine Sexuality* (New York: Norton, 1983).

2. Ben Agger, *Gender, Culture, and Power: Toward a Feminist Postmodern Critical Theory* (Westport, CT: Praeger, 1993), p. 4.

3. Patricia M. Longermann and Jill Niebrugge-Brantley, "Contemporary Feminist Theory," in George Ritzer (ed.), *Sociological Theory*, 5th ed. (New York': McGraw-Hill, 2000), pp. 443–489.

4. Octavio Paz, *The Labyrinth of Solitude*, translated by Lysander Kemp (New York: The Grove Press, 1962).

5. Robert C. Williamson, "Shifts in Feminism and Women's Rights in Latin America," *MACLAS Latin American Essays*, 1999, 12, 137–156.

6. Jeane Delaney, "Making Sense of Modernity: Changing Attitudes toward the Immigrant and the Gaucho in Turn-of-the-Century Argentina," *Journal for Comparative Study of Society and History*, 1996, 39, 434–459.

7. Matthew C. Gutmann, *The Meanings of Macho: Being a Man in Mexico City* (Berkeley: University of California Press, 1996), pp. 223–242.

8. Richard G. Parker, *Bodies, Pleasures, and Passions: Sexual Culture in Contemporary Brazil* (Boston: Beacon Press, 1991), p. 44.

9. Roger N. Lancaster, *Life is Hard: Machismo, Danger, and the Intimacy of Power in Nicaragua* (Berkeley: University of California Press, 1992), pp. 236–237.

10. Alma Guillermoprieto, *The Heart That Bleeds* (New York: Vantage Books, 1994), p. 111 cited.

11. Gilberto Freyre, *The Masters and the Slaves*, translated by Samuel Putnam (New York: Knopf, 1956), p. 70.

12. Geraldo Reichel-Dolmatoff and Alicia Reichel-Dolmatoff, *The People of Aritama* (Chicago: University of Chicago Press, 1961), p. 178.

13. Sarah LeVine, Dolor y Alegría: *Women and Social Change in Mexico* (Madison: University of Wisconsin Press, 1993), pp. 36–46, 86–87.
14. Fernando Peñalosa, "Mexican Family Roles," *Journal of Marriage and Family*, 1968, 30, 680–689.
15. Robert C. Williamson, "Differential Perception of the Parental Relationship: A Chilean Sample of Deviant and 'Normal' Population," *Journal of Comparative Family Studies*, 1973, 4, 239–253.
16. Lancaster, *Life is Hard*, pp. 42–43.
17. Noel F. McGinn, Ernest Harburg, and Gerald P. Ginsburg, "Dependency Relations with Parents and Affiliative Responses in Michigan and Quadalajara," *Sociometry*, 1965, 28, 305–321.
18. Alberto Hernández Medina et al., *Cómo Somos los Mexicanos?* (Mexico, DF: Centro de Estudios Educativos, 1987), p. 60.
19. Kathleen Logan, *Hacienda Pueblo: The Development of a Guadalajara Suburb* (Tuscaloosa: University of Alabama Press, 1984), pp. 67–73.
20. Patricia Van Dorp, *Algunos Aspectos de la Familia Chilena* (Santiago: Centro Bellarmino, 1984), p. 64.
21. Nena Delpino, *Salieindo a Flota: La Jefa de la Familia Popular* (Lima: Fundación Naumann, 1990), p. 121.
22. Gutmann, *The Meaning*, pp. 58–64, 83–91.
23. Virginia Gutiérrez de Pineda, *El Gamín* (Bogotá: UNICEF, 1977).
24. Claudio Stern, *El Papel del Trabajo Materno en la Salud Infantil* (Mexico, DF: El Colegio de México, 1996), pp. 335–340.
25. Anne Bar-Din, *Los Niños de Santa Ursula* (Mexico, DF: UNAM, 1991).
26. Francisco Zarama, *La Familia Hoy en América Latina* (Bogotá: Indo-American Press Service, 1980).
27. William Rowe and Vivian Schelling, *Memory and Modernity: Popular Culture in Latin America* (London: Verso, 1991), pp. 98–100, 185–188.
28. Sonya Lipsett-Rivera, "Latin America and the Caribbean," in Teresa A. Meade and Merry E. Wiesner-Hanlks (eds), *A Companion to Gender History* (Malden, MA: Blackwell Publishers, 2004), pp. 477–491.
29. Reichel-Dolmatoff and Reichel-Dolmatoff, *The People*, pp. 89–95.
30. William W. Stein, Hualcan: Life in the Highlands of Peru (Ithaca, New York: Cornell University Press, 1961), 211–213.
31. Marvin Harris, *Town and Country in Brazil* (New York: Columbia University Press, 1956), pp. 158–159.
32. Roger N. Lancaster, "On Homosexualities in Latin America (and Other Places)," *American Ethnologist*, 1997, 24, 193–202.
33. Joseph Varrier, *Se Los Otros: Intimacy and Homosexuality Among Mexican Men* (New York: Columbia University Press, 1995), pp. 188–195.
34. G. Nuñez Noriega, *Sexo Entre Varones: Poder y Residencia en el Campo Sexual* (Hermosillo: Universidad de Sonora, 1994), pp. 318–327.
35. Alejandro D. Marroqúin, *San Pedro Nonualco: Investigación Sociológica* (San Salvador: Editorial Universitaria, 1962).
36. Carlos Monsiváis, "*La Mujer en la Cultura Mexicana*" as cited in Debra A. Castillo, *Easy Women*(Minneapolis: University of Minnesota Press, 1998), p. 39.
37. Robert C. Williamson, "Society, Women, and Mores in Nineteenth-Century Colombia," unpublished manuscript, 1994.
38. Elsa M. Chaney, *Supermadre: Women in Politics in Latin America* (Austin: University of Texas Press, 1978), p. 8.

39. Talcott Parsons, *The Social System* (New York: Free Press, 1951).
40. Olga Grau et al., *Discurso, Género, Poder* (Santiago: LOM-ARCIS, 1997).
41. Lynn Stephen, "Women's Rights are Human Rights: The Merging of Feminine and Feminine Interests among El Salvador's Mothers of the Disappeared (CO-MADRES)," *American Ethnologist*, 1995, 22, 807–822.
42. Brenda Rosenbaum, *With Our Heads Bowed: The Dynamics of Gender in a Maya Community* (Albany: State University of New York Press, 1993), pp. 7–9.
43. Sueli Carneiro, "Identidade Feminina," in Helenieth Saffioti and Monica Muñoz-Vargas (eds), *Muhler Brasileira e Assim* (Rio: Fundo de Naçoes Unidos, 1994), pp. 187–194.
44. Lynn Stephen, "The Construction of Diagnosis Suspects: Militarization and the Gendered and Ethnic Dynamics of Human Rights Abuses in Southern Mexico," *American Ethnology*, 1999, 26, 822–826.
45. K. Lynn Stoner, "Directions in Latin American Women's History, 1977–1985," *Latin American Research Review*, 1987, 22(2), 101–134.
46. Florence E. Babb, "Women and Men in Vicos, Peru: A Case of Unequal Development," in William W. Stein (ed.), *Peruvian Contexts of Change* (New Brunswick, NJ: Transaction Books, 1985), pp. 163–210.
47. Deborah Brandt, *Tangled Routes: Women, Work, and Globalization on the Tomato Trail* (Boulder, CO: Rowman & Littlefield, 2002).
48. Graciela Morgade, "Aproximaciones a la Docencia como Trabajo Feminina," *Revista Paraguaya de Sociología*, 1992, 29(83), 43–54.
49. Helenieth I. B. Saffioti, "Technological Change in Brazil: Its Effect on Men and Women in Two Firms," in June Nash and Helen Safa (eds), *Women and Change in Latin America* (South Hadley, MA: Bergin & Garvey, 1985), pp. 109–135.
50. Julia Tuñon, *Mujeres en México: Una Historia Olvidada* (Mexico, DF: Planeta, 1987).
51. Ana I. García and Enrique Gomáriz, "Tendencias Estructurales: Información Estatística por Sexo," in FLACSO, *Mujeres Centroamericanas ante la Crísis, la Guerra y el Proceso de la Paz* (San José: CSUCA, 1989).
52. Sarah Hamilton, "Neoliberalism, Gender, and Property Rights in Rural Mexico," *Latin American Research Review*, 2000, 37(1), 119–143.
53. Luis Abramo and Marianela Armijo, "Casmbio Técnológico y el Trabajo de Mujeres," *Estudos Feministas*," 1997, 5(1), 31–67.
54. Helen Collinson, *Women and Revolution in Nicaragua* (London: Zen Books, 1990), p. 64.
55. Marilyn Thompson, *Women of El Salvador: The Price of Freedom* (Philadelphia: Institute for the Study of Human Issues, 1986), p. 19.
56. Saffioti, "Technological Change," pp. 109–135.
57. Cristina Bruschini, "O Trabalho da Mulher no Brasil," in Saffioti and Muñoz-Vargas, *Muhler Brasileira*, pp. 15–44.
58. Eliana Chávez, "Mujer y Trabajo Informal," in Ana Ponce et al. (eds), *Hogar y Familia en el Peru* (Lima: Universidad Católica del Peru, 1985), pp. 135–167.
59. Philip L. Russell, *Mexico Under Salinas* (Austin, TX: Mexico Resource Center, 1994), p. 275.
60. Kerry L. Preibisch et al., "Defending Food Security in a Free-Market Economy: The Gendered Dimensions of Restructuring in Rural Mexico," *Human Organization*, 2002, 61, 68–77.

61. Lourdes Benería and Martha Roldán, *The Crossroads of Class and Gender: Industrial Homework, Subcontracting and Household Dynamics in Mexico City* (Chicago: University of Chicago Press, 1987).
62. Instituto Interamericano de Cooperación para la Agricultura, *Producturas de Alimentos: Frente a las Mujeres Productoras de Alimentos en América Latina y el Caribe* (San José: IICA, 1994).
63. Tuñon, *Mujeres*, pp. 37–39.
64. Verena Redkau, *"La Fama" y la Vida: Una Fábrica y sus Obreras* (Tlapán, Mexico, DF: Centro de Investigaciones y Estudios Superiores en Antropología Social, 1984).
65. Altha J. Gravey, *Women and Work in Mexico's Maquiladoras* (Lanham, MD: Rowman & Littlefield, 1998), pp. 85–91.
66. Marie-Dominique de Suremein, "Women's Involvement in the Urban Economy of Colombia," in Joycelin Massiah (ed.), *Women in Developing Countries: Making Visible the Invisible* (Paris: UNESCO, 1993), pp. 195–279.
67. Lois M. Smith and Alfred Padula, *Sex and Revolution: Women in Socialist Cuba* (New York: Oxford University Press, 1966), pp. 122–130.
68. La Mujer Interamericana, "Women of Cuba: An Overview," unpublished manuscript, 2003.
69. Judith Adler Hellman, "Making Women Visible: New Works on Latin American and Caribbean Women," *Latin American Research Review*, 1992, 27(1), 182–204.
70. Susan Triano, "Women and Industrial Development in Latin America," *Latin American Research Review*, 1986, 21, 151–170.
71. Lynne Phillips, "Rural Women in Latin America: Directions for Future Research," *Latin American Research Review*, 1990, 25(3), 89–107.
72. Ricardo Romero Aceves, *La Mujer en la Historia de México* (Mexico, DF: UNAM, 1982).
73. Latin American and Caribbean Women's Collective, *Slave of Slaves: The Challenge of Latin American Women* (London: Zed Press, 1980), pp. 23–24.
74. Suzy Bermúdez Q., *Hijas, Esposas y Amantes: Género, Etnica y Edad en la Historia de América Latina* (Bogotá: Ediciones Uniandes, 1992).
75. Gillian McGillivray, "Evita's Legacy?: Argentine Women and the First Peronist Administration," *Entrecaminos*, 1998, 3(Spring), 71–81.
76. Olga S. Opfell, *Women Prime Ministers and Presidents* (Jefferson, NC: McFarland Publishers, 1993).
77. Tim Hodgon, "*Fem*: A Window onto the Cultural Coalescence of a Mexican Feminist Politics of Sexuality," *Mexican Studies*, 2000, 16(1), 79–105.
78. Lady E. Repetto, "Women Against Violence Against Women," in Gaby Küppers, (ed), *Companeras: Voices from the Latin American Women's Movement*, (New York: Random House, 1994), pp. 126–137.
79. Susan K. Besse, "Engendering Reform and Revolutionary Change in Latin America and the Caribbean," in Teresa A. Meade and Merry E. Wiesner-Hanks (eds), *A Companion to Gender History* (Malden, MA: Blackwell Publishers, 2004), pp. 568–585.
80. Jane S. Jaquette (ed.), *The Women's Movement in Latin America: Feminism and the Transition to Democracy* (Boston: Unwin Hyman, 1989), pp. 4–6.
81. Miguel A. Mudarra, *Aproximación a la Mujer Venezolana* (Caracas: Editorial Buchivacas, 1990), pp. 172–174.

82. Beatriz Manz, *Paradise in Ashes: A Guatemalan Journey of Courage, Terror, and Hope* (Berkeley: University of California Press, 2004), pp. 100–102, 201–204.
83. Tommie S. Montgomery, "Constructing Democracy in El Salvador," *Current History*, February 1997, 96(607), 61–67.
84. Latin American Center, *Statistical Abstract of Latin America* (Los Angeles: University of California, 2000), p. 105.
85. Susan Franceschet, " 'State Feminism' and Women's Movement: The Impact of Chile's Servicio Nacional de la Mujer on Women's activism." *Latin American Research Review*, 2003, 38(1), 9–40.
86. Maritsa Villavicencio, "The Feminist Movement and the Social Movement: Willing Partners," in Küppers, *Compañeras*, pp. 59–70.
87. Joane Martin, "Motherhood and Power: The Production of a Woman's Culture of Politics in a Mexican Community," *American Ethnolgist*, 1990, 17, 470–483.
88. Alejandra Massola, *Por Amor y Coraje: Mujeres en Movimientos Urbanos de la Ciudad de México* (Mexico, DF: El Colegio de México, 1992).
89. Helena Theodoro, *Mito e Esperitualidade: Muhleres Negras* (Rio: Pallas Editores, 1996); Roser Solà, *Ser Madre en Nicaragua: Testimonios de una Historia No Escrita* (Barcelona: Icaria Editorial, 1988); Mzya Diana, *Mujeres Guerrillas* (Buenos Aires: Editorial Planeta, 1997).
90. Linda Lobao, "Women in Revolutionary Movements: Changing Patterns of Latin America Guerrilla Struggle," in Guida West and Rhoda L. Blumberg (eds), *Women and Social Protest* (New York: Oxford University Press, 1990), pp. 180–207.
91. Jennifer G. Schirmer, " 'Those Who Die for Life Cannot be Called Dead,' Women and Human Rights in Latin America," in Marjorie Agosin (ed.), *Surviving Beyond Fear: Women, Children and Human Rights* (Fredonia, NY: White Pine Press, 1993), pp. 31–57.
92. Claudia N. Laudano, *Las Mujeres en los Discursos Militares: Un Análisis Semiótico, 1976–1983* (La Plata: Editorial de la Universidad de La Plata, 1995), p. 80.
93. Jennifer Schirmer, "The Claiming of Space and the Body Politic within National-Security States: The Plaza de Mayo Madres and the Greenham Common Women," in Jonathan Boyerin (ed.), *Remapping Memory: The Politics of Time/Space* (Minneapolis: University of Minnesota Press, 1994), pp. 185–220, 215 cited.
94. Patricia M. Chuchrik, "Subversive Mothers: The Women's Opposition to the Military Regime in Chile," in Agosin, *Surviving*, pp. 86–97.
95. Ximena Bunster, "Surviving Beyond Fear: Women and Torture in Latin America," in Agosin, *Surviving*, pp. 98–125, 109 cited.
96. Rosario Ibarra, "The Search for Disappeared Sons: How it Changed the Mothers," in Küppers *Compañeras*, pp. 116–119.
97. Heleieth Saffioti, "Movimentos Sociais: Face Feminina," in Nanci Valadaras de Carvalho (ed.), *A Condição Feminina* (São Paulo: Vértice, 1988), pp. 143–178.
98. Kristin Herzog, *Finding Their Voice: Women's Testimonies of War* (Valley Forge, PA: Trinity Press International, 1993).
99. Shannon Speed and Melissa Forbis, "Embodying Alternative Knowledges: Everyday Leaders and the Diffusion of Power in Zapatista Autonomous Regions," *LASA Forum*, 2005, 36(1), 17–18.

100. Sonia E. Alvarez et al., "Encountering Latin American and Caribbean Feminisms," *Signs: Journal of Women in Culture and Society*, 2002, 28, 537–580.

Chapter Seven   Marriage and the Family

1. Emmanuel Todd, *The Explanation of Ideology* (Oxford: Basil Blackwell, 1985), p. 196.
2. Wesley R. Burr et al., "Symbolic Interaction and the Family," in Wesley R. Burr et al. (eds), *Contemporary Theories about the Family*, vol. 2 (New York: Free Press, 1979), pp. 42–111.
3. Tessa Cubitt, *Latin American Society* (New York: Wiley-Longman, 1988), p. 106.
4. Keith Farrington and Ely Certok, "Social Conflict Theories of the Family," in Pauline G. Boss et al. (eds), *Sourcebook of Family Theories and Methods* (New York: Plenum Press, 1993), pp. 357–382.
5. David H. Olson and Hamilton I. McCubbin, *Families: What Makes Them Work?* (Newbury Park, CA: Sage, 1983).
6. Jack Knight, *Institutions and Social Conflict* (New York: Cambridge University Press, 1992), pp. 136–137.
7. Hélène Tremblay, *Families of the World* (New York: Farrar, Straus, and Giroux, 1988), pp. 31–32, 61–62.
8. Rogelio Díaz Guerrero, *El Ecosistema Sociocultural y la Calidad de la Vida* (Mexico, DF: Editorial Trillas, 1986), p. 47.
9. William J. Goode, *World Changes in Divorce Patterns* (New Haven, CT: Yale University Press, 1993), p. 189.
10. María A. Fauné, "Cambios de las Familias en Centro América," in Ediciones de la Mujer 20: *Familias Siglo 21* (Santiago: ISIS Internacional, 1994), pp. 107–150.
11. Tremblay, *Families*, pp. 108–109.
12. Cristina Bruschini, "O Trabalho da Mulher no Brasil," in Helenieth Saffioti and Monica Muñoz-Vargas (eds), *A Mulher Brasileira e Assim* (Rio: Fundo de Naçcoes Unidos, 1994), pp. 15–44.
13. Goode, *World Changes*, p. 194.
14. Debra A. Castillo, *Easy Women: Sex and Gender in Modern Mexican Fiction* (Minneapolis: University of Minnesota Press, 1998), p. 219.
15. Paul L. Doughty, *Huaylas: An Andean District in Search of Progress* (Ithaca, NY: Cornell University Press, 1968), p. 31.
16. Lucero Zamudio and Norma Rubiano, *La Nupcialidad en Colombia* (Bogotá: Universidad Externada de Colombia, 1991), p. 129.
17. Asuncion Lavrin, "Sexuality in Colonial Mexico: A Church Dilemma," in Asuncion Lavrin (ed.), *Sexuality and Marriage in Colonial Latin America* (Lincoln: University of Nebraska Press, 1989), pp. 47–95.
18. Diana Balmori, Stuart F. Voss, and Miles Wortman, *Notable Family Networks in Latin America* (Chicago: University of Chicago Press, 1984), p. 163.
19. Thomas McCorkle, *Fajardo's People* (Los Angeles: Latin American Center, University of California, 1965), pp. 75–77.
20. Gerardo Reichel-Dolmatoff and Alicia Reichel-Dolmatoff, *The People of Aritama* (Chicago: University of Chicago Press, 1961), pp. 111–114.

21. David Yaukey and Timm Thorsen, "Differential Female Age at First Marriage in Six Latin American Cities," *Journal of Marriage and the Family*, 1972, 34, 375–379; Latin American Center, *Statistical Abstract of Latin America* (Los Angeles: University of California, 2001), p. 154.

22. *World Population Data Sheet* (Washington, DC: Population Reference Bureau, 1999).

23. Ana Ponce, *Hogar y Familia en el Peru* (Lima: Universidad Católica del Peru, Facultad de Ciencias Sociales, 1985), p. 15.

24. Matthew C. Gutmann, *The Meanings of Macho: Being a Man in Mexico City* (Berkeley: University of California Press, 1996) pp. 70–73.

25. Richard Pace, *The Struggle for Amazon Town: Gurupá Revisited* (Boulder, CO: Rienner Publishers, 1988), p. 143.

26. Sarah LeVine, *Dolor y Alegría: Women and Social Change in Urban Mexico* (Madison: University of Wisconsin Press, 1993), pp. 90–91.

27. Verena Stolcke, *Coffee Planters, Workers and Wives* (New York: St. Martin's Press, 1988), pp. 221–239.

28. Robert M. Levine and José Bom Meily, *The Life and Death of Carolina Maria de Jesus* (Albuquerque: University of New Mexico Press, 1995).

29. Latin American Center, Statistical Abstract of Latin America (Los Angeles: University of California, 2001), 155.

30. Michael L. Conniff and Frank D. McCann, *Modern Brazil: Elites and Masses in Historical Perspective* (Lincoln: University of Nebraska Press, 1989), p. 276.

31. Mariana Aylwin and Ignacio Walker, *Familia y Divorcio: Razones de una Posición* (Santiago: Editorial Los Andes, 1996).

32. Comisión Económica para Latinamérica y el Caribe, *Cambios en el Perfil de la Familia* (Santiago: United Nations, 1993), p. 315.

33. Fundación Andes, *Familias Populares: Historia Cotidana y Intervención Social* (Santiago: ECO, 1997), pp. 164–168.

34. Cathy A. Rokpwski. "Women as Political Actors: The Move from Maternalism to Citizenship Rights and Power," *Latin American Research Review*, 2003, 38(2), 180–193.

35. Gutmann, *The Meanings*, pp. 74–88.

36. Richard G. Parker, *Bodies, Pleasures, and Passions: Sexual Culture in Contemporary Brazil* (Boston: Beacon Press, 1991), pp. 33–35.

37. Alex H. Westfried, "The Emergence of the Democratic Brazilian Middle-Class Family: A Mosaic of Contrasts with the American Family (1960–1994)," *Journal of Comparative Family Studies*, 1997, 28, 25–54.

38. Pierre L. van den Berghe and George P. Primov, *Inequality in the Peruvian Andes: Class and Ethnicity in Cuzco* (Columbia: University of Missouri Press, 1977), p. 162.

39. Claudia B. Isaac, "Class Stratification and Cooperative Production among Rural Women in Central Mexico," *Latin American Research Review*, 1995, 30(2), 123–150.

40. Oscar Lewis, *Five Families: Mexican Case Studies in the Culture of Poverty* (New York: Basic Books, Inc., 1959).

41. Cathy A. Rakowski, "Women as Political Actors: The Move from Maternalism to Citizenship Rights and Power," *Latin American Research Review*, 2003, 38(2), 180–194.

42. Gabriel Careaga, *Mitos y Fantasías de la Clase Media en México* (Mexico, DF: Cal y Aena, 1990), chapters 3 and 4.

43. Reichel-Dolmatoff and Reichel-Dolmatoff, *The People*, pp. 184–185.
44. Rolando Amcs C., *Familia y Violencia en el Peru de Hoy* (Lima: Comite Peruana de Bienestar Social, 1986), pp. 36–40.
45. June E. Hahner, "Recent Research on Women in Brazil," *Latin American Research Review*, 1985, 22(3), 163–179.
46. Ronald M. Sabatelli and Constance L. Shehan, "Exchange and Resource Theories," in Boss, *Sourcebook*, pp. 385–411.
47. Waldo Ansaldi, "Fragmentados, Excluidos, Famelicos y Como si Eso Fuese, Violentos y Corruptos," *Revista Paraguaya de Sociología*, 1997, 34(98), 7–36.
48. Susan Vincent, "Flexible Families: Capitalist Development and Crisis in Rural Peru," *Journal of Comparative Family Studies*, 2000, 26, 155–170.
49. Susan K. Besse, "Engendering Reform and Revolutionary Change in Latin America and the Caribbean," in Teresa A. Meade and Merry E. Wiesner-Hanks (eds), *A Companion to Gender History* (Malden, MA: Blackwell Publishers, 2004), pp. 568–585.
50. Anna Nygren, "Violent Conflicts and Threatened Lives: Nicaraguan Experiences of Wartime Displacement and Postwar Distress," *Journal of Latin American Studies*, 2003, 35, 391–409.
51. R. S. Oropesa, "Development and Marital Power in Mexico," *Social Forces*, 1997, 75, 1291–1317.
52. Gutmann, *The Meanings*, pp. 193–199.
53. Caroline Moser and Cathy McIlwaine, *Violence in a Post-Conflict Context* (Washington, DC: World Bank, 2001), pp. 57–63.
54. Christina MacCulloch, "Domestic Violence: Private Pain, Public Issue," *IDB Special Report* (Washington, DC: Inter-American Development Bank, 1998).
55. Fauné, "Cambios de," pp. 107–109.
56. Instituto de la Mujer, *Como Les Ha Ido a las Mujeres Chilenas en la Democracia? Balance y Propuestas Mirando al 2000* (Santiago, 1996), p. 3.
57. Nancy E. Dowd, *In Defense of Single-Parent Families* (New York: New York University Press, 1997).
58. Lorraine Davies, William R. Avison, and Donna D. McAlpine, "Significant Life Experiences among Single and Married Mothers," *Journal of Marriage and the Family*, 1997, 59, 294–308.
59. Roberto DaMatta, *Carnivals, Rogues, and Heroes: An Interpretation of the Brazilian Dilemma* (Notre Dame, IN: University of Notre Dame Press, 1991), pp. 189–190.
60. Darcy Ribeira, *The Brazilian People* (Gainesville: University Press of Florida, 2000), pp. 49–51.
61. Susan Eckstein, *The Poverty of Revolution* (Princeton, NJ: Princeton University Press, 1977), p. 75.
62. Andrew H. Whiteford, *An Andean City at Mid-Century* (East Lansing: Michigan State University, 1977), pp. 174–177.
63. William W. Stein, *Hualcan: Life in the Highlands of Peru* (Ithaca, NY: Cornell University Press, 1961), p. 177f.
64. McCorkle, *Fajardo's People*, p. 83.
65. Pace, *The Struggle*, pp. 140–143.
66. Carl Kendall, "Urban–Rural Differences in the Selection Strategies of Compadrazgo: A Controlled Comparison," in John M. Hunter et al. (eds), *Population Growth and Urbanization in Latin America* (Cambridge, MA: Shenkman Publishing Co., 1983), pp. 281–290.

67. UNICEF, *Qué Piensan los Niños Latinoamericanos sobre la Familia?* (La Paz: Defensa de los Niños, 1993).

68. Larissa Adler Lomnitz and Marisol Perez-Lisaur, "The Solidarity of Mexican Grand-Families," in Elisabeth Jelin (ed.), *Family, Household and Gender Relations in Latin America* (London: Kegan Paul International, 1991), pp. 123–132.

69. Gilberto Freyre, *The Masters and the Slaves*, translated by Samuel Putnam (New York: Knopf, 1956), p. 356.

70. Larissa Adler Lomnitz and Marisol Perez-Lisaur, *A Mexican Elite Family, 1820–1980, Kinship, Class and Culture* translated by C. Lomnitz (Princeton, NJ: Princeton University Press, 1987), pp. 232–238.

71. Alex M. Saragoza, *The Monterrey Elite and the Mexican State, 1880–1940* (Austin: University of Texas Press, 1988).

72. Eni de Mesquita Samara, *A Familia Brasileira* (São Paulo: Editorial Brasiliense, 1983).

73. Piers Armstrong, " 'The Brazilianists' Brasil' Interdisciplinary Portraits of Brazilian Society and Culture," *Latin American Research Review*, 2000, 35(1), 227–241.

74. Leo A. Despres, *Manaus: Social Life and Work in Brazil's Free Trade Zone* (Albany: State University of New York Press, 1991), pp. 156–157.

75. Susan de Vos, "Is There a Socioeconomic Dimension to Household Extension in Latin America?" *Journal of Comparative Family Studies*, 1993, 24, 21–34.

76. Robert C. Williamson and Georgene H. Seward, "Concepts of Social Sex Roles Among Chilean Adolescents," *Human Development*, 1971, 14, 184–194.

77. Frank Bonilla, *The Failure of Elites* (Cambridge, MA: M.I.T. Press, 1970), pp. 116–126.

78. C. H. Browner, "Gender Roles and Social Change: A Mexican Case Study," *Ethnology*, 1985, 26, 89–104.

79. Consejo Consultivo del Programa Nacional de Solidaridad, *El Combate a la Pobreza: Lineamientos Programáticos* (Mexico, DF: El Nacional, 1990), pp. 102–105.

80. Richard Maclure and Melvin Sotelo, "Children's Rights and the Tenuousness of Local Conditions: A Case Study of Nicaragua," *Journal of Latin American Studies*, 1994, 36, 81–108.

81. Carlos García Martínez, *La Puericultua y la Mitología Popular* (Guerétaro: Universidad Autónoma de Querétaro, 1989).

82. Leoncio Barios, "Television, Telenovelas, and Family Life in Venezuela," in James Lull (ed.), *World Families Watch Television* (Newbury Park, CA: Sage, 1988), pp. 48–79.

83. Keista E. Van Fleet, "Adolescent Ambiguities and the Negotiation of Belonging in the Andes," *Ethnology*, 2003, 42, 349–363.

84. LeVine, *Dolor y Alegría*, p. 193.

85. Paz Covarrubias, Monica Muñoz, and Carmen Reyes, *Crisis en la Familia?* (Santiago: Instituto de Sociología, Universidad Cátolica, 1983).

86. Beverly Raphael, *The Anatomy of Bereavement* (New York: Basic Books, 1983), p. 37.

87. Robert W. Shirley, *The End of a Tradition: Culture, Change and Development in the Municipio of Cunha* (New York: Columbia University Press, 1971), pp. 38–39.

Chapter Eight   Religion at the Crossroads

1. Peter L. Berger, *The Sacred Canopy* (New York: Doubleday, 1967), p. 45.
2. Gisela Cónepa Koch, "Identidad Regional y Migrantes en Lima: La Fiesta de la Virgen del Carmen de Paucartango," in Bernd Schmalz and N. Ross Crumarine (eds), *Estudios sobre el Sincretismo en América Central y en los Andes* (Bonn: Hoxlos Verlag, 1996), pp. 153–185.
3. J. Lloyd Mecham, *Church and State in Latin America*, revised ed. (Chapel Hill: University of North Carolina Press, 1969), pp. 34–35.
4. John F. Schwaller, *Origins of Church Wealth in Mexico* (Albuquerque: University of New Mexico Press, 1985), pp. 173–182.
5. Alberto Armani, *Ciudad de Dios y Ciudad del Sol* (Mexico, DF: tomdo de Cultura Económica, 1987).
6. Robert N. Bellah and Phillip E. Hammond, *Varieties of Civil Religion* (New York: Harper & Row, 1980), pp. 57–58.
7. John M. Hart, *Revolutionary Mexico* (Berkeley: University of California Press, 1987), pp. 346–347.
8. Roderic Al Camp, "The Cross in the Polling Booth: Religion, Politics, and the Laity in Mexico," *Latin American Research Review*, 1994, 29(3), 69–100.
9. Lisandro Pérez, "The Catholic Church in Cuba: A Weak Institution," in Jay P. Dolan and Jaime R. Vidal (eds), *Puerto Ricans and Cubans in the U.S., 1900–65* (Notre Dame, IN: University of Notre Dame Press, 1994), pp. 147–157.
10. Thomas C. Bruneau, *The Church in Brazil: The Politics of Religion* (Austin: University of Texas Press, 1982), pp. 13–14.
11. Robert M. Levine, *Father of the Poor? Vargas and His Era* (Cambridge, UK: Cambridge University Press, 1998), pp. 35–36.
12. Jon S. Bincent, *Culture and Customs of Brazil* (Westport, CT: Greenwood Press, 2003), pp. 74–75.
13. Souza Barros, *Messianismo e Volência de Massa em Brasil* (São Paulo: Instituto Nacional do Livro, 1986).
14. David Martin, *Tongues of Fire: The Explosion of Protestantism in Latin America* (Cambridge, MA: Basil Blackwell, 1990), p. 69.
15. Lindsay L. Hale, "Preto Velho: Resistance, Redemption, and Engendered Representations of Slavery in a Brazilian Possession–Trance Religion," *American Ethnologist*, 1997, 24, 392–414.
16. Alma Guillermoprieto, *The Heart That Bleeds: Latin America Now* (New York: Vintage Books, 1994), p. 153.
17. David J. Hess, *Spirits and Scientists: Ideology, Spiritism, and Brazilian Culture* (University Park: Pennsylvania State University Press, 1991), p. 163.
18. Kasey O. Dolin, "Yoruban's Religious Survival in Brazilian Candoblé, *MACLAS Latin American Essays*, 2002, 15, 69–82.
19. Rachel E. Harding, *A Refuge in Thunder: Candomblé and Alternative Spaces of Blackness* (Bloomington: University of Indiana Press, 2000).
20. Hebe Vessuri, "Brujos y Aprendices de Brujos en una Comunidad Rural de Santiago del Estero," *Revista Latinamericana de Sociologia*, September–December, 1976, 443–458.
21. Ralph della Cava, *Miracle at Joaseiro* (New York: Columbia University Press, 1970).

22. Miguel Barnet, *Afro-Cuban Religions* (Kingston, Jamaica: Wycliff Publishers, 2001), pp. 1–7, 120–125.

23. David Rock, *Argentina 1516–1987: From Spanish Colonialism to Alfonsín* (Berkeley: University of California Press, 1987), pp. 120–131.

24. Latin American Center, *Statistical Abstract of Latin America*, vol. 38 (Los Angeles, CA: University of California, 2002), p. 379; Judith Laiken Elkin, *Jews of the Latin American Republics* (Chapel Hill: University of North Carolina Press, 1989) 239 cited, Victor A. Mirelman, *Jewish Buenos Aires, 1890–1930: In Search of an Identity* (Detroit: Wayne State University Press, 1990).

25. Beatriz Manz, *Paradise in Ashes: A Guatemalan Journey of Courage, Terror, and Hope* (Berkeley: University of California Press, 2004), pp. 25–26.

26. Maureen E. Shea, *Culture and Customs of Guatemala* (Westport, CT: Greenwood Press, 2001), pp. 33–35.

27. William Rowe and Vivian Schelling, *Memory and Modernity: Popular Culture in Latin America* (London: Verso, 1991), p. 50.

28. Daniel H. Levine, *Popular Voices in Latin American Catholicism* (Princeton, NJ: Princeton University Press, 1992), pp. 81–82.

29. Robert C. Williamson, *El Estudiante Colombiano y sus Actitudes*, in *Monografías Sociológicas*, no. 13 (Bogotá: Universidad Nacional de Colombia, 1962), pp. 31–32.

30. Enrique Luengo González, *La Religión y los Jovenes: El Desgaste de una Religión*. vol. 3, *Cuadernos de Cultura y Religión* (Mexico, DF: Universidad Interamericana, 1993), p. 215.

31. Sarah LeVine, *Dolor y Alegría: Women and Social Change in Urban Mexico* (Madison: University of Wisconsin Press, 1993), pp. 106–108.

32. Latin American Center, *Statistical Abstract of Latin America* (Los Angeles: University of California, 2001), p. 293.

33. Thomas C. Bruneau, "Brazil: The Catholic Church and Basic Christian Communities," in Daniel H. Levine (ed.), *Religion and Political Conflict in Latin America* (Chapel Hill: North Carolina University Press, 1986), pp. 106–123.

34. Guillermo de la Peña, *A Legacy of Promises* (Austin: University of Texas Press, 1981), pp. 226–230.

35. Ivan Vallier, "Religious Elites in Latin America," in Seymour M. Lipset and Aldo Solari (eds), *Elites in Latin America* (New York: Oxford University Press, 1967), pp. 221–252.

36. Charles A. Reilly, "Latin America's Religious Populists," in Daniel H. Levine (ed.), *Religion and Political Conflict in Latin America* (Chapel Hill: University of North Carolina Press, 1986), pp. 42–57.

37. Levine, *Popular Voices*, p. 217.

38. Phillip Berryman, "What Happened at Puebla?" in Daniel H. Levine (ed.), *Churches and Politics in Latin America* (Newbury Park, CA: Sage, 1982), pp. 55–86.

39. Gustavo Gutiérrez, *Una Teología de Liberación* (Lima: CEP, 1971).

40. Paul E. Sigmund, "The Development of Liberation Theology: Continuity or Change?" in Richard L. Rubenstein and John K. Roth (eds), *The Politics of Latin American Liberation Theology* (Washington, DC: Washington Institute Press, 1988), pp. 21–47.

41. Hugo José Sáurez, *Laberinto Religioso: Sociedad, Iglesia y Religión en América Latina* (La Paz: Umbrales CIDES, UMSA, 1996), pp. 129–130.

42. Roger N. Lancaster, *Thanks to God and the Revolution* (New York: Columbia University Press, 1988), pp. 73–79.
43. Roger N. Lancaster, *Life is Hard: Machismo, Danger, and the Intimacy of Power in Nicaragua* (Berkeley: University of California Press, 1992), pp. 170–174.
44. Scott Mainwaring and Alexander Wilde, "The Progressive Church in Latin America: An Interpretation," in Scott Mainwaring and Alexander Wilde (eds), *The Progressive Church in Latin America* (Notre Dame, IN: University of Notre Dame Press, 1989), pp. 1–37.
45. Manuel María Marzal, *Estudios sobre la Religión Campesina* (Lima: Universidad Católica, Fondo Editorial, 1977), p. 298.
46. José L. González, *La Religión Popular en el Peru: Informe y Diagnóstico* (Cuzco: Instituto de Pastoral Andina, 1987), p. 29.
47. Delir Brunelli, *A Mulher Religiosa: Presença e Atuacão* (Rio: Publicações CRB, 1990).
48. Phillp Berryman, *Liberation Theology* (Philadelphia: Temple University Press, 1987), pp. 172–175.
49. W. E. Hewitt, "Myths and Realities of Liberation Theology: The Case of Basic Christian Communities in Brazil," in Rubenstein and Roth, *The Politics*, pp. 135–155.
50. Hannah Stewart-Gambino, "Introduction: New Games, New Rules," in Edward L. Cleary and Hannah Stewart-Gambino (eds), *Conflict and Competition: The Latin American Church in a Changing Environment* (Boulder, CO: Rienner Publishers, 1992), pp. 1–19, p. 2 cited.
51. J. Amando Robles, *Religión y Paradigmas: Ensayos Sociólogos* (Heredia, CR: Editorial Fundación UNA, 1995), pp. 119–141.
52. Kristen Norget, "Progressive Theology and Popular Religiosity in Oaxaca, Mexico," *Ethnology*, 1997, 36, 67–83.
53. Stephen D. Glazier, "Latin American Perspectives on Religion and Politics," *Latin American Research Review*, 1995, 30(1), 247–255.
54. Miguel A. Sobrino, "Reflexiones Marginmales en Torno a dos Problemas de la Iglesia Católica en América Latina: El Centrismo del Poder y la Corrupción, *Anuario de Estudios Latinoamericanos*, 1999, 31, 39–54.
55. Barbara Bode, *No Bells to Toll: Destruction and Creation in the Andes* (New York: Scribner's, 1989), pp. 191–232.
56. George Priestley, "Squatters, Oligarchs and Soldiers in San Miguelito, Panama," in Matthew Edel and Ronald G. Hellman (eds), *Cities in Crisis: The Urban Challenge in the Americas* (New York: Bildner Center, 1989), pp. 127–156.
57. David Stoll, *Is Latin America Turning Protestant? The Politics of Evangelical Growth* (Berkeley: University of California Press, 1992), p. 8; *Statistical Abstract*, 2001, p. 306.
58. Hannah Stewart-Gambino, "Church and State in Latin America," *Current History*, 1994, 93(March), 129–133.
59. Susan Hawley, "Protestantism and Indigenous Mobilization: The Moravian Church among the Miskitu Indians of Nicaragua," *Journal of Latin American Studies*, 1997, 29, 111–129.
60. Christian Gros, "Evangelical Protestantism and Indigenous Populations," *Bulletin of Latin American Research*, 1999, 18, 175–197.
61. Scott Mainwaring, *The Catholic Church and Politics in Brazil, 1916–1985* (Stanford, CA: Stanford University Press, 1986), p. 38.

62. Phillip Berryman, "Is Latin America Turning Pluralist? Recent Writings on Religion," *Latin American Research Review*, 1995, 30(3), 107–122.

63. Jean-Pierre Bastian, "The Metamorphosis of Latin American Protestant Groups: A Sociohistorical Perspective," *Latin American Research Review*, 1993, 28(2), 33–61.

64. Karsten Paerregaard, "Conversion, Migration, and Social Identity: The Spread of Protestantism in the Peruvian Andes," *Ethnos*, 1994, 59(3–4), 163–185.

65. Mark Thurner, "Peasant Politics and Andean Haciendas in the Transition to Capitalism: An Ethnographic History," *Latin American Research Review*, 1993, 28(3), 41–82.

66. Anne Motley Hallum, "Taking Stock and Building Bridges: Feminism, Women's Movements and Pentecostalism in Latin America," *Latin American Research Review*, 2003, 38(1), 169–182.

67. Latin American Center, *Statistical Abstract of Latin America* (Los Angeles: University of California, 2000), p. 285.

68. José L. Pérez Guadalupe, *Las Sectas en el Peru: Estudio sobre los Principales Movimientos en Nuestro País* (Lima: Conferencia Episcopal Peruana, 1991), p. 55.

69. Juan G. Prado Ocaranza, *Sectas Juveniles en Chile* (Santiago: Editorial Covadonga, 1984), p. 79.

70. Patricia Birman and David Lehman, "Religion and Media in a Battle for Ideological Hegemony: The Universal Church of the Kingdom of God and TV Globo in Brazil," *Bulletin of Latin American Research*, 1999, 18, 145–164.

71. David L. Parkyn, "Pentecostal Faith and National Politics: The Presidency in Contemporary Guatemala," MACLAS, *Latin American Essays*, vol. 6, 1993, pp. 65–82.

72. Shea, *Culture and Customs*, p. 94.

73. Stoll, *Is Latin*, p. 307.

74. Germán Guzmán, *Camilo Torres*, translated by John D. Ring (New York: Sheed and Ward, Inc., 1969), pp. 130–131.

75. Penny Leroux, "The Journey from Medellín and Puebla: Conversion and Struggle," in Edward L. Cleary (ed.), *Born of the Poor: The Latin American Church since Medellín* (Notre Dame, IN: University of Notre Dame Press, 1990), pp. 45–63.

76. Patricia Birman, "Modos Perifieicos de Crença," in Pierre Saanchis (ed.), *Catolocismo: Unidade Religiosa e Pluralismo Cultural* (São Paulo: Edições Loyola, 1992), pp. 167–195.

77. John M. Kirk, *Politics and the Catholic Church in Nicaragua* (Gainesville: University Press of Florida, 1992), pp. 184–187.

78. Kevin Neuhouser, "The Radicalization of the Brazilian Catholic Church in Comparative Perspective," *American Sociological Review*, 1989, 54, 233–244.

79. Michael Fleet and Brian H. Smith, *The Catholic Church and Democracy in Chile and Peru* (Notre Dame, IN: University of Notre Dame Press, 1997), pp. 274–278.

80. Penny Lernoux, *Cry of the People* (New York: Penguin Books, 1982), p. 488.

81. Sarah Cline, "Competition and Fluidity in Latin American Christianity," *Latin American Research Review*, 2000, 35(2), 244–251.

82. Mark Chaves, "Secularization as Declining Religious Authority," *Social Forces*, 1994, 72, 749–774.

83. Fleet and Smith, *The Catholic Church*, pp. 282–286.

84. Scott Mainwaring, "Democrastization, Socioeconomic Disintegration, and the Latin America Church After Puebla," in Edward L. Clearly (ed.), *Born of the*

Poor: The Latin America Church since Medellin (Notre Dame, IN: University of Notre Dame Press, 1990), 163.

85. MarcoLar Klahr, "Mercadotécnia y Angelomania: Ecos del Esoterismo de la New Age a Fin del Melenio," *Revista Académoca para Esudios de las Religiones*, 2000, 3, 309–324.

Chapter Nine Education:
A Continuing Challenge

1. Ingemar Fägerlind and Lawrence J. Saha, *Education and National Development*, 2d ed. (Oxford: Pergamon Press, 1989), pp. 16–17.
2. John Hawkins, *Inverse Images: The Meaning of Culture, Ethnicity and Family in Postcolonial Guatemala* (Albuquerque: University of New Mexico Press, 1984), p. 42.
3. Orlando Fals-Borda, "La Educación en Colombia," *Monografías Sociológicas*, no. 11 (Bogotá: Universidad Nacional de Colombia, 1962).
4. Thomas S. Popkewitz, *A Political Sociology of Educational Reform* (New York: Teachers College, Columbia University, 1991), pp. 58–59.
5. Carlos Newland, "The *Estado Docente* and its Expansion: Spanish American Elementary Education, 1900–1950," *Journal of Latin American Studies*, 1994, 26, 449–467.
6. Robert M. Levine, *Father of the Poor? Vargas and his Era* (Cambridge: University of Cambridge Press, 1998), pp. 124–126.
7. Marta M. Chagas de Carvalho, "Quando a História de Educação é a História da Disciplina e da Higienação da Pessoa," in Marcos Cezar de Freitas (ed.), *História da Infância no Brasil* (São Paulo: Cortez Editora, 1997), pp. 269–288.
8. David S. Brown, "Democracy, Authoritarianism and Education Finance in Brazil," *Journal of Latin American Studies*, 2002, 34, 115–141.
9. Jacques Hallak, *Investing in the Future: Setting Educational Priorities in the Developing World* (Oxford: Pergamon Press, 1990), p. 21.
10. Latin American Center, *Statistical Abstract of Latin America*, vol. 1 (Los Angeles: University of California, 2000), pp. 210–212.
11. Concha Delgado-Gaitan, "Traditions and Transitions in the Learning Process of Mexican Children: An Ethnographic View," in George Spindler and Louise Spindler (eds), *Interpretive Ethnography of Children* (Hillsdale, NJ: Erlbaum Publishers, 1987), pp. 333–359.
12. Charles E. Bidwell and Noah E. Friedkin, "The Sociology of Education," in Neal J. Smelser (ed.), *Handbook of Sociology* (Newbury Park, CA: Sage Publications, 1988), pp. 449–472.
13. *Statistical Abstract*, 1999, p. 263.
14. Fernando Reimers, "The Impact of Economic Stabilization and Adjustment on Education in Latin America," *Comparative Education Review*, 1991, 35, 319–354.
15. Daniel A. Morales-Gómez and Carlos A. Torres, *Education, Policy, and Social Change: Experiences from Latin America* (Westport, CT: Praeger, 1992), p. 5.
16. Emilio Parrado, "Expansion of Schooling, Economic Growth, and Regional Inequalities," *Comparative Education Review*, 1998, 42(3), 223–235.

17. Martin Carnoy, "National Voucher Plans in Chile and Sweden: Did Privatization Make for Better Education," *Comparative Education Review*, 1998, 42, 309–332.
18. Luís E. Valcárcel, "La Educación del Campesino," in Eilio Barrentos (ed.), *Ensayos sobre Educaciíon Peruana* (Lima: Universidad Palma, 1999), pp. 143–169.
19. Altamirano Rua et al., *Necesidades y Demanda para un Cambio Edducativo* (Magdalena: Foro Educativo, 1996).
20. Nelly P. Stromquist, "Gender Delusions and Exclusions in the Democratization of Schooling in Latin America," *Comparative Education Review*, 1996, 40, 404–425.
21. *Statistical Abstract*, 2000, pp. 181–184.
22. René López Camaclho, *La Irrazonable Educación Mexicana* (Mexico, DF: Universidad Autónoma del Estado de México, 1994), pp. 77–78.
23. George Psacharopoulos, Carlos Rojas, and Eduardo Velez, "Achievement Evaluation of Colombia's *Escuela Nueva*: Is Multigrade the Answer?" *Comparative Education Review*, 1993, 37, 263–276.
24. Gilbeert A. Valverde, "Curriculum Convergence in Chile: The Global and Local Context of Reforms in Curriculum Policy," *Comparative Education Review*, 2004, 18, 180–207.
25. Juan Prawda, *Logros, Irregularidades y Retos del Futuro del Sistema Educativo Mexicano* (Mexico, DF: Editorial Grijalbo, 1989), p. 273.
26. Gilberto Guevara Niebla, *La Catástrofe Silenciosa* (Mexico, DF: Fondo de Cultura Económica, 1992), pp. 17–20.
27. Minerva Vincent and Josefina Pimental, *Concepciones, Procesos y Estructuras Educativas* (Santo Domingo: FLACSO, 1996).
28. Rosa M. Torres, *Los Achaques de la Educación* (Quito: Instituto Fronesis, 1995), pp. 252–254.
29. Jorge Alegria et al., "Chile," in Walter Wickremasinghe (ed.), *Handbook of World Education* (Houston: American Collegiate Service, 1991), pp. 159–165.
30. "Brazil," in *World Education Encyclopedia*, vol. 1 (Detroit: Gale Group, 2001), p. 163.
31. Alfredo Traversoni and Diosma Piotti, *Nuestro Sistema Educativo Hoy* (Montevideo: Ediciones de la Banda Oriental, 1984), pp. 104–108.
32. Peter A. Easton and Simon M. Fass, "Monetary Consumption Benefits and the Demand for Primary Schooling in Haiti," *Comparative Education Review*, 1989, 33, 176–193.
33. Frans J. Schryer, *Ethnicity and Class Conflict in Rural Mexico* (Princeton, NJ: Princeton University Press, 1990), pp. 163, 298.
34. Juan Carlos Palafox, Juan Prawda, and Eduardo Vélez, "Primary School Quality in Mexico," *Comparative Education Review*, 1994, 38, 167–180.
35. T. Neville Postlethlwaite (ed.), *The Encyclopedia of Comparative Education and National Systems of Education* (New York: Pergamon, 1988), p. 158.
36. Nariw–Andrée Somers, Patrick J. McEvan, and J. Douglas Williams, "How Effective Are Private Schools in Latin America?" *Comparative Education Review*, 2004, 48, 48–67.
37. Isolda Rodríguez Rosales, "Educación Asimiladora en el Litoral Atlántico: Nicaragua 1893–1909," in Msargsarita Vanninmi and Frances Kinloch (eds), *Política, Cultura y Sociedad en Centroamérica Siglos XVIII–XX* (Managua: Universidad Centroamericana, 1998), pp. 123–135.

38. Luis A.Galeano, "Escenarios Socio-demográficos y Reforma Educativva," *Revista Paraguaya de Sociología*, 2001, 38, 51–82.
39. Bradley A. Levinson, " '*Una etapa seimpre díficil*': Concepts of Adolescence and Secondary Education in Mexico," *Comparative Education Review*, 1999, 43(2), 120–140.
40. Octavio A. Pescador, "Education in Mexico: Historical Evolution and Ethnographic Perspectives," *Comparative Education Review*, 2002, 46, 509–518.
41. Carlos A. Torres, "The State, Nonformal Education, and Socialism in Cuba, Nicaragua, and Grenada," *Comparative Education Review*, 1991, 35, 110–130.
42. Carlos A. Torres, "Paulo Freire as Secretary of Education for the Municipality of São Paulo," *Comparative Education Review*, 1994, 38, 181–214.
43. Paulo Ghiradelli, "Pedagogia e Infância em Tempos Liberales," in Celestino A. da Silva et al. (eds), *Infância e Educação* (São Paulo: Cortez Editora, 1996), pp. 11–41.
44. Adoja F. Jones de Almeida, "Unveiling the Mirror: Afro-Brazilian Identity and the Emergence of a Community School Movement," *Comparative Education Review*, 2003, 47, 41–63.
45. *Statistical Abstract*, 2000, p. 219.
46. Ernesto Toro Balart, "Calidad de la Educación: Concepto, Medición y Estrategias de Mejoramiento," *Revista Paraguaya de Sociología*, 1994, 31(90), 201–206.
47. Thomas J. La Belle, *Nonformal Education in Latin America and the Caribbean: Stability*, Reform or Revolution? (New York: Praeger, 1986), p. 23.
48. Everett M. Rogers and William Herzog, "Functional Literacy Among Colombian Peasants," *Economic Development and Cultural Change*, 1966, 14(January), 190–203.
49. Ricardo Morales Basadra et al., *Educación: Retos y Esperanzas* (Lima: Centro de Estudios y Promoción del Desarrollo, 1995), pp. 37–38.
50. Anna F. Bosch, "Popular Education, Work Training, and the Path of Women's Empowerment in Chile," *Comparative Education Review*, 1998, 42, 163–179.
51. Gonzalo Retamal, "The Politics of Education in Central America: 1950–1980," in Colin Brock and Donald Clarkson (eds), *Education in Central America and the Caribbean* (New York: Routledge, 1990), pp. 174–226.
52. Carlos Rodrigues Bandão, *O Ardil da Ordem: Caminhos e Armadilhos da Educaçáo Popular* (Campinas, SP: Papirus Livaria e Editora, 1983).
53. António Teodoro, "Paulo Freire, or Pedagogy as the Space and Time of Possibility," *Comparative Education Review*, 2003, 47, 328.
54. Rolland G. Paulston, "Ways of Seeing Education and Social Change in Latin America," *Latin* American Research Review, 1992, 27(3), 177–202.
55. Carlos A. Torres, *The Politics of Nonformal Education in Latin America* (New York: Praeger, 1990), pp. 6–16.
56. Liliana Vaccaro, "Transferencia y Apropiación en Intervencias Comunitarias: Un Marco de Referencia para su Análisis," *Harvard Educational Review*, 1990, 60, 79–95.
57. Deborah Brandt, "Popular Education," in Thomas W. Walker (ed.), *Nicaragua: The First Five Years* (New York: Praeger, 1985), pp. 317–345.
58. Robert Arnove, *Education and Revolution in Nicaragua* (New York: Praeger, 1986), p. 40.

59. George Psacharopoulos, "Ethnicity, Education, and Earnings in Bolivia and Guatemala," *Comparative Education Review*, 1993, 37, 9–20.
60. Gonzálo Portocarrero M., "Peru: Education for National Identity—Ethnicity and Andean Nationalism," in Morales-Gómez and Torres, *Education*, pp. 69–82.
61. La Belle, *Nonformal*, pp. 210–214.
62. Judith A. Weiss, *Latin American Popular Theater* (Albuquerque: University of New Mexico Press, 1993).
63. *Statistical Abstract*, 2000, pp. 203–206.
64. Maltilde Zimmermann, *Sandinisita: Carlos Fonseca and the Nicaraguan Revolution* (Durham, NC: Duke University Press, 2000), pp. 28–49.
65. Carlos Ornelas and David Post, "Recent University Reform in Mexico," *Comparative Educational Review*, 1992, 36, 278–297.
66. Dale A. Anderson, *Management and Education in Developing Countries: The Brazilian Experience* (Boulder, CO: Westview Press, 1987), as reported in Paulston, "Ways of Seeing."
67. Antonio Muniz de Rezende, *Crise Cultural e Subdesenvolvimento* (Campinas, SP: Papirus, 1983), pp. 59–64.
68. Augustino Pérez Lindo, *La Batalla de la Inteligenia* (Buenos Aires: Cántoro Editores, 1989), pp. 101–102.
69. Gilberto Guevara N., *La Democracia: Crónica del Movimiento Estudiantil Mexicano* (Mexico, DF: Siglo Veinte, 1988), as reported in Paul L. Haber, "Identity and the Political Process," *Latin American Research Review*, 1996, 31, 171–187.
70. Robert C. Williamson, "Modernism and Related Attitudes: An International Comparison among University Students," *International Journal of Comparative Sociology*, 1970, 11 (June), 130–145.
71. Gabriel Careaga, *Mitos y Fantasías de la Clase Media en México* (Mexico, DF: Cal y Arena, 1990), pp. 140–144.
72. Leo A. Despres, *Manaus: Social Life and Work in Brazil's Free Zone* (Albany: State University of New York Press, 1991), pp. 196–197.
73. George Psacharopoulos and Eduardo Velez, "Educational Quality and Labor Market Outcomes: Evidence from Bogotá, Colombia," *Sociology of Education*, 1993, 66, 130–145.
74. Jorge Millas, *Idea y Defensa de la Universidad* (Santiago: Editorial del Pacífico, 1981).
75. David Post, "Political Goals of Peruvian Students: The Foundations of Legitimacy in Education," *Sociology of Education*, 1988, 61, 178–190.
76. Orlando Albornoz, *La Educatión bajo el Signo de la Crisis* (Mérida: Universidad de los Andes, 1987).
77. Jorge Méndez, "El Divorcio entre el Tratamiento de los Grandes Temas Nacionales y la Universidad: El Caso Especial del Desarrollo Industrial," in *Memorias del Semnario sobre Ciencia y Tecnología* (Bogotá: Universidad Nacional, 1989), pp. 159–165.
78. Armanda Dias Mendes, "O Papel da Universidade no Desenvolvimento Científico e Tecnológico da Região Amazónica," in Luis E. Araganez (ed.), *Universidade e Desenvolvimento Amazónico* (Belem: Universidade Federal de Pará, 1988), pp. 9–17.
79. David Post, "Student Movements and User Fees: Trends in the Effect of Social Background and Family Income on Access to Mexican Higher Education, 1984–1996," *Mexican Studies*, 2000, 161, 11–159.

80. David E. Lowey, *The Rise of the Professions in Twentieth-Century Mexico* (Los Angeles: Latin American Center, University of California, 1992),
81. Pamela Constable and Arturo Valenzuela, *A Nation of Enemies: Chile under Pinochet* (New York: Norton, 1991), p. 267.
82. Louis A. Pérez, *Cuba: Between Reform and Revolution* (New York: Oxford University Press, 1988), pp. 360–361.
83. Torres, *The Politics*, pp. 119–120; John A. Britton (ed.), *Molding the Hearts and Minds: Education, Communications, and Social Change in Latin America* (Wilmington, DE: Scholarly Resources, 1994).

Chapter Ten   Communication, Science, and Technology

1. Javier Esteinou Madrid, *La Comujnicación y las Culturas Nacionales en los Tiempos del Libre Comercial* (Mexico, DF: Fundaciíon Manuel Buendía, 1993), pp. 218–220.
2. Elizabeth Mahan, "Media, Politics, and Society in Latin America," *Latin American Research Review*, 1995, 30 (3), 138–162.
3. Carlos Catalan, "Mass Media and the Collapse of a Democratic Tradition in Chile," in Elizabeth Fox (ed.), *Media and Politics in Latin America* (Newbury Park, CA: Sage Publications, 1988), pp. 45–65.
4. Judy Polumbaum, "Free Society, Repressed Media: The Chilean Paradox," *Current History*, 2002, 101(452), 66–71.
5. Maria E. Díaz, "The Satiric Penny Press for Workers in Mexico,1900–1910," in John A. Briton (ed.), *Molding the Hearts and Minds: Education, Communications, and Social Change in Latin America* (Wilmington, DE: SR Books, 1994), pp. 65–91.
6. Marvin Alisky, *Latin American Media: Guidance and Censorship* (Ames: Iowa State University Press, 1991), p. 35.
7. Silvio Waisbord, *Watchdog Journalism in South America: News, Accountability, and Democracy* (New York: Columbia University Press, 2000), pp. 47–48.
8. David Wood, "The Peruvian Press under Recent Authoritarian Regimes, with Special Reference to the *Autogolpe* of President Fujimori," *Bulletin of Latin American Research*, 2000, 19, 17–31.
9. Roque Faraone and Elizabeth Fox, "Communication and Politics in Uruguay," in Fox, *Media*, pp. 148–156.
10. Antonio Menéndez Alarcon, *Power and Television in Latin America: The Dominican Case* (Westport, CT: Praeger, 1992) as cited in Mahan, "Media, Politics," p. 155.
11. James Schwoch, *The American Radio Industry and its Latin American Activities, 1900–1939* (Urbana: University of Illinois Press, 1990), as cited in Mahan, "Media, Politics," p. 143.
12. José Marques de Melo, *Teoria da Comunicação: Paradimas Latino-Americanos* (Petropolis: Editores Vozes, 1998), pp. 388–401.
13. Ariel Dorfman and Armand Mattelart, *Para Leer al Pato Donald* (Valparaíso: Ediciones Univesitarias, 1972).
14. Robert Buckman. "Cultural Agenda of Latin American Newspapers and Magazines: Is U.S. Domination a Myth?" *Latin American Research Review*, 1990(2), 134–155.

15. Louise Montgomery, "The Role of Women in Latin American Mass Media," in Richard R. Cole (ed.), *Communication in Latin America: Journalism, Mass Media, and Society* (Wilmington, DE: SR Books, 1996), pp. 37–49.
16. Latin American Center, *Statistical Abstract of Latin America* (Los Angeles: University of California, 2000), p. 59.
17. David Kunzle, "Roger Sánchez's 'Humor Erótico' and *Semana Crónica*: A Sexual Revolution in Sandinista Nicaragua," *Latin American Perspectives*, 1998, 25(101), 89–120.
18. Robert T. Buckman, "Present Status of the Mass Media in Latin America," in Cole, *Communication*, pp. 3–36.
19. Ola Guedes, "Environmental Issues in the Brazilian Press," *Gazette*, 2000, 62(6), 537–554.
20. Waisbord, *Watchdog*, p. 155.
21. Alisky, *Latin American Media*, pp. 138–162.
22. Waisbord, *Watchdog*, pp. 248–252.
23. Jacques A. Weinberg, "Telecommunication Revolution in Brazil and its Impact in Self-Image Perceptions," *Convergencia*, 2001, 8(25), 141–191.
24. Cecilia Krphling Peruzzo, *Comunicação e Culturas Populares* (São Paulo: INTERCOM, 1995), p. 53.
25. Cythia J. Miller, "The Social Impact of Televised Media among the Yucatec Maya," *Human Organization*, 1998, 57, 207–312.
26. Pataricia Aufderheide, "Grassroots Video in Latin America," in Chon A. Noriega (ed.), *Visible Nations: Latin American Cinema and Video* (Minneapolis: University of Minnesota Press, 2000), pp. 219–238.
27. Taylor C. Boas, "Television and Neopopulism in Latin America: Media Effects in Brazil and Peru," *Latin American Research Review*, 2005, 40(2), 27–49.
28. Alicia Fraerman, *Identidad y Nuevos Medios: La Comunicación en Iberoamérica* (Madrid: Editorial Comjhnica, 1885), p. 102.
29. Michael Morgan and Jamwa Shanahan, *Democracy Tango: Television, Adolescents, and Authoritarian Tensions in Argentina* (Cresskill, NJ: Hampton Press, 1995).
30. Heidi N. Nariman, *Soap Operas for Social Change* (Westport, CT: Praeger, 1993).
31. Rafael Ahumada Barajas, *Análisis de la Imagen Teelevisa* (Mexico, DF: UNAM, 1999), pp. 21–27.
32. Constantin von Barloewen, *Cultural History and Modernism in Latin America: Technology and Culture in the Andes Region* (Providence, RI: Berghahn Books, 1995), pp. 107–116.
33. Eduardo Estrella, "Medicina Autóctono Precolombina," in Juan J. Saldaña (ed.), *Historia Social de las Ciencias en América Latina* (Mexico, DF: UNAM, 1996), pp. 43–69.
34. Ferenando de Azevedo, *As Ciências no Brasil*, vol. 2 (São Paulo: Edições Melhoramientos, 1963), pp. 153–154.
35. Simon Collier, Thomas E. Skidmore and Harold Blakemore, *The Cambridge Encyclopedia of Latin America and the Caribbean* (New York: Cambridge University Press, 1992), pp. 451–460.
36. Collier, *The Cambridge*, pp. 453–456.
37. Elias Trabulse, *Historiia de la Ciencia y la Tecnología* (Mexico, DF: Colegio de México, 1994), pp. 218–229.
38. Juan J. Saldaña, "Ciencia y Libertad: La Ciencia y la Tecnologia como Política de los Nuevos Estados Americanos," in Saldaña, *Historia*, pp. 283–297.

39. Luis Carranza, *La Ciencia en el Peru en el Siglo XIX* (original ed., 1890, Lima: Editorial Eddili, 1988).

40. David J. Hess, *Spirit and Scientists: Ideology, Spiritism, and Brazilian Culture* (University Park: Pennsylvania State University Press, 1991), pp. 84–88, 156.

41. Ricardo Israel, *Educación, Ciencia y Tecnología: Reflexiones de Fin de Milenio* (Santiago: LOM Editores, 1998), pp. 74–76.

42. Augusto Ruíz Zeballos, *Psiquiatras y Locos entre la Moderniización Contra los Andes y el Nuevo Proyecto de Modernidad, Peru, 1850–1930* (Lima: Instituto Pasado y Presentre, 1994), as reported in Ann Zulawski, "New Trends in Studies of Science and Medicine in Latin America," *Latin American Research Review*, 1999, 34(3), 241–251.

43. Jorge Cañizaraes-Esguerra and Marcos Gurto, "Latin American Science: The Long View," *NACLAS Report on the Americas*, 2002, 35(5), 18–52.

44. Silvia Segal, *Intelectuales y Poder en la Década del Sesenta* (Buenos Aires: Editorial Punto Sur, 1991), pp. 260–283.

45. José Leite Lopes, *Ciência, e Desenvolvimento* (Niteroi: Universidade Federal Fluminense, 1987), p. 64.

46. Zulawski, "New Trends," p. 250.

47. C. A. Hooker, *A Realistic Theory of Science* (Albany: State University of New York Press, 1987), p. 209.

48. Jacqueline Fortes and Larissa A. Lomnitz, *Becoming A Scientist in Mexico* (University Park: Pennsylvania State University Press, 1994), pp. 75–90.

49. Roberto Hidalgo and Guillermo Monge, *El Futuro Cercamo y la Capacidad Tecnológica Costariccense* (San José: Editorial de la Universidad de Costa Rica, 1989), pp. 20–27.

50. Fortes and Lomnitz, *Becoming*, pp. 134–164.

51. Susana García and Lorissa Lomnitz, "Evaluacíon de la Ciencia y la Tecnología: Ambiguiuedades y Discrepenciasen el Sistema Nacional," in Miguel Campos and Jaime Jiménez (eds), *El Sistema de Ciencia y Tecnología en México* (Mexico, DF: UNAM, 1991), pp. 167–262.

52. Maario de Lima Bezera (ed.), *Ciência e Tecnologia para o Desenvolvimento Sustentável* (Brasilia: Ministério do Meio Ambiente, 2000), pp. 178–180.

53. George Zarur, *A Arena Ciéntífica* (Brasilia, DF: FLACSO, 1994), pp. 180–186.

54. Darío Saráchaga, *Ciencia y Tecnología en Uruguay: Una Agenda Hacia el Futuro* (Montevideo: CONICYT, Ediciones Trilece, 1997), pp. 93–97.

55. Ana María Fernandes et al., *Colapso de Ciência e Tecnologia no Brasil* (Rio: Relume Dumara, 1994).

56. Carlos Dávila and Rory Miller (eds), *Business History in Latin America* (Liverpool: Liverpool University Press, 1999).

57. Horacio Godoy, "La Ciencia, la Politica: La Ciencia Política," in Hernando Rua Suárez et al. (eds), *La Investigación Científica en Colombia Hoy* (Bogotá: Editora Guadalupa, 1979), pp. 363–379.

58. Roberto Briceño León and Heinz Sonntag, *Pueblo, Epoca, y Desarrollo: La Sociología en América Latina* (Caracas: Nueva Sociedad, 1998), pp. 57–60.

59. Pablo González Casanova et al., *Las Ciencias Sociales en los Años Noventa* (Mexico, DF: UNAM, 1993), pp. 11–24.

60. Roberto J. González, "From *Indigenismo* to *Zapatismo*: Theory and Practice in Mexican Anthropology," *Human Organization*, 2004, 63, 141–149.

61. Jacques Gaillard, *Scientists in the Third World* (Lexington: University Press of Kentucky, 1991), pp. 20–32.

62. Ricardo Arrellano Castro, *Estado, Ciencia-Tecnología y Desarrollo en México* (Mexico, DF: UNAM, 1996); Jimena F. Beltrão, *Ciência e Tecnologia: Desafio Amazonico* (Belem: Editorial Universitario UFPA, 1992).
63. Leopold de Meis and Jaquelin Leta, *O Perfil da Ciência Brasileira* (Rio: Editora URI, 1996), pp. 95–96.
64. Simon Schwartzman, *A Space for Science: The Development of the Scientific Community in Brazil* (University Park: Pennsylvania State University Press, 1991), pp. 200–201.
65. Tony Reichhardt, "Brazil's Space Age Comes of Age," *Nature*, April 1999, 398, A19.
66. Luis E. Aragón et al., *Science Development, Environment and Development in Brazil* (Stockholm: Stockholm University Press, 1995), pp. 156–164.
67. Christopher Roper and Jorge Silva (eds), *Science and Technology in Latin America* (New York: Longman, 1983), p. 209.
68. Hector Hianay Escobar, "Política Científica y Tecnológica en el Peru: Antrecendtes, Situación Actual y Perspectivas," paper presented at AlAS meeting, Miami 2000.
69. Marshall C. Eakin, "The Origins of Modern Science in Costa Rica: The Instituto Físico-Geográfico Nacional, 1867–1904," *Latin American Research Review*, 1999, 34(1), 123–149.
70. Iván Molina Jiménez and Steven Palmer, *La Voluntud Radiante: Cultura, Empresa, Magia y Medicina en Costa Rica (1897–1932)* (San José: Editorial Porvenir, 1996), pp. 109–122.
71. Gaillard, *Scientists*, pp. 69, 109, 119.
72. Roper and Silva, *Science and Technology*, p. 95.
73. Dimlus D. James, "Science, Technology, and Development," in James L. Dietz and Dilmus D. James (eds), *Progress Toward Development in Latin America* (Boulder, CO: Rienner Publishers, 1990), pp. 159–176.
74. William J. Stover, *Information Technology in the Third World* (Boulder, CO: Westview Press, 1984), pp. 61–71.
75. Guilleromo Guajardo, "Tecnología y Campesinos en la Revolución Mexicana," *Mexican Studies*, 1999, 14, 291–320.
76. Arnold Facey, *Meaning in Technology* (Cambridge, MA: MIT Press, 1999), pp. 86–91.
77. Devon G. Peña, *The Terror of the Machine: Technology, Work, Gender, and Ecology on the U.S.–Mexico Border* (Austin: University of Texas Press, 1997).
78. Ibid., pp. 191–193.
79. Patricia Atlas and Fiona Wilson, *La Aguja y el Surco: Cambio Regional, Consumo y Relaciones de Género en la Industria de Ropa en México* (Guadalajara: Universidad de Guadalajara, 1997).
80. Nadya Araujo Castro, "Impacto Social das Mudanças Tecnológicas, Organização Idnuustrial," in Simon Schwartzman (ed.), *Ciênçia e Tecnologia no Brasil: Política Industrial, Mercado de Trabajo, e Instituçoes de Apoio* (Rio: Fundação Getulio Vargas, 1995), pp. 207–212.
81. Henri Coing and Etienne Henry, "Balance y Perspecrivas de los Servicios Urbanos," *Ciudades*, 1992, 11, 9–14.
82. Peña, *The Terror*, pp. 214–215.
83. Manfred Nitsch, "The Biofuel Programme PROALCOHOL within the Brazilian Energy Strategy," in Aragon et al., *Science, Development*, pp. 43–63.

84. Emanuel Adler, *The Power of Ideology: The Quest for Technological Autonomy in Argentina and Brazil* (Berkeley: University of California Press, 1987), pp. 327–331.
85. Rafael Urrelo, *Capital y Conocimiento: Ciencia y Tecnologgía para el Desarrollo* (Lima: Edicioones del Congreso, 2000), pp. 127–130.
86. Francisco J. Pichón, Jorge E. Uquillas, and John Frechione, *Traditional and Modern Natural Resources Management in Latin America* (Pittsburgh: University of Pittsburgh Press, 1999), pp. 21–23.
87. Betty B. Paust, *Mexican Rural Development and the Plumed Serpent: Technology and Mayan Cosmology in the Tropical Forest of Campeche, Mexico* (Westport, CT: Bergin and Garvey, 1998), pp. 163–164.
88. Paul H. Gelles, *Water and Power in Highland Peru: The Cultural Politics of Irrigation and Development* (New Brunswick, NJ: Rutgers University Press, 2000), pp. 156–161.
89. Nigel J. H. Smith, "Biodiversity and Agroforestry along the Amazon Floodplain," in Pichón et al., *Traditional*, pp. 232–260.
90. William M. Loker, "Sowing Discord, Planting Doubts, Rhetoric and Reality in an Environment and Development Project in Honduras," *Human Organization*, 2000, 59(3), 300–310.
91. Wency Call, "Plan Puebla-Panama," *MACLAS Report on the Americas*, 2002, i5, 224–226.
92. Francisco R. Sagasti, *Ciencia, Tecnología y Desarrollo Lationmoamericano* (Mexico, DF: Fondo de Cultura Económica, 1997).
93. Margaret Everett. "Latin America On-Line: The Internet, Development, and Democratization," *Human Organization*, 1998, 57, 385–393.
94. Robert K. Heldman, *Information Telecommunications* (New York: McGraw-Hill, 1993), pp. 252–257.
95. Rainer Randolph, "Vivendo e Apreendo no Cyberespação Urbana do Rio de Janeiro," paper presented at the Latin American Studies Association meeting, Miami, March 2000.
96. Ibid.
97. Sean Cubitt, *Digital Aesthetics* (Thousand Oaks, CA: Sage Publications, 1994), pp. 148–151.
98. María Saantos Corral, *Cien Mil Llamadas por el Ojo de una Aguja* (Mexico, DF: UNAM (Aragan), 2000), pp. 193–203.
99. Jane Robinett, *This Rough Magic: Technology in Latin American Fiction* (New York: Peter Lang, 1994), pp. 74, 80, 221–229.
100. Philip Scranton, "Determinism and Indeterminism in the History of Technology," in Merritt R. Smith and Leo Marx (eds), *Does Technology Drive History?* (Cambridge, MA: MIT Press, 1995), pp. 143–167.

Chapter Eleven   Social Structure and Change: Latin America and the World

1. Jonathan H. Turner, *The Structure of Sociological Theory* (Homewood, IL: Dorsey Press, 1974), pp. 17–19.
2. Roland I. Perusse, *Haitian Democracy Restored, 1991–1995* (Lanham, MD: University Press of America, 1995), pp. 174–197.

3. Carlos Moneta, "The Latin American Economic System as a Mechanism to Control Conflicts," in Michael A. Morris and Victor Millán (eds), *Controlling Latin American Conflicts: Ten Approaches* (Boulder, CO: Westview Press, 1983), pp. 99–116.
4. Ben R. Schneider, "Why is Mexican Business So Organized?" *Latin American Research Review*, 2002, 37(1), 77–117.
5. Lewis A. Coser, *The Functions of Social Conflict* (New York: Free Press, 1966).
6. Victor Vich, *El Discurso de la Calle* (Lima: Instituto de Estudios Peruanos, 2001), pp. 169–186.
7. David R. Mares, *Violent Peace: Militarized Interstate Bargaining in Latin America* (New York: Columbia University Press, 2001), pp. 132–148, 201–203.
8. Randall Collins, *Conflict Sociology: Toward an Explanatory Science* (New York: Academic Press, 1975), p. 551.
9. Frank Safford and Marco Palacios, *Colombia: Fragmented Land, Divided Society* (New York: Oxford University Press, 2002), pp. 347–355.
10. Peter D. Scott, *Drugs, Oil, and War* (Boulder, CO: Rowman & Littlefield, 2003).
11. Raymond Boudon, *Theories of Social Change: A Critical Approach*, translated by J. C. Whitehouse (Berkeley: University of California Press, 1986).
12. Trevor Noble, *Social Theory and Social Change* (New York: St Martin's Press, 2001), pp. 5–10.
13. Everett Hagen, *On the Theory of Social Change* (Homewood, IL: Dorsey Press, 1962), pp. 353–384.
14. Jack Knight, *Institutions and Social Conflict* (New York: Cambridge University Press, 1992), pp. 28–34.
15. Adolfo Gilly, "América Latina: Abajo y Afuera," in Coloquio de Inverno (ed.), *Las Américas en el Horizonte del Cambio*, vol. 2 (Mexico, DF: UNAM, 1992), pp. 106–118.
16. Eric Selbin, *Modern Latin American Revolutions* (Boulder, CO: Westview Press, 1993), pp. 2–4.
17. Jan K. Black, "Participation and Political Process: The Collapsible Pyramid," in Jan K. Black (ed.), *Latin America: Its Problems and Its Promises* (Boulder, CO: Westview Press, 1998), pp. 203–231.
18. Susan Eckstein (ed.), *Power and Popular Protest* (Berkeley: University of California Press, 1989), pp. 1–8.
19. Peter E. Newell, *Zapata of Mexico* (Montreal: Black Rose Books, 1997), pp. 138–146.
20. Black, "Participation," pp. 219–234.
21. Hugh Byrne, *El Salvador's Civil War: A Study of Revolution* (Boulder, CO: Rienner Publishers, 1996), pp. 197–206.
22. David W. Dent, *The Legacy of the Monroe Doctrine* (Westport, CT: Greenwood Press, 1999), pp. 170–179.
23. May B. Warren, "Interpreting *La Violencia* in Guatemala: Shapes of Mayan Silence and Resistance," in Kay B. Warren (ed.), *The Violence Within Cultural and Political Opposition in Divided Nations* (Boulder, CO: Westview Press, 1993), pp. 25–36.
24. Hernando de Soto and Deborah Orsiini, "Overcoming Underdevelopment," in Uner Kirdar (ed.), *Change: Threat or Opportunity*, vol. 4 (New York: United Nations, 1992), pp. 79–89.

25. David Harvey, Globalization in Question," in Johannes D. Schmidt and Jacques Hersh (eds), *Globalization and Social Change* (New York: Routledge, 2000), pp. 20–36.
26. Anna Magoles, "Nuevo Programa Bracero en la Era de Globalización," *Revista Paraguaya de Sociología*, 2000, 37(107), pp. q56–q168.
27. Josaignacio Morenmo Leo, *América Latina: Del Realismo a la Sociedad Global* (Caracas: CERET, 1994), pp. 48–52.
28. John Peeler, *Building Democracy in Latin America* (Boulder, CO: Rienner Publishers, 1998), pp. 142–144.
29. Guillermo Palacios, *La Pluma y el Arado: Los Intelectuales y la Construción del Problema Campesino en México* (Mexico, DF: Colegio de México, 1999), pp. 233–235.
30. World Bank's 1990 *World Development Report* as reported in Duncan Green, "A Trip to the Market: The Impact of Neoliberalism in Latin America," in Julia Buxton and Nicola Phillips (eds), *Developments in Latin American Political Economy* (Manchester: Manchester University Press, 1999), pp. 13–31.
31. Mauricio González Gómez, "Crisis and Economic Change in Mexico," in Susan K. Powell and Luis Rubin (eds), *Mexico under Zedillo* (Boulder, CO: Rienner Publishers, 1998), pp. 37–66.
32. Thomas C. Wright, *Latin America in the Era of the Cuban Revolution*, revised ed. (Westport, CT: Praeger, 2001), pp. 192–194.
33. Michael Snodgrass, "Assessing Everyday Life in Post-Soviet Cuba," *Latin American Research Review*, 2001, 36(3), 204–220.
34. Emilio Pereira Alegía, "Las Reformas Institucionales y su Ippacto en la Política Económica Social," in Alfredo Solorzano et al. (eds), *Administración Política y Democracia* (Managua: Instituto Nicaragaense de Administración, 1997), pp. 69–79.
35. Wright, *Latin*, pp. 180–182.
36. Albert Fishlow, "Is the *Real* Plan for Real?" in Susan K. Purcell and Riordan Roett (eds), *Brazil under Cardoso* (Boulder, CO: Rienner Publishers, 1997), pp. 43–62.
37. Michael Shaafter and Vinay Iaavahar, "Latin America's Populist Turn," *Current History*, February 2005, 104, 51–57.
38. Peeler, *Building*, p. 153.
39. George Philip, "Institutions and Democratic Consolidation in Latin America," in Buxton and Phillips, *Developments*, pp. 33–48.
40. Carlos Franco, *Acerca del Modo de Pensar la Democracia en América Latina* (Lima: Friedrich Ebert Stiftung, 1998).
41. Miriam Kornblith and Daniel H. Levine, "Venezuela: The Life and Times of the Party System," in Scott Mainwaring and Timothy R. Scully (eds), *Building Democratic Institutions: Party Systems in Latin America* (Stanford, CA: Stanford University Press, 1995), pp. 37–71.
42. Philip, "Institutions," pp. 36–38.
43. Luis E. González, "Continuity and Change in the Uruguayan Party System," in Mainwaring and Scully, *Building*, pp. 138–163.
44. Ab ril Trigo, "*El tamaño de la esperanza*: Some Reflections on the Elections in Uruguay," *LASA Forum*, 2005, 35(1), 13–14.
45. Eduardo Morales Pérez, *Una Nueva Visión de la Política Internacional de México* (Mexico, DF: Fondo de Cultura Economica, 1997), pp. 177–179.

46. Gerald K. Haines, *The Americanization of Brazil: A Study of U.S. Cold War Diplomacy in the Third World, 1945–1954* (Wilmington, DE: SR Books, 1989).
47. T. David Mason, "The Civil War in El Salvador: A Retrospective Analysis," *Latin American Research Review*, 1999, 34(3), 179–196.
48. Dent, *The Legacy*, pp. 75–77.
49. Cheryl L. Eschbach, "Explaining U.S Policy toward Latin America and the Caribbean," *Latin American Research Review*, 1990, 25(2), 204–216.
50. Robert Trudeau and Luis Choutz, "Guatemala," in Morris J. Bachman et al. (eds), *Confronting Revolution* (New York: Pantheon Books, 1986), pp. 23–49.
51. Philip Zwerling and Connie Martin, *Nicaragua: A New Kind of Revolution* (Chicago: Lawrence Hill Books, 1985).
52. Raay M. Marini, *América Latina, Dependência e Integração* (São Paulo: Editora Brasil Urgente, 1992), pp. 141–149.
53. Josefina Zoraída Vásquez and Lorenzo Meyer, *The United States and Mexico* (Chicago: University of Chicago Press, 1985), pp. 90–101.
54. Martha L. Cottam, *Images and Intervention: U.S. Policies in Latin America* (Pittsburgh: University of Pittsburgh Press, 1994), pp. 178–181.
55. Kenneth E. Boulding, *Ecodynamics: A New Theory of Societal Evolution* (Thousand Oaks, CA: Sage Publications, 1978), pp. 264–265.
56. Claarence Zwekas, Jr., "The Honduran Poverty Reduction Strategy," MACLAS *Latin American Essays*, 2003, 16, 1–31.
57. Alfredo Jocelin-Holt, *El Chile Perplejo* (Santiago: Planeta/Ariel, 1998), pp. 272–285.
58. Black, "*Participation*," pp. 610–611.
59. Jay Olson, "Terrorism: Stop Inflating the Concept," *Cross-Currents*, (Washington Office on Latin America, June 2004), pp. 1–3.

Index

Printed in the United States
67931LVS00001B/85-264

9 781403 968869